FOCUS ON NEUROCHEMISTRY RESEARCH

FOCUS ON NEUROCHEMISTRY RESEARCH

ROBERT M. COLEMAN
EDITOR

Nova Science Publishers, Inc.
New York

Library of Congress Cataloging-in-Publication Data
Available upon request

ISBN 1-59454-417-4

Published by Nova Science Publishers, Inc. ✣ New York

Contents

Preface

As reported in Chapter I, each of the two main cell types in the CNS (neurons and glial cells) is made up by subgroups (neurons utilizing different transmitters, the glial cells astrocytes, oligodendrocytes and microglia), and each individual cell type displays a variety of subcellular structures, including cell processes that are intricately intertwined with cell processes from other cells. An extracellular space between different cells and cell processes constitutes an important communication channel, where the concentrations of neuroactive compounds like the potassium ion (K^+) and glutamate constantly vary, are regulated by cellular release and subsequent reuptake, especially in astrocytes, and exert metabolic effects. The stimulation of glucose metabolism during functional activity mainly occurs in the neuropil, perhaps in dendrites carrying excitatory impulses and in adjacent astrocytes. A large fraction of this energy expenditure serves the purpose of restoring resting distribution between intracellular and extracellular K^+ and Na^+ following its disturbance by neuronal excitation.

Ethanol is a small molecule acting on several neurotransmitter systems in the brain. Accumulating evidences suggest that the primary excitatory – i.e. the glutamatergic – neurotransmitter system is a particularly important site of ethanol's action. Several studies showed that ethanol is a potent and selective inhibitor of the N-methyl-D-aspartate (NMDA) receptors and prolonged ethanol exposition leads to a compensatory "up-regulation" of these receptors resulting in enhanced NMDA receptor mediated functions after removal of ethanol. These alterations are supposed to contribute to the development of ethanol tolerance, dependence as well as the acute and delayed signs of ethanol withdrawal and is discussed in Chapter II.

Benzodiazepines are widely prescribed anxiolytic, anticonvulsant and sedative drugs. Their pharmacological effects are mediated by a variety of mechanisms that converge in the alteration of synaptic transmission in the central nervous system. Initially, benzodiazepines were thought to modulate synaptic transmission by interacting with postsynaptic neurotransmitter receptors. High-affinity benzodiazepine binding sites were described on neuronal $GABA_A$ receptors. Chapter III focuses on the newly identified benzodiazepine targets in the modulation of neurotransmission.

Microglia are the principal immune cells in the central nervous system (CNS) and have a critical role in host defense against invading microorganisms and neoplastic cells. These cells are known to interact with peripheral immunocompetent cells via production of cytokines and co-stimulatory molecules. In fact, like macrophage, microglia have an array of cellular defense mechanisms, including secretion of inflammatory cytokine, production of complement proteins and receptors. The authors of Chapter IV activated in vitro human fetal microglia either by infection with Mycobacterium avium (SmD, SmT) and Mycobacterium tuberculosis (Myc) or by stimulation with myelin basic protein (MBP) and estrogens. The results suggest that microglia inflammatory response as well as their interaction with immunocompetent cells can be modulated by pathogenic factors such as bacteria or MBP (multiple sclerosis). They also help in understanding the mechanism(s) of brain damage in inflammatory disease.

Oxidative stress plays an important role in neuron death associated with various neurodegenerative conditions. Ceramide, the product of sphingomyelin hydrolysis, is reported to be a key mediator of the cell death induced by reactive oxygen species (ROS). Antioxidants that prevent the detrimental consequences of ROS are considered to be a promising approach to neuroprotection. Crocin is a pharmacologically active component of Crocus sativus L. (saffron). Described in Chapter V is the in vitro effect of crocin on the oxidative stress-induced death of neuronally differentiated rat pheochromocytoma (PC-12) cells. The authors suggest that crocin is a unique and potent antioxidant that combats ischemic stress-induced neuron death by increasing GSH levels, and prevents the activation of the JNK pathway, which is reported to have a role in the signaling cascade downstream ceramide.

The molecular basis for developing epilepsy remains ill defined. In epileptic patients and several experimental models of epileptogenesis, the integrity of dendritic spines and the molecular organization of the postsynaptic density, both essential for normal neurotransmission and controlled neuronal excitability, are thought to be dramatically altered. The authors of Chapter VI have also investigated the effect of protein kinase and protein phosphatese inhibitors on experimental epileptic seizures. They found that ascomycin (100□M), a specific calcineurin inhibitor, prevents picrotoxin-induced seizures. H-9 dihydrochloride (100□M), an inhibitor of several protein kinases, also shows antiepileptic effect. The results show that in vivo neurochemical research methods such as chronic microdialysis are a good tool to explore the dynamic molecular changes associated with neuronal plasticity and epileptogenesis, complementing in vitro studies.

5-Aminolevulinic acid (ALA) is an haem precursor that accumulates in acute intermittent porphyria and lead poisoning. It has been proposed that ALA overload is involved in the neurological dysfunctions associated with those disorders. Chapter VII was aimed at investigating the effects of ALA on the uptake of amino acid neurotransmitters by rat brain synaptic vesicles. The results provide evidence that ALA may disturb the vesicular uptake of amino acids, which is an important event in the storage and release of these neurotransmitters by synaptic vesicles.

Methylmercury (MeHg) is a significant environmental contaminant that will continue to pose great risk to human health. Considerable attention in the scientific and health policy fora has focused on the question of whether MeHg intake from a diet high in fish is associated with aberrant CNS function. A number of studies suggest that fetal exposure at levels attained by mothers regularly eating fish during pregnancy are associated with neurological deficits in their offspring. In Chapter VIII The authors present data in support of the hypotheses that activation of the cytosolic phospholipase A_2 ($cPLA_2$) and the ensuing hydrolysis and release of arachidonic acid (AA) are mediators for glutamate release upon exposure to MeHg.

Neurons strongly depend on oxygen supply and ATP production. Short periods of hypoxia or anoxia occurring e.g. in stroke or trauma can result in necrosis/apoptosis and damaged axons. Attendant with energy depletion is the formation of purine nucleosides adenosine, inosine, and guanosine. In Chapter IX, hypoxia was induced with rotenone, a mitochondrial complex I inhibitor, which leads to loss of viable cells and neurite formation. In the authors study, neuroprotective effects of adenosine and inosine were apparently supported by A1 receptor-mediated signaltransduction, whereas guanosine-mediated protection seemed to be A1 receptor independent. Their data supports current hope of the possible use of purine nucleosides for the neuronal regeneration of primary brain cells following hypoxic insult.

In: Focus on Neurochemistry Research
Editor: Robert M. Coleman, pp.1-60

ISBN 1-59454-417-4

Chapter I

The Neuronal-Glial Brain: Fine Structure, Ion Homeostasis and Glucose Metabolism

*Leif Hertz**
College of Basic Medical Sciences, China Medical University, Shenyang, P.R. China

Abstract

Each of the two main cell types in the CNS (neurons and glial cells) is made up by subgroups (neurons utilizing different transmitters, the glial cells astrocytes, oligodendrocytes and microglia), and each individual cell type displays a variety of subcellular structures, including cell processes that are intricately intertwined with cell processes from other cells. An extracellular space between different cells and cell processes constitutes an important communication channel, where the concentrations of neuroactive compounds like the potassium ion (K^+) and glutamate constantly vary, are regulated by cellular release and subsequent reuptake, especially in astrocytes, and exert metabolic effects. The stimulation of glucose metabolism during functional activity mainly occurs in the neuropil, perhaps in dendrites carrying excitatory impulses and in adjacent astrocytes. A large fraction of this energy expenditure serves the purpose of restoring resting distribution between intracellular and extracellular K^+ and Na^+ following its disturbance by neuronal excitation. Energy requiring extrusion of calcium ions accumulated into nerve terminals in connection with transmitter release and in astrocytes during depolarization and transmitter exposure, and reuptake of transmitters may account for smaller parts of the excess energy demand during CNS stimulation. Glucose metabolism in astrocytes is also required for synthesis of precursors for the neurotransmitters glutamate and GABA, which cannot be generated in neurons, lacking

* Please send correspondence to: Dr. L. Hertz, RR 2, Box 245. Gilmour, ON, K0L 1W0, Canada. Phone: 1 613 474 0537; Fax: 1 613 474 0538; e-mail: lhertz@northcom.net

specific enzymes which are found in astrocytes. Neuronal glutamate release affects not only postsynaptic neurons but also adjacent astrocytes, which can either reinforce and prolong synaptic activity of the glutamate-releasing neuron after incoming stimulation has ceased, or transfer the signal in the form of a calcium wave through an astrocytic syncytium and astrocytic release of glutamate or adenosine to excite or inhibit other neurons. Such a process constitutes a potentially flexible and plastic neuronal-astrocytic-neuronal impulse transmission system complementing synaptic transmission. After its release glutamate is preferentially accumulated in astrocytes where it can be returned to neurons for continued transmitter use, released by astrocytes or oxidatively degraded.

Introduction

In order to obtain an understanding of the unique way in which the central nervous system (CNS) operates during its normal function and how these functions are disrupted in disease states one must realize (i) the extreme complexity of this organ with respect to its cellular and subcellular anatomical, biochemical and functional organization; (ii) the extent of functional and metabolic interactions between the different structures of which it is composed; and (iii) the autonomy of the brain in relation to the systemic circulation. The purpose of this chapter is to provide the reader with some appreciation of each of these three points.

In the CNS, fine structure, metabolism and function are connected to an extent that is only presently becoming unraveled. It is becoming evident that metabolic activities differ not only according to the functional state of the CNS, but also between different types of cells and their subcellular constituents (neurons versus non-neuronal cells; one type of neurons versus another type of neurons; one glial cell type versus another type of glial cell; one neuronal constituent versus other neuronal constituents; one astrocytic compartment versus another compartment), and that metabolic intermediates travel not only between different subcellular structures within the same cell (for example from cytosol to mitochondria during glucose metabolism) but also between different cell types (for example from astrocytes to neurons during biosynthesis of transmitter glutamate). The complicated interaction between metabolic events in cells or cell constituents with different functional characteristics reflects the extraordinarily complex function of the CNS.

The metabolic interactions are to a large extent regulated by compounds released from cells to the extracellular (interstitial) space and affecting metabolic and physiological processes in adjacent cells. Such agents include simple ions like potassium ions (K^+), which exit in large amounts from excited neurons; transmitters, which are not always released in the intimacy of a synapse conveying private information between a presynaptic and a postsynaptic neuron; and metabolites, needed in one cell type or subcellular constituent, which is unable to synthesize them and therefore must import them from other cells or parts of cells. We can only begin to understand function (and dysfunction) of the CNS if we possess information about such interactions, which, in turn, requires knowledge about structural organization.

In the concept of the CNS sketched above, the extracellular space functions as an important information channel. For this purpose it is essential that it is isolated from the systemic circulation. This is achieved by the blood-brain-barrier and other barriers separating the CNS *as an organ* from the remainder of the body. It is commonly believed that the main purpose of the blood-brain-barrier is to prevent toxic compounds from getting access to the brain, but it has been pointed out by Cserr and Bundgaard (1986) that a much more important physiological role of this barrier is to surround CNS cells with a microenvironment, which is unperturbed by exchange with the systemic circulation. This applies to the composition of the interstitial space with respect to such neuroactive compounds as glutamate, which shows little or no net transport from the systemic circulation to the CNS and limited transport from CNS to blood (O'Kane et al., 1999; Smith 2000), and K^+, which crosses the blood-brain-barrier exceedingly slowly compared to the great rapidity with which it is transferred across the membranes of CNS cells through ion channels and carriers (reviewed by Walz and Hertz, 1983; Hertz, 1986). Glucose is transported very efficiently across the blood-brain-barrier and cell membranes by aid of specific carriers (Vannucci et al., 1997), which is a major reason that the intact brain normally utilizes glucose as virtually its only metabolic substrate. Less active carriers are found for essential amino acids (Smith, 2000). Like glutamate, most other non-essential amino acids do not rapidly cross the blood-brain-barrier and are not accumulated into the brain from the systemic circulation. The gases oxygen, carbon dioxide (CO_2) and ammonia diffuse freely across the blood-brain-barrier and cell membranes.

The other barriers isolating the CNS from the systemic circulation include the *glia limitans*, a multi-layered structure made up of astrocytic processes and covering the surface of the brain towards the pia mater of the meninges, as well ependyma cells and tanycytes lining the walls of the ventricles (Peters et al., 1991; Szele and Szuchet, 2004). The areas covered by these barriers are much smaller than the total area involved in the blood-brain-barrier. Both of them separate brain tissue from cerebrospinal fluid. A special barrier system with very active transport carriers is localized in the *choroid plexus*, which is the major site of production of cerebrospinal fluid (Weaver et al., 2004). Evidence is accumulating that the function of the different non-neuronal cells controlling access to the brain (endothelial cells, subpial astrocytes, ependymal cells and choroid plexus epithelial cells) may be co-ordinated with that of parenchymal astrocytes in order to regulate the composition of extracellular fluid in brain (Hertz et al., 2002).

In other parts of the body, such functions as energy metabolism, ion permeability and vascular perfusion are regulated by the autonomic nervous system, utilizing transmitters like noradrenaline and acetylcholine, which have direct effects on, e.g., metabolic activities, ion channels or muscle contraction. It was, for many years, a dogma that the CNS, in contrast to all other organ systems, does not possess an autonomic nervous system. The main reason for this belief was the presence of the blood-brain-barrier, which makes receptors located on the abluminal side of the endothelial cells and on other target cells within the CNS inaccessible for most systemically administered receptor agonists. However, it is now well established that noradrenergic and serotonergic fibers ascend from, respectively, *locus coeruleus* and the *raphe nuclei* in the brain stem and supply cells all over the cerebral and cerebellar cortex as well as in the hippocampus with afferent adrenergic and serotonergic input (reviewed by Hertz et al., 2004). A similar "globally afferent" system of cholinergic neurons extends from the *nucleus basalis magnocellularis* in the basal forebrain. These nuclei are part of a supraspinal system in the brain stem, which also regulates autonomic output in the peripheral nervous system. Several peptides also have transmitter activity; some of these compounds (vasopressin and oxytocin) can after their synthesis in specific hypothalamic nuclei be released as hormones, a release that to a large extent is controlled by adjacent astrocytes (Hatton, 1997; Salm et al., 2004).

Fine Structure

Cell Types

Neurons

The most important general characteristic of nerve cells is their excitability. Excitation triggers "downhill" transport of K^+ and sodium (Na^+) ions along their electrochemical gradients through ion channels (initially Na^+ channels leading to an influx of Na^+ and accompanying depolarization of the negative membrane potential and slightly later, K^+ channels, leading to an efflux of K^+ and accompanying repolarization). In most neurons, these passive fluxes of Na^+ and K^+ constitute the biophysical basis of the action potential, but in Purkinje cells depolarization is partly mediated by calcium (Ca^{2+}) entry (e.g., Llinas and Sugomori, 1980; Pouille et al., 2000). These ionic movements are not energy dependent, whereas the subsequent restoration of the resting low extracellular K^+ concentration and a low intracellular Na^+ concentration requires active transport, catalyzed by the Na^+,K^+-ATPase, and accounting for a sizeable fraction of the brain's energy demand.

Nerve cells consist of a cell body (soma) from which two types of processes extend, the input receiving dendrites and the output conducting axons (Figure 1). In most neurons, the volume and, especially the surface area, of the processes are much larger than those of the cell body. The axons can be extremely long, and they exchange essential substrates with the cell body by axonal transport. The cell bodies often contain very conspicuous granules, the Nissl bodies, which mainly consist of ribonucleic acid, reflecting an intense protein synthesis in nerve cells. The cellular composition and neuronal connections have been especially well studied in the cerebellum, but the general principles of neuronal function are similar in different regions of the CNS.

The entire cellular distribution and neuronal connectivity within the cerebellum is beyond the scope of the present review. However, two simple, and well studied nerve cells in the cerebellum, the excitatory, glutamatergic cerebellar granule cell and the inhibitory, GABAergic Purkinje cell with their dendrites, somata and axons are schematically illustrated in Figure 1. The Purkinje cells are relatively large, compared to the smaller granule cells, and their dendritic network (which for graphical reasons is only partly shown in the Figure) is much more extensive. Neurons communicate chemically one-to-one at synapses by the aid of transmitters released from the presynaptic neuron, and affecting its postsynaptic neuron; however, crosstalk can in some situations occur between adjacent glutamatergic

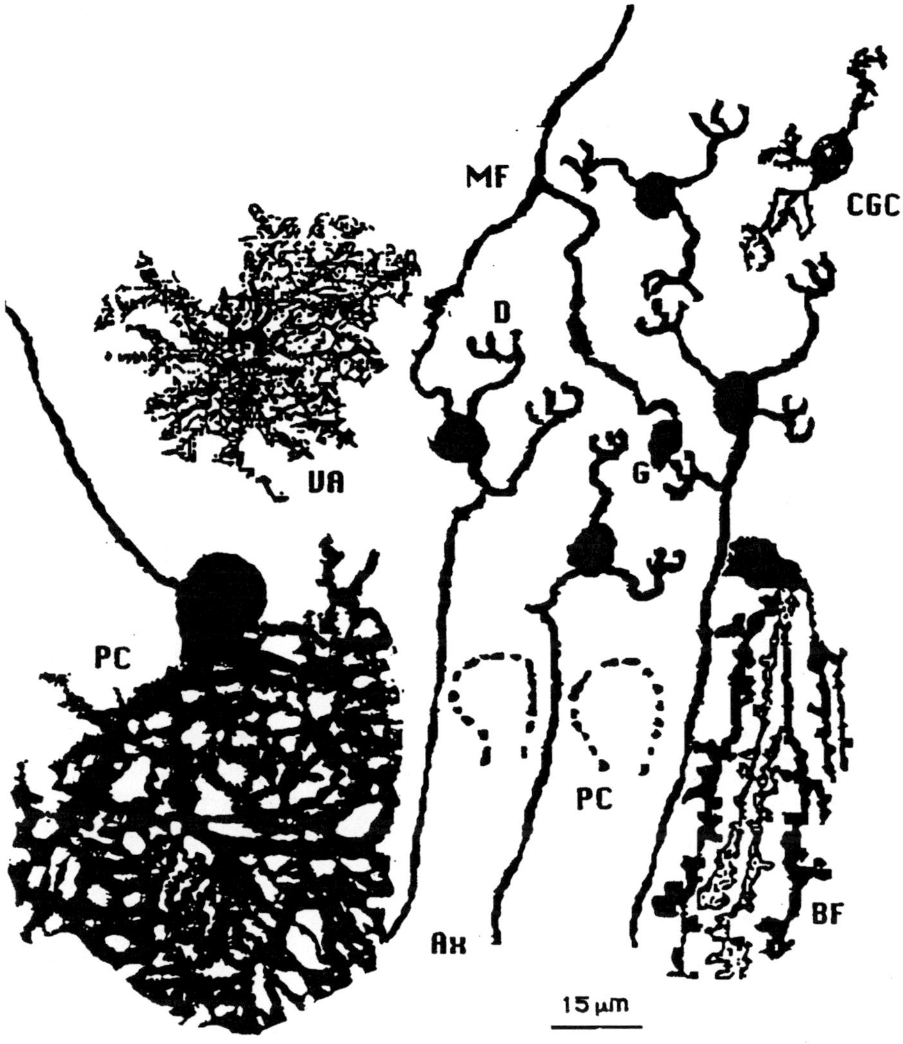

Figure 1. Schematic presentation of two cerebellar types of neurons, cerebellar granule cells (CGC) and Purkinje cells (PC) with surrounding astrocytes. One of the 5 cerebellar granule cells (CGC) is a camera lucida drawing, as are the two astrocytes or parts of astrocytes: a velate astrocyte (VA), and Bergmann fibers (BF) from an epithelial Golgi cell. The arborization of the dendritic network of the Purkinje cell was visualized by intracellular dye injection, but for graphical reasons only a minor part of this network is shown here. The cerebellar granule cells are innervated by glutamatergic mossy fibers (MF) from the brain stem, synapsing at the very end of the granule cell dendrites (D) in cerebellar glomeruli (G). Cerebellar granule cell axons (Ax) extend through the layer of Purkinje cells (PC) to synapse with different cerebellar cell types, including Purkinje cells. The axon of the Purkinje cell returns towards the deep cerebellar nuclei. The Bergmann fibers make close contacts with cerebellar granule cell axons, and the terminal parts of the axons make synaptic connections, which are densely covered by astrocytic processes. The granule cell bodies are enshrouded in veils from velate astrocytes. Composed, and slightly retouched, from several different figures in Palay and Chan-Palay (1974).

synapses and be enhanced by inhibition of glutamate uptake (Isacsson, 1999). Once the axonal action potential reaches the synapse, a depolarization of synaptic terminals is induced and transmitter substance is released, contingent upon a depolarization induced entry of calcium ions (Ca^{2+}) along their concentration gradient and upon adequate supply of transmitter or transmitter precursors. Subsequently, accumulated Ca^{2+} must be removed by active transport out of the cell. The synapses between the cerebellar granule cells and their presynaptic input, the mossy fibers are very conspicuous and morphologically somewhat unconventional. They form the so-called cerebellar glomeruli, each of which consists of a single mossy fiber axon and many granule cell dendrites (Peters et al., 1991). In turn, the axons of the cerebellar granule cells synapse, in morphologically conventional synapses, with their postsynaptic neurons, the Purkinje cells. Most of the synapses occur on specific structures of the dendrites, called spines. An example of such a synapse is shown in Figure 2.

Neurons are often classified according to the transmitter they release during synaptic transmission, e.g., glutamate, γ-aminobutyric acid (GABA), norepinephrine, acetylcholine or a neuropeptide, but the receptor types they express are equally important for determination of neuronal circuitry. Glutamatergic and GABAergic neurons are found in about equal numbers in the mammalian brain cortex and by far outnumber monoaminergic, cholinergic and peptidergic neurons. For all practical purposes, postsynaptic glutamatergic neurotransmission is always depolarizing and excitatory, and GABAergic neurotransmission always hyperpolarizing (due to opening of chloride [Cl^-] channels) and inhibitory. Glutamatergic and GABAergic neurons innervate each other, and glutamatergic innervation of a GABAergic neuron (e.g., stimulation of Purkinje cells by cerebellar granule cells) obviously leads to increased inhibition. Most neurons (including Purkinje cells) receive both excitatory and inhibitory input, and inhibition as well as sudden cessation of inhibitory activity (disinhibition), play a major role in normal CNS function. Effects of monoaminergic transmitters like norepinephrine and of neuropeptides on neuronal excitability are more variable, maybe because they often act on second messenger systems located in different cell types rather than causing a local depolarization or hyperpolarization of a postsynaptic neuron.

Dendrites emerge from the cell body like the branches of a tree (hence their name). The shape of the dendrites vary between different types of neurons, reflecting the distribution of the area(s) from which they receive their afferent input and the specific characteristics of the neuronal type. Only at very few locations are dendrites covered with myelin, and in these places the myelin sheath is thin (Peters et al., 1991). During their course the dendrites extend side branches (Figure 2), which in the case of many dendrites are studded with dendritic spines. The density of the spines often increases with the distance from the cell body, and in some neurons the spines constitute a very large fraction of the total volume of the dendritic tree. In contrast to a high mitochondrial density in the dendrites proper the spines only rarely contain mitochondria. Synapses are often found on spines (Figure 2), but can also be formed with other structures on the postsynaptic neurons. In some synapses, the released transmitters act not only on the postsynaptic neurons but also on "autoreceptors" at the presynaptic neuron, an action, which may result in a reduction of transmitter release.

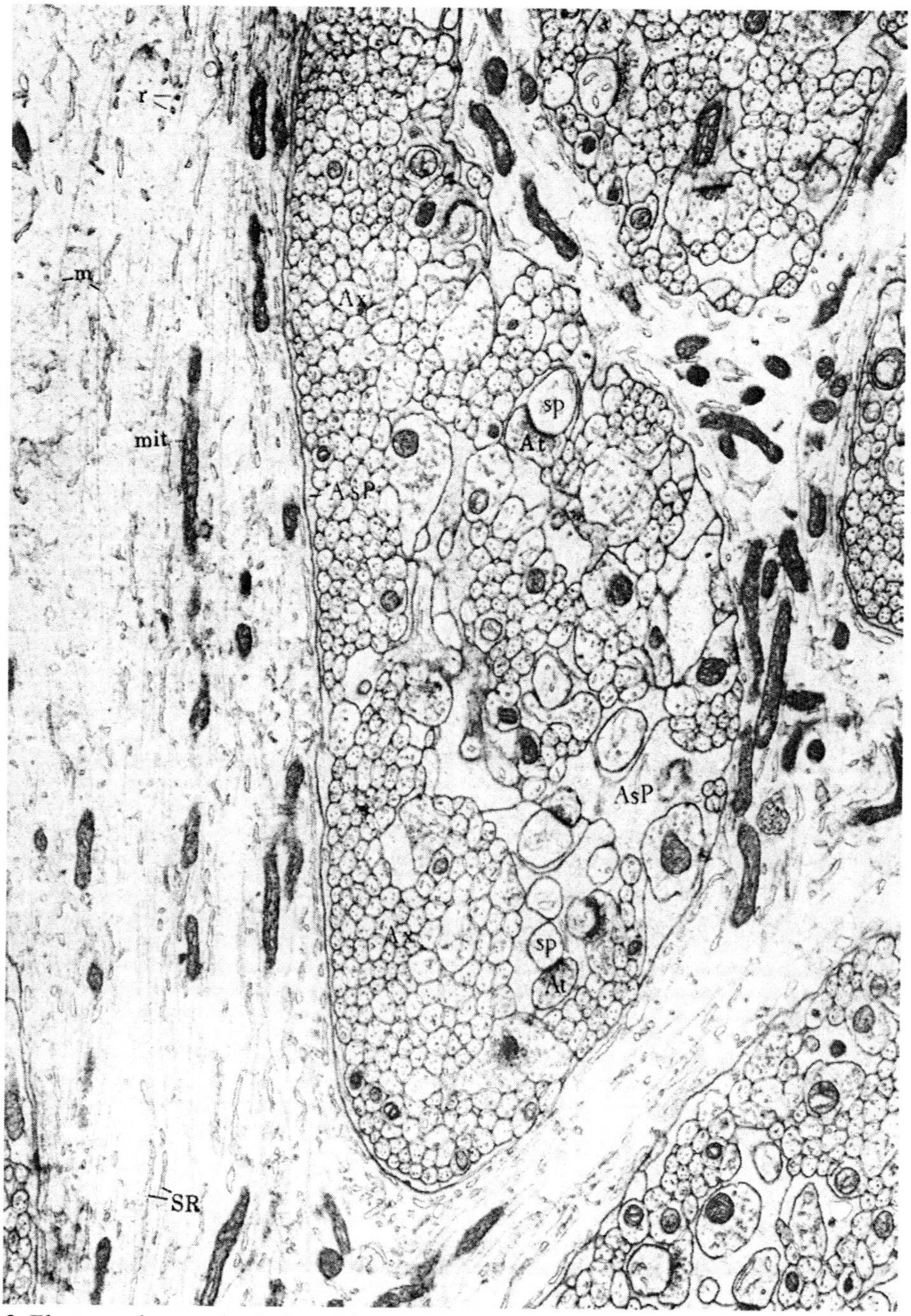

Figure 2. Electron micrograph of a branching dendrite of a Purkinje cell from the cerebellar cortex of a rhesus monkey. A large dendrite ascends on the left side of the field and extends a thinner dendrite, which forms the spiny branchlets occupying the upper right portion. Organelles in the dendrite include mitochondria (mit), microtubules (m), aggregates of ribosomes (r) and an agranular reticulum (SR), With the exception of the branchlets both branches of the dendrite are covered by astrocytic processes (AsP). The neuropil between the two dendrites contain granule cells axons (Ax), intermingled with astrocytic processes (AsP), and forming synapses with the spines (sp) of the Purkinje cell. From Peters et al., 1991.

Synapses mediating excitation, i.e., expressing receptors for excitatory transmitters (generally one of the glutamate receptors) are often located distally at the dendrites, a prime example being the receptors on cerebellar granule cells, which are innervated by the glutamatergic mossy fibers (Figure 1). In contrast, receptors for inhibitory, hyperpolarizing transmitters are often located closer to the cell body, and are thus able to block incoming activity. Inhibitory synapses might even be found on axons, where release of the inhibitory transmitter GABA causes presynaptic inhibition (Miyata and Otsuka, 1975). Most synapses are formed by glutamatergic or GABAergic neurons, and show typical synaptic morphology (Figure 3). Some synapses are terminations of the monoaminergic or cholinergic neurons ascending from the more basal areas of the brain. However, many ascending monoaminergic fibers (e.g., serotonergic, dopaminergic and noradrenergic) show a structural anomaly from other nerve fibers in the CNS, i.e., they do not terminate in true synapses but in "varicosities", from which transmitters reach their target cells (neurons, glial cells, microvessels) by diffusion (see Hertz, 1992; Hertz et al., 2004b). This is an arrangement similar to that in the peripheral part of the autonomic nervous system.

It was previously assumed that the postsynaptic depolarization caused by exposure to an excitatory transmitter is of insufficient magnitude to elicit an action potential but, instead, creates an excitatory postsynaptic potential (EPSP), that spreads electrotonically along the dendrite and the cell body. The passive ion fluxes involved in the EPSP's may be relatively modest as long as depolarization is not strong enough to evoke an action potential, and eventually some EPSP's may be extinguished due to the decremental character of electrotonic conduction. Other EPSP's reach the axonal hillock (at the base of the axon), where spatial and/or temporal summation of EPSP's, facilitated by a high local density of Na^+ channels, evokes an action potential. However, it is now well established that many dendrites display action potentials, which can either move towards the cell body or represent a backpropagation of the action potential into the dendrites (Johnston et al., 1996; Häusser et al., 2000; Vetter et al., 2001). Propagation of dendritic action potentials will create a very considerable energy demand for restoration of resting levels of intra- and extracellular K^+ and Na^+ (and in the case of Purkinje cell dendrites also Ca^{2+}) by active cellular re-accumulation and re-extrusion following the action potential, because of the lack of insulation by myelin in the big majority of dendrites. Accordingly action potential propagation must be contiguous. The density of Na^+ and K^+ channels are often high on dendrites and even Purkinje cell dendrites, which have a low density of Na^+ channels (compensated for by Ca^{2+} channels), have a high density of voltage-dependent K^+ channels, allowing large release of K^+ during the repolarizing phase of the membrane potential (Martina et al., 2003).

A single axon exits from the cone-shaped axonal hillock. In contrast to dendrites, many axons, including all axons extending between different regions (Peters et al., 1991), are covered with myelin (generated in an interaction between axons and non-neuronal satellite cells (oligodendrocytes in the CNS and Schwann cells in the periphery). The myelin sheath isolates the axon from its neighbours except at the nodes of Ranvier (the boundary zones between adjacent satellite cells) and provides the possibility for saltatory conduction from one node of Ranvier to the next, rather than contiguous conduction of an action potential. This arrangement confines most ion movements accompanying excitation to the nodes of Ranvier and thereby reduces the amount of energy required for subsequent re-establishment of resting distribution between intracellular and extracellular ions. Therefore, conduction of action potentials along axons might be much less energy requiring than conduction of action potentials along dendrites. At the same time, myelination increases the speed of impulse conduction.

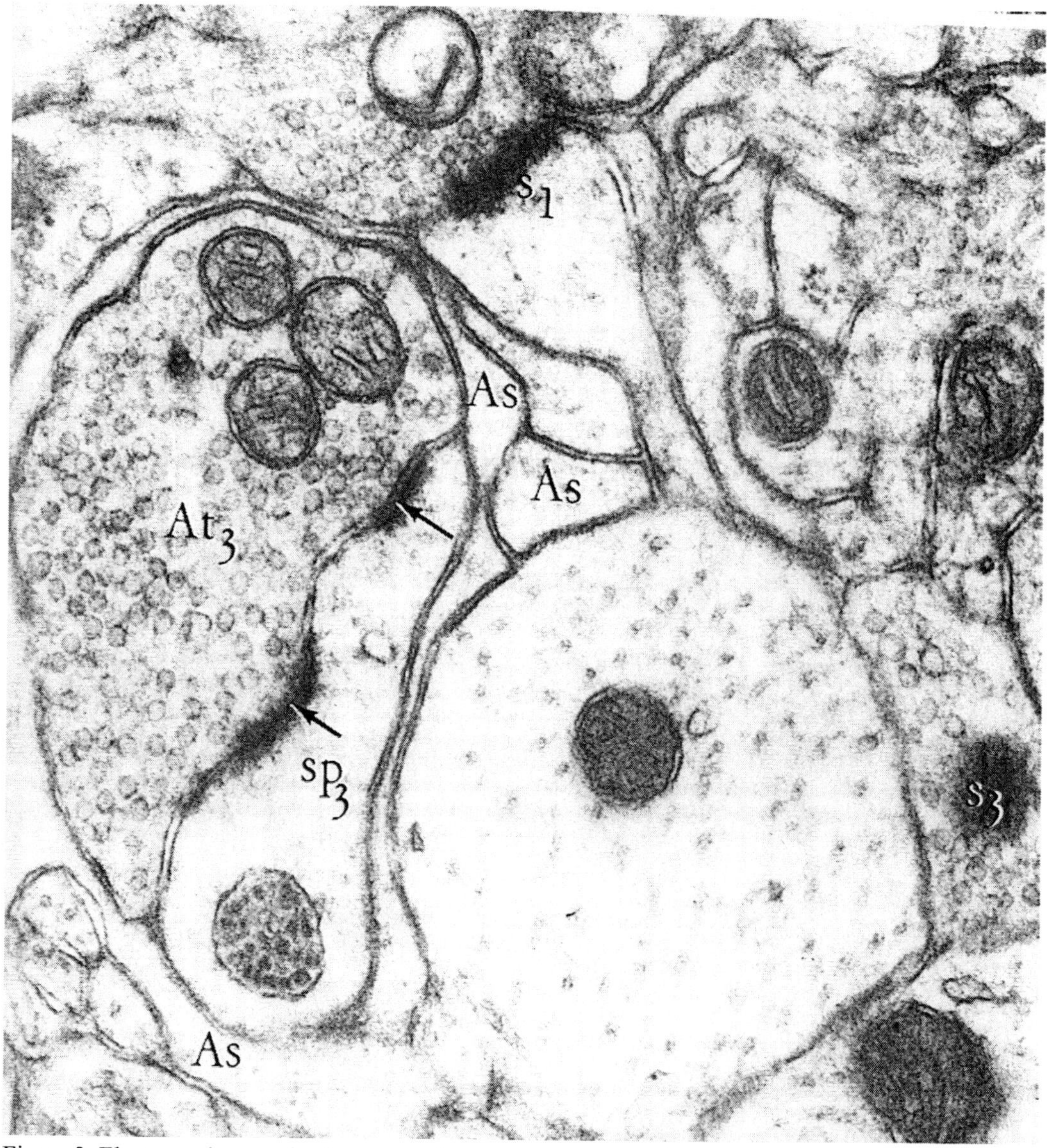

Figure 3. Electron micrograph of synapses from rat visual cortex. An axon terminal (AT_3) containing synaptic vesicles and 3 mitochondria synapses with a large spine (sp_3), and the synaptic cleft is enclosed by astrocytic processes (As). Further up in the picture another spine (sp1) forms another synapse, and at the lower right corner a third synapic junction (s_3) has been sectioned parallel to the plane of the participating membranes. From Peters et al., 1991.

Glutamatergic and GABAergic neurons generally either project over short distances locally (interneurons) or extend across longer distances (e.g., as association fibers between different regions in the cerebral cortex and hippocampus, as thalamocortical and corticothalamic fibers, or as corticospinal fibers), and both glutamate and GABA are formed locally from glucose as described later in this chapter. Differences exist between different neuronal cell types with respect to the presence of the machinery for production of specific transmitters. Thus, glutamate decarboxylase (GAD), the enzyme catalyzing production of GABA, is restricted to GABAergic neurons. It is unknown to what extent there might also be more widespread metabolic differences between neurons releasing different types of transmitters and/or between neurons expressing receptors for different types of transmitters. However, there appears to be different modes of pH homeostasis in the two cell types, as will be discussed later. Some individual neurons release more than one type of transmitter, for example a monoamine transmitter together with a peptide transmitter, which may "modify" the action of the monoamine.

Glial Cells

These cells were originally named neuroglia, i.e., nerve "glue" and were in the not too distant past regarded as being of little functional importance. This point of view has been challenged during recent decades, and glial cells should nowadays be considered as equal partners of neurons in CNS function. Glial cells are traditionally divided into protoplasmic and fibrous astrocytes (on the basis of cell morphology), oligodendrocytes and microglia. In large vertebrates, astrocytes and oligodendrocytes together outnumber nerve cells by a factor of 5-10, and in many gray matter areas astrocytes are more abundant than oligodendrocytes. Oligodendrocytes are the myelinating cells of the CNS, although non-myelinating oligodendrocytes also exist (Szuchet and Seeger, 2004). Most oligodendrocytes are found in white matter, but these cells are also found close to neurons in gray matter. Microglia primarily perform phagocytic and immunological roles and play a large role in many infectious and neurodegenerative diseases (Rock et al., 2004). Astrocytes are the glial cells that will receive most attention in this chapter. Much information about astrocytes originates from experiments using cultured cells. In order to deduct meaningful information from cells grown and differentiating *in vitro*, it is essential as far as possible to secure that the cultured cells express the characteristics of their *in vivo* counterparts (Kimelberg et al., 2000). During recent years many conclusions drawn from tissue culture experiments have been confirmed in the brain in vivo (e.g., the role of astrocytes in provision of glutamate precursors for neurons; a high rate of oxidative metabolism; intense active uptake of glutamate and of K^+ from the extracellular space).

Astrocytes account for 15-25 % of the total cellular volume in brain cortex, perhaps somewhat more in man (Bass et al., 1991; Pope, 1978; Williams et al., 1980; Wolff and Chao, 2004), with regional and laminar differences. The predominant astrocytes in gray matter are the protoplasmic astrocytes, whereas fibrous astrocytes are found mainly in white matter. Astrocytes are crucial in regulation of transmitter, ion and metabolic substrate concentrations in the neuronal-glial microenvironment, and they function in collaboration with neurons. They also participate in a non-synaptic transcellular signaling system, by reacting to neuronal release of glutamate (or to certain other stimuli) with an increase in free cytosolic Ca^{2+} concentration ($[Ca^{2+}]_i$), which spreads in a waveform across an astrocytic syncytium and eventually may excite the neurons it encounters by release of astrocytic glutamate or inhibit other neurons by release of adenosine (see "Calcium Homeostasis"). The syncytium is made up by astrocytes coupled by gap junctions, which are permeable for second messengers such as inositol trisphosphate (IP_3), K^+ and some dyes. However, gap junctions are also formed between astrocytes and oligodendrocytes and between these cells and ependyma cells, giving rise to the concept of a "panglial syncytium" (Rash et al., 1997).

Astrocytes originally received their name on account of a sometimes stellate appearance, but nowadays the presence of a specific protein, glial fibrillary acidic protein (GFAP), is conventionally used as an indication of the astrocytic character of a CNS cell. Ironically, the precise function of this protein is not known. It is probably of importance for the maintenance of the cytoskeleton and it is metabolically active, capable of undergoing rapid, transmitter induced phosphorylation (Wofchuk and Rodnight, 1995).

Astrocytes are located at strategic positions, e.g., around synapses (Peters et al., 1991; Derouiche et al., 2002), where they virtually enclose the synaptic cleft (Figure 3), giving rise to the expression "the tripartite synapse" (Volterra et al., 2002); they also ensheath the larger part of many dendrites, including those of Purkinje cells (Figure 2). Bergmann fibers are longitudinal astrocytic processes following the axons of cerebellar granule cells, with which they interact (Figure 1). Most of the surface of brain capillaries are covered by astrocytic processes; astrocytes surround larger vessels with astrocytic endfeet (Wolff and Chao, 2004) and they send perinodal processes to nodes of Ranvier (Butt et al., 1994). The *glia limitans* on the surface of the brain is made up by subpial astrocyte processes.

As can be seen in Figure 4, most astrocytes extend a large number of finely branched processes (filopodia) and thin veils from their primary processes (Derouiche and Frotcher, 2001; Bushong et al., 2002; Derouiche, 2004), and astrocytes have an extremely large surface/volume ratio (Wolff and Chao, 2004). Glycogen stores, which are found in astrocytes but, with a few exceptions, normally not in neurons, have been demonstrated in even the thinnest processes (Peters et al., 1991). The peripheral astrocytic processes are of highly irregular shape, they form a special compartment, expressing proteins not found in other locations of the astrocytes (Derouiche and Frotcher, 2001; Derouiche 2004), their average length is ~3 μm, their thickness 0.2 μm, and they account for 80% of the astrocytic surface area (Wolff and Chao, 2004). The very complex "appendages" extended from Bergmann fibers (Figure 5) have an extremely high surface-to-volume ratio (Grosche et al., 1999).

The density of astrocytes in a given brain region is important for the ability of individual astrocytes to interact with each other in the astrocytic syncytium. In layers 5 and 6 of the parieto-occipital part of the neocortex of the adult rat the astrocytic density is relatively low, so that the mean distance of evenly distributed astrocyte cell bodies is ~50 μm. However, there are cell clusters in which the average distance between the cell bodies is reduced to ~30 μm. This increases the mean radial overlap between processes from adjacent astrocytes substantially and therefore greatly enhances the possibilities for intercellular interactions (Wolff and Chao, 2004). As illustrated in Figure 4 it is only the peripheral astrocyte processes (filopodia), which communicate with other astrocytes through gap junctions (Bushong et al., 2002; Scemes and Spray, 2004). Gap junctions can also be formed between different

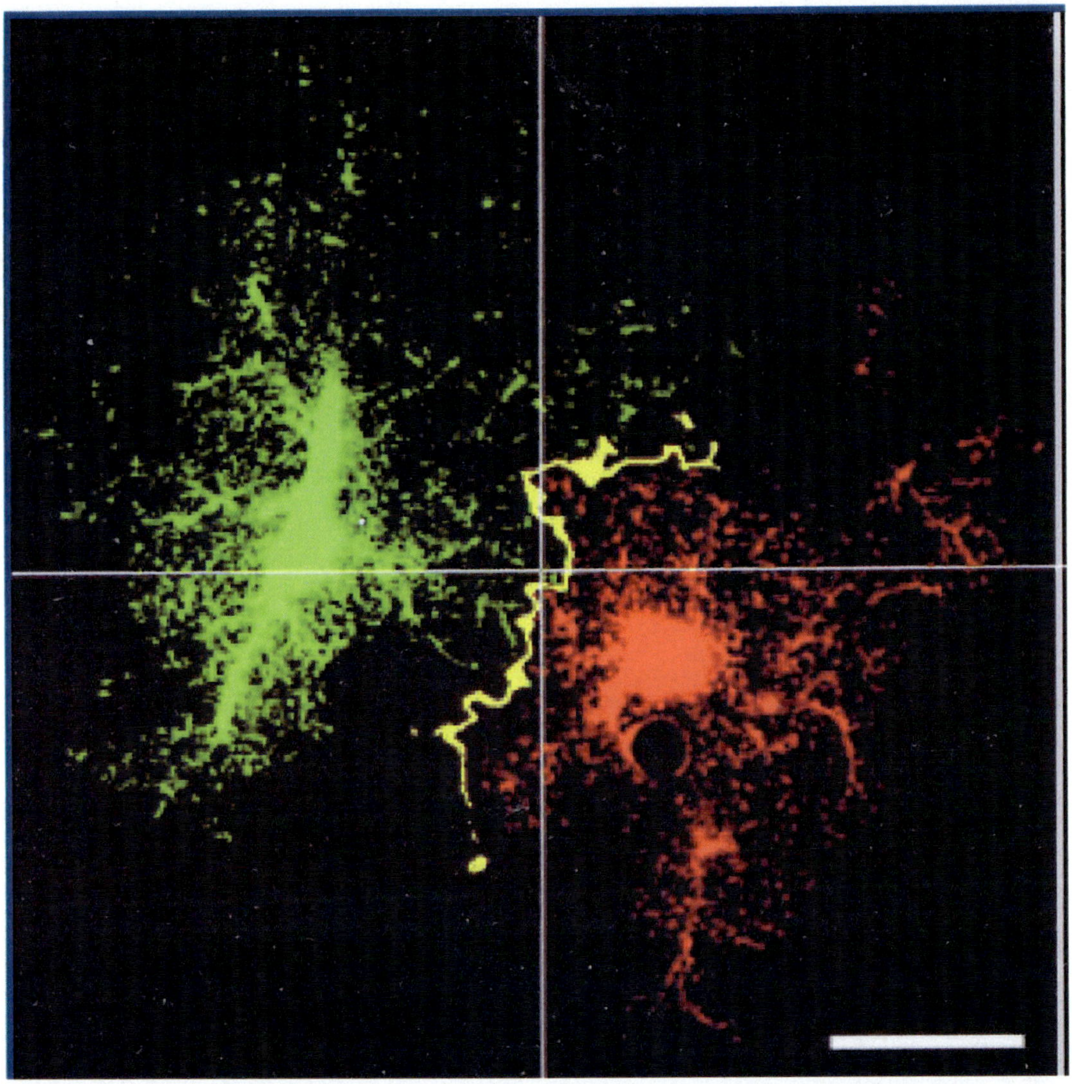

Figure 4. Two protoplasmic rat astrocytes in the CA 1 *stratum radiatum* have been filled with, respectively, a green and a red dye. Only pixels containing both red and green correspond to the overlapping astrocytic processes and have been pseudocolored in bright yellow, which forms a very narrow band between the two cells. Scale bar: 20 μm. From Bushong et al., 2002.

processes on the same astrocyte (or oligodendrocyte). Similar gap junctions between different parts of a single oligodendrocyte may be important for the nutrition of the innermost layers of the developing myelin sheath.

In contrast to neurons, astrocytes are not excitable cells (Barres et al., 1990; Sontheimer, 1994). This difference has the functionally important consequence that CNS excitation will lead to entry of Na^+ into neurons but not into astrocytes. Nevertheless, astrocytes resemble neurons in expressing receptors for a multitude of transmitters (Hösli and Hösli, 2000;

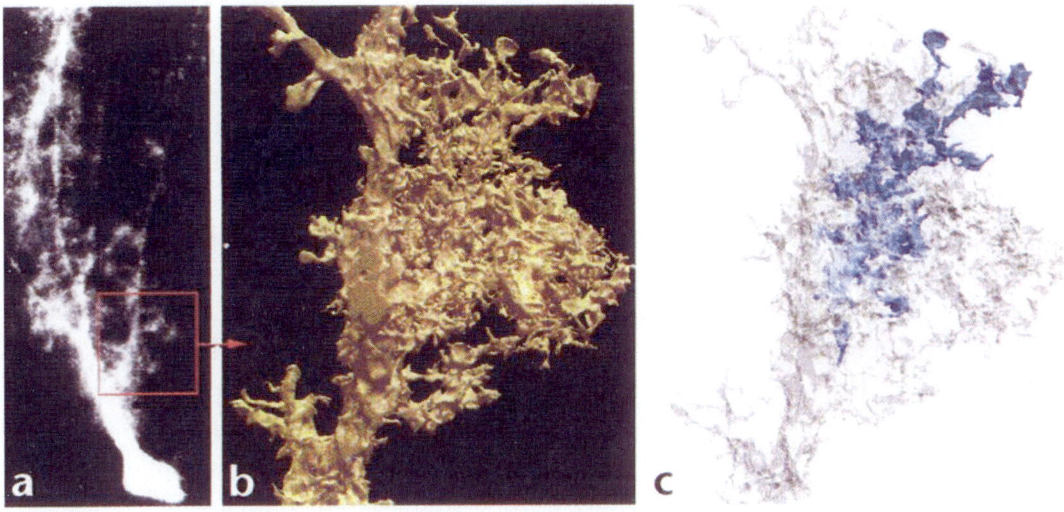

Figure 5. An astrocytic appendage extends from a dye-injected Bergmann fiber (a); the red square (20 μm x 20 μm) indicates a section that was reconstructed from electron micrographs of serial ultrathin sections to show one of the lateral appendages, with other side branches omitted for clarity (b); one of these side branches or "glial microdomains" is shown in blue, extending from the appendage in (c). From Grosche et al., 1999.

Hansson and Rönnbäck, 2004). It came as a surprise to many neuroscientists that astrocytes (and oligodendrocytes) displayed such "neuronal characteristics". However, it should be kept in mind that receptor expression is not a neuronal prerogative but shared by a multitude of cell types, e.g., muscle cells and epithelial cells, in which receptor activation regulates functions like energy metabolism and ion transport. There is a large diversity between individual astrocytes with respect to receptor expression. Certain receptors are expressed on cerebral but, for example, not on spinal astrocytes, or on striatal but not on cerebrocortical astrocytes. Adrenergic, and especially ß-adrenergic, receptors seem to be expressed on virtually all cerebral astrocytes, whereas other receptors, e.g., serotonergic and certain adrenergic subtypes (such as the α_2-adrenergic receptor) are found only on subsets of astrocytes (Nilsson, et al., 1991; Hertz et al., 2004).

Like neurons, astrocytes constitute a morphologically and functionally heterogeneous cell population (Zhou and Kimelberg, 2000; Wallraff et al., 2004), displaying not only pronounced regional differences, but also differences between individual astrocytes within each region. A detailed and elegant morphological study was performed of glial cells in the cerebellum by Chan-Palay and Palay (1974). In the cerebellar glomeruli, where cerebellar granule cells receive their afferent input from the glutamatergic mossy fibers, there are only few astrocytic processes. However, outside the glomeruli granule cell dendrites are intimately surrounded and covered by astrocytic processes. The cell bodies of the granule cells are almost completely ensheathed, and separated, by astrocytic veils, which extend from "velate" astrocytes (Figure 1) and form honeycomb-like compartments, in which individual cell bodies or groups of cell bodies are lodged. The Bergmann fibers are studded with lateral protrusions ("domains"), where they interact with these neurons (Fig. 5, Grossche et al.,

1999). Dendrites of Purkinje cells are also covered by a thin astrocytic sheath except at the very end (Figure 2).

The barrier systems surrounding individual cells can be functionally extremely important, both by enabling neuronal-astrocytic interactions, e.g., in K^+ homeostasis (see below) and by insulating neuronal constituents from the extracellular space. It has thus been demonstrated, that at least some types of glutamatergic neurons in the isolated and incubated retina are not directly accessible for glutamine or glutamate in the incubation medium, because they are completety invested with a glial (Müller cell) sheath (Robinson and Dreher, 1990). These neurons must accumulate glutamate, needed for neurotransmission, *via* non-neuronal cells, converting glutamate to glutamine, which can then be transferred to the neurons and hydrolyzed to glutamate as described later.

Neuropil

After the volumes occupied by nerve cell and glial cell bodies and their easily recognizable, larger processes have been added together, a major fraction of the tissue volume in gray matter remains to be accounted for. The main part of this fraction is made up by the so-called neuropil, seen between the larger dendritic branches shown in Figure 2. The neuropil is composed of an intricately interwoven network of minute neuronal and glial processes. Although the distant (apical) parts of dendrites, and the associated spines, may constitute the largest fraction of the neuropil, both nerve endings and glial cell processes account for a considerable volume. The neuropil is the part of the tissue where virtually all functional interactions occur between individual neurons, between neurons and glial cells and between individual glial cells. Synaptic profiles are abundant in the neuropil and this is the location where most gap junction-mediated astrocytic contacts occur. Since dendrites are not covered by myelin, interactions may be especially pronounced between glial cells and dendritic elements, separated by an extremely narrow interstitial space. It is a reflection of the immense surface area and the intensity of the active, energy requiring ion movements required for re-establishment of resting intra- and extracellular concentrations of Na^+ and K^+ ions after neuronal excitation, that it is the energy metabolism in the neuropil, not in nerve cell bodies, which is increased during functional activity (Sokoloff,1992).

Interstitial Space

At one time it was believed that there is no true interstitial or extracellular space in the CNS, but electron microscopic studies by van Harreveld showed that the apparent absence of an extracellular space in the CNS was an artifact due to *post mortem* cell swelling (e.g., Van Harreveld and Steiner, 1970). It is now generally recognized that an extracellular space of ~15 % is normally present in CNS tissue (Peters et al., 1991). Pronounced cell swelling, which may occur as a result of cell damage, leads to a drastic reduction of its magnitude, but minor changes are evoked even by normal CNS function (Sykova 1992; Sykova and Chvatal, 1993; Hertz et al., 2002). The interstitial space is shared by neurons and glial cells, and it is of critical importance as a communication channel between cells of identical or different types during functional activity in the brain. As described above, this space is isolated from the rest of the body by barriers, and it is to a large extent localized in the narrow extracellular clefts penetrating the neuropil. At many locations, the extracellular space is compartmentalized by astrocytic veils, isolating groups of cells. An elegant study in the hypothalamic supraoptic nucleus (SON), where the extent of neuronal investiture with astrocytic membranes is altered during lactation (Salm et al., 2004) has shown that morphological remodeling greatly reduces coverage of SON neurons by astrocytes and increases diffusion in the tissue; this enhances heterosynaptic depression of GABAergic transmission by glutamate diffusing from its release site to adjacent GABAergic terminals where it stimulates presynaptic metabotropic glutamate receptors that inhibit GABA release (Piet et al., 2004).

The importance of alterations in the composition of the extracellular neuronal-glial microenvironment has only been realized during the last 30 years. Aided by newly developed microelectrodes that can be inserted into the extracellular space, it was shown that extracellular K^+ rises dramatically during functional activity or pathological conditions: a few millimolar (e.g., from 3 to 4 mM) during normal nerve cell activity, to 6-8 mM during intense afferent stimulation (Figure 6), to a "ceiling level" of 10-12 mM during seizures, and to exceedingly high levels (more than 50 mM) during energy failure and 'spreading depression' (reviewed by Walz and Hertz, 1983; Sykova, 1992). In most other tissues such alterations in extracellular ion concentrations would rapidly be reduced or abolished by diffusion into the vascular system. This does not occur in the CNS due to the presence of the blood-brain-barrier. Therefore, the extracellular concentrations of K^+ and other neuroactive compounds are regulated by release and subsequent uptake into adjacent cells.

Ion Homeostasis

Potassium and Sodium Homeostasis

Active K^+ Uptake

It is of crucial importance for CNS function and metabolism that the increases of K^+ in the interstitial space during neuronal function and pathophysiological conditions are rapidly reversed, whereas correction of the corresponding intracellular loss of K^+ is less urgent. One reason for this is the low extracellular concentration of K^+ and the small volume of the extracellular space, compared to the intracellular water space, rendering the relative alteration in ion concentrations especially marked for extracellular K^+. The alteration of the K^+ gradient across the cell membrane affects cellular membrane potentials, which are largely determined by the ratio between extra- and intracellular K^+ concentrations. Although the effect on membrane potential is more pronounced in astrocytes than in neurons (because neurons are not very good K^+ electrodes at low K^+ concentrations), the effects of minor alterations in extracellular K^+ concentration on neuronal function are more pronounced and widespread

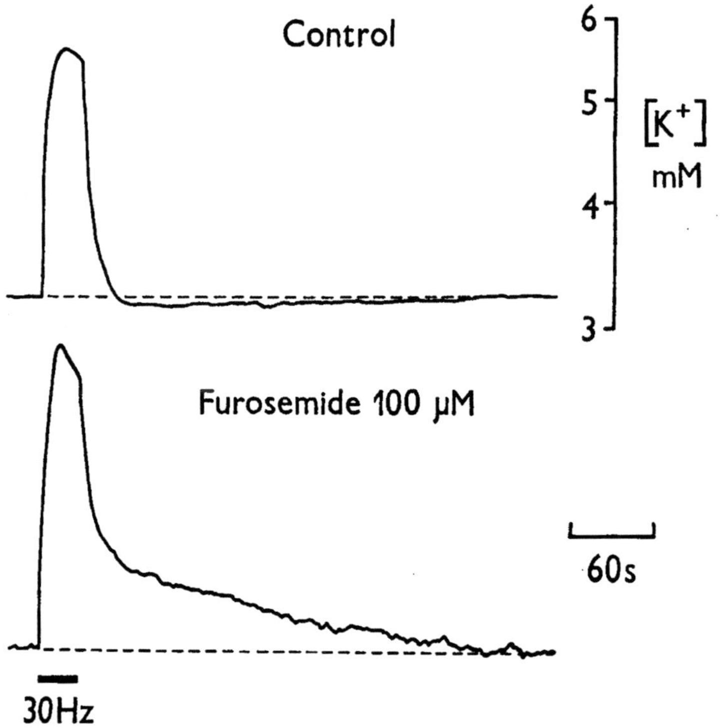

Figure 6. Increase in extracellular K^+ concentration and subsequent decline due to K^+ removal in the frog spinal cord after sensory stimulation under control conditions (top) and in the presence of 100 μM furosemide (bottom). Note the slowing of the reduction of extracellular K^+ and the abrogation of the "undershoot" in the presence of furosemide. From Hertz et al. (1996).

than might be expected on the basis of a relatively small neuronal depolarization; slightly elevated K^+ concentrations may, for example, influence pathway selection at bifurcations or lead to conductance block (Walz and Hertz, 1983).

An energy requiring subsequent restoration of the previous balance between intra- and extracellular concentrations of K^+ is therefore of critical importance for regulation of normal signaling activities, and stimulation of active ion transport accounts for a major fraction of the total energy demand in the CNS and for most of its increase during increased activity. The finding that the increase in energy metabolism resulting from excitation by afferent stimulation occurs in the neuropil (Sokoloff 1992) reflects that this is where most of the active transport of K^+ and Na^+ occurs, reflecting the large surface area of the neuronal and glial processes constituting the neuropil. Although it a priori would appear logical that K^+ released from neurons during their excitation was re-accumulated in neurons, an additional, quantitatively very significant and highly energy-requiring uptake in astrocytes was suggested 40 years ago (Hertz, 1965). Such an uptake has now been confirmed (Walz and

Wuttke, 1999; Walz 2004), and it must secondarily be followed by release from astrocytes and reuptake in neurons.

It has been known since the discovery of the Na^+,K^+-ATPase by Skou in the late 1950's (Skou, 1957, 1998) that this enzyme is responsible for active transport of Na^+ and K^+ across cell membranes fueled by conversion of ATP to ADP. In accordance with classical concepts of metabolic regulation the elevation of ADP and decrease in ATP trigger an increase in energy metabolism (see, e.g., Hertz and Dienel, 2002). Na^+,K^+-ATPase activity is present in both neurons and astrocytes. In neurons, an excitation-induced increase in intracellular Na^+ concentration stimulates the Na^+, K^+-ATPase at its intracellular Na^+-sensitive site, leading to Na^+ extrusion and K^+ accumulation and to an increase in oxidative metabolism, but the affinity of the neuronal Na^+, K^+-ATPase for extracellular K^+ is so high that it is not further stimulated when the extracellular K^+ concentration increases above its normal level (Grisar et al., 1979; Hajek et al., 1996).

In astrocytes no stimulus-induced increase occurs in intracellular Na^+ concentration, because they are non-exitable. However the astrocytic Na^+, K^+-ATPase is stimulated at its extracellular K^+ sensitive site by an increase in extracellular K^+ concentration above its resting level (Figure 7). This stimulation is reflected by an increase in glucose phosphorylation, which is abolished by ouabain, an inhibitor of the Na^+, K^+-ATPase (Peng et al., 1994, 1996). The difference between the effect of the extracellular K^+ concentration on Na^+, K^+-ATPase activity in neurons and astrocytes is shown by a steep increase in intracellular K^+ content in cultured astrocytes, but not in cultured neurons, when the extracellular K^+ concentration is raised from 4 to 10 mM (White et al., 1992), an increase, which is too small to lead to neuronal excitation. Thus, after neuronal excitation the resulting increase in interstitial K^+ concentration must lead to an enhanced uptake of K^+ into astrocytes but not into neurons, at the same time as the excitation-induced increase in intracellular Na^+ concentration stimulates Na^+,K^+-ATPase activity and thus also K^+ uptake in neurons. At an extracellular K^+ concentration of ~6 mM *active* uptake of K^+ in cultured astrocytes amounts to ~400 nmol/min per mg protein or, based on a water content of 4-5 µl/mg protein (Chen et al., 1992) ~80 µmol/ml cell water (80mM) per min (Hertz, 1986, 1990); the uptake rate into neurons appears to be somewhat lower, although no less than one half of this value. These values are fully compatible wuth the measured clearance rate of extracellular K^+ after the neuronal stimulation shown in Figure 6, since concentration changes in the extracellular space, which is approximately 6-8 times smaller than the intracellular water space, must be correspondingly more pronounced. The importance of Na^+, K^+-ATPase activity for K^+ clearance from the extracellular space in brain slices is indicated by the observation that ouabain reduces the rate at which extracellular K^+ falls after induction of seizure-like activity (Xiong and Stringer, 2000). Strophantidin has a similar effect in the intact rat optic nerve (Ransom et al., 2000).

Many cell types, including astrocytes in primary cultures, express a cotransport system, accumulating or extruding in conjunction one K^+, one Na^+ and two Cl^- and inhibited by furosemide, ethacrynic acid or bumetanide (Walz and Hertz, 1984; Kimelberg and Frangakis, 1985; Tas et al., 1987). In astrocytes the cotransport system moves K^+, Na^+ and Cl^- into the cells, energetically driven by the Na^+ gradient, which allows Na^+ to enter passively and to

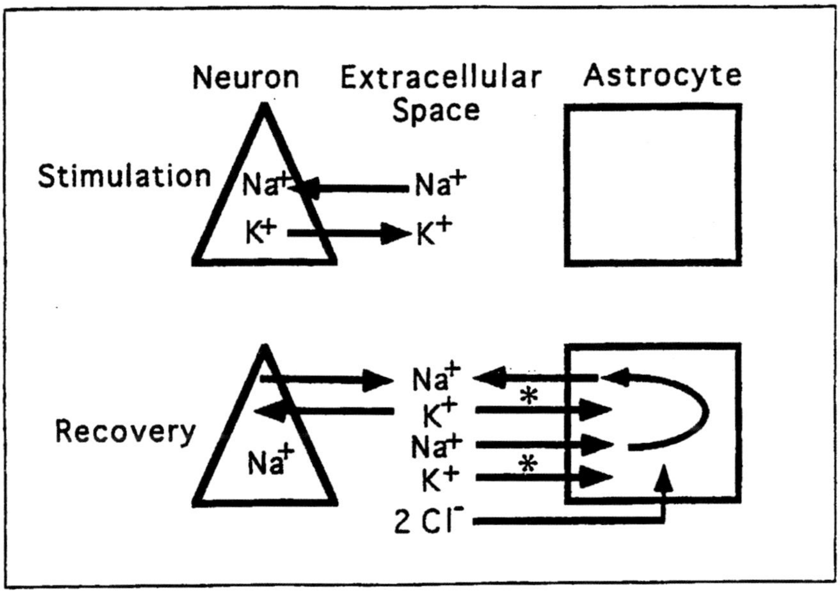

Figure 7. Diagram of sodium (Na^+), potassium (K^+) and chloride (Cl^-) ion fluxes during neuronal stimulation (passive Na^+ uptake and K^+ release in neurons; no ion movements in astrocytes) and subsequent recovery by two active transport mechanisms, the Na^+,K^+-ATPase, exchanging extracellular K^+ with intracellular Na^+, and a cotransport system, transporting in conjunction 1 Na^+, 1 K^+ and 2 Cl^- into the cell. In astrocytes, the Na^+,K^+-ATPase is stimulated at its extracellular K^+-sensitive site by the increase in extracellular K^+ concentration resulting from neuronal excitation (indicated by an asterisk) and in neurons at its intracellular Na^+-sensitive site by the stimulus-induced increase in intracellular Na^+ concentration. Uptake of Na^+, K^+ and Cl^- by the cotransport system is energetically driven by the Na^+ gradient. The cotransport system is active in astrocytes, and its operation stmimulated by entry of calcium ions (Ca^{2+}) through voltage-sensitive Ca^{2+} channels by K^+-mediated depolarization (indicated by an asterisk). The joint operation of the Na^+,K^+-ATPase and the cotransport system leads to a reaccumulation of K^+, together with Cl^-, but without any concomitant uptake of Na^+, because accumulated Na^+ after its cellular accumulation is extruded by stimulation of the intracellular, Na^+-sensitive site of the astrocytic Na^+,K^+-ATPase. If intracellular Na^+ is still elevated in neurons (as a result of the Na^+ entry during the previous stimulation) after complete clearance of excess K^+ from the extracellular space, neuronal Na^+,K^+-ATPase activity will remain stimulated and lead to an "undershoot" in the extracellular concentration of K^+. The abrogation of the "undershoot" in the presence of furosemide (Figure 6) is consistent with this concept.

bring with it the two other ion species. Once inside the cell, the accumulated Na^+ stimulates Na^+, K^+-ATPase activity at the intracellular, sodium sensitive site of the enzyme (Walz and Hinks, 1986), leading to an active, energy dependent extrusion of Na^+ and a further uptake of K^+ (Figure 7). In the CNS, the functional implication of the joint operation of the cotransport system and the Na^+, K^+-ATPase is a cellular uptake of K^+ together with Cl^-, and a corresponding removal of interstitial K^+ (and Cl^-) without any net movement of Na^+.

In cultured astrocytes and rat optic nerve the cotransport system is stimulated by elevated extracellular K^+ concentrations (Walz and Hertz, 1984; MacVicar et al., 2002), secondary to an increase in $[Ca^{2+}]_i$, mediated by voltage dependent L-channels for Ca^{2+} (Su et al., 2000). Thus, this system is activated by extracellular K^+ concentrations high enough to cause an opening of L-channels (Hertz et al., 2004). The importance of the cotransporter for clearance of excess extracellular K^+ after neuronal excitation in intact brain tissue is indicated by the finding that the extracellular K^+ concentration in slices of dentate gyrus reaches a higher level during induced epileptic activity when the cotransporter is inhibited by furosemide (Xiong and Stringer, 2000) and that clearance of a stimulus-induced increase in extracellular K^+ concentration is delayed by furosemide (Figure 6). Ethacrynic acid, another cotransporter inhibitor, also inhibits a Ca^{2+}-dependent increase in oxygen uptake in brain slices evoked by K^+ concentrations high enough to cause an increase in $[Ca^{2+}]_i$ (Hertz et al., 2004); such K^+ concentrations also stimulate oxidative metabolism in astrocytes (Hertz, 1966; Hertz et al., 1973; Hertz and Hertz, 1979). It is consistent with the concept that activation of the cotransporter system stimulates oxidative metabolism that other interventions increasing intracellular Na^+ concentration in astrocytes, e.g., exposure to glutamate which is accumulated together with Na^+, or to the Na^+/H^+ exchange enhancer monensin stimulate oxidative metabolism in astrocytes (Eriksson et al., 1995; Peng et al., 2001).

Due to the intensity of the K^+ stimulated K^+ transport mechanisms operating across the astrocytic membrane, extracellular K^+ tends to return to its normal resting level, at least after massive neuronal excitation, before intracellular, and thus intraneuronal, Na^+ concentration has been normalized (Herreras and Somjen 1993). Continuous stimulation of neuronal Na^+, K^+-ATPase activity after this point by elevated intracellular Na^+ will lead not only to Na^+ extrusion from neurons but also to a coupled active uptake of K^+ from the extracellular space into neurons (Figure 7), i.e., to an "undershoot" of extracellular K^+ concentration. Such a mechanisms links at least one mechanism involved in decreasing the extracellular post-stimulus K^+ concentration below its resting level to the intensity of the astrocytic uptake of K^+. It is in agreement with such a mechanism that administration of furosemide not only reduces the rate of K^+ removal from the extracellular space after excitation but also abolishes the "undershoot" of extracellular K^+ (Figure 6).

The importance of astrocytes for removal of excess extracellular K^+ in neocortex and hippocampus in vivo has been demonstrated in animals injected with the astrocyte-specific toxins, fluoroacetate or α-aminoadipate (Largo et al., 1996; Xiong and Stringer, 1999; Lian and Stringer, 2004). However, these experiments do not prove that extracellular K^+ is *actively* accumulated by astrocytes, since passive redistribution or uptake mechanisms have also been described (see below). The importance of active transport for removal of K^+ from the extracellular space has been demonstrated in brain slices or brain tissue treated with ouabain, but the latter experiments are unable to distinguish between uptake in neurons and astrocytes (Balestrino et al., 1999; Xiong and Stringer, 2000; D'Ambrosio et al., 2002). The most clear-cut evidence in favor of an active uptake of K^+ in astrocytes in the brain in vivo is illustrated in Figure 8, which shows the intracellular K^+ concentration in hippocampal astrocytes in gliotic tissue, measured by K^+-sensitive microelectrodes (Walz and Wuttke, 1999; Walz, 2000). Even in the presence of an inhibitor of Cl^- channels (preventing channel-mediated uptake of K^+ plus Cl^-), an increase in the extracellular K^+ concentration to 10 mM causes an increase in intracellular K^+ of almost 20 mM. This increase is completely abolished in the presence of ouabain.

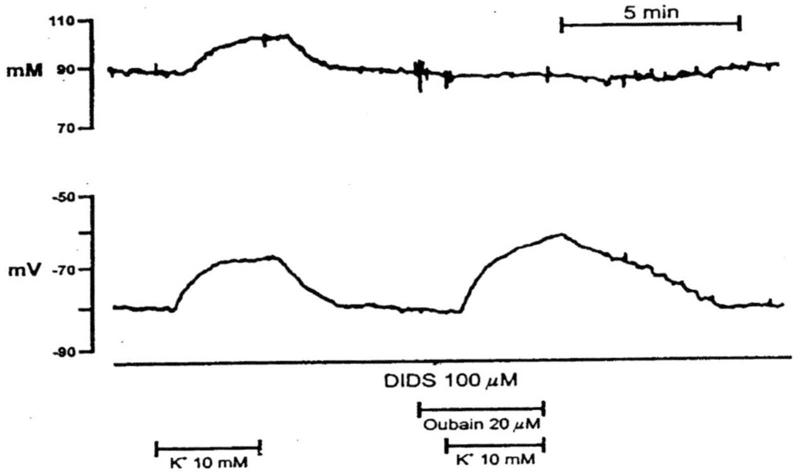

Figure 8. Registration of intracellular K^+ concentration (measured with a K^+-sensitive electrode) and membrane potential in a single astrocyte in hippocampal tissue made partly gliotic by previous administration of kainic acid. During the entire experiment the cell was exposed to 100 µM DIDS, an inhibitor of Cl^- channels and thus of channel-mediated uptake of K^+ and Cl^-. During the periods indicated, the extracellular K^+ concentration was raised from 3 to 10 mM, the second time in the additional presence of 20 µM ouabain, a blocker of the Na^+,K^+-ATPase. Note the complete inhibition by ouabain of the increase in intracellular K^+ concentration following the increase in extracellular K^+ concentration, and the increased amplitude of the K^+-induced depolarization, reflecting the absence of a simultaneous increase in intracellular K^+ concentration. From Walz and Wuttke, 1999.

Channel-mediated K^+ Redistribution or Uptake

In addition to the active uptake of extracellular K^+ into astrocytes channel-mediated mechanisms for K^+ homeostasis have been suggested. A "spatial buffer" mechanism was first suggested by Kuffler et al. (1966), based on experiments in amphibian nervous systems. The large size of some glial cells in these species allowed the performance of intracellular recordings, which were used to investigate membrane characteristics. These studies showed characteristics of glial cells, which might allow a *local* increase in extracellular K^+ to be re-distributed through a glial syncytium to different extracellular locations by a current-carried redistribution of K^+. These characteristics are: i) selective K^+ permeability of the glial cell membrane even at low extracellular K^+ concentrations (i.e., a much higher conductance for K^+ than for any other ion); and ii) electrical coupling of glial cells into a functional syncytium. Both characteristics have been confirmed in mammalian astrocytes. The electrical coupling is a consequence of the gap junction-mediated astrocytic syncytium described above. A high K^+ conductance of the astrocytic cell membrane has also repeatedly been demonstrated and an apparent "sub-Nernstian" behaviour, i.e., a smaller decline in membrane potential than theoretically expected when the extracellular K^+ concentration is increased, has been shown to reflect the simultaneous increase in intracellular K^+ (Walz et al., 1984b; Walz, 2000). There is also recent evidence that hippocampal astrocytes in addition to expressing inward rectifying K^+ channels (facilitating K^+ entry) also express outward rectifiers (facilitating K^+ exit) both in situ and in culture (Emmi et al., 2000; Zhou and Kimelberg, 2000).

Because of the membrane characteristics in astrocytes, one would expect a *local* increase in extracellular K^+, evoked by localized neuronal firing, to depolarize the glial cell membrane *at this location*. Initially, an unaltered membrane potential in the remainder of the glial syncytium creates electrical currents moving towards the depolarized region, primarily carried by a flux of K^+ in the opposite direction. The electrical circuit is closed through the extracellular space by the transport of Cl^- in the same direction, or Na^+ in the opposite direction. K^+ is the only ion, which can freely cross the glial membrane and thus continue to travel from the extracellular space through the intracellular space and out again, *and it will only do so as long as a localized increase in K^+ persists*. The whole process leads to redistribution of local K^+, but it does *not* involve any measurable cellular *uptake* of K^+, and it ceases once extracellular K^+ has evenly re-equilibrated. It is driven by the membrane potential of the glial cells (which decreases in the process), but it does not require expenditure of metabolic energy. Transfer of K^+ by the spatial buffer is much faster and can reach over much longer distances than K^+ transfer by simple diffusion.

The spatial buffer plays a major role in the non-mammalian CNS and in the retina. In the retina, it rapidly "siphons" K^+ released from photoreceptors (modified neurons) along the interior of Muller cells, which span the entire width of the retina, into the *corpus vitreum*, which due to its large volume serves as a "potassium sink" (Newman et al., 1984; Reichenbach et al., 1993; Bringmann et al., 2004). It is occasionally suggested that a similar "siphoning" is of importance for clearance of excess extracellular K^+ in brain by transporting K^+ through astrocytic endfeet into the systemic circulation, but a very slow exchange of ^{42}K across the blood-brain-barrier shows that such a process cannot play a quantitatively significant role for extracellular K^+ homeostasis in the mammalian brain (reviewed by Walz and Hertz, 1983). However, the relatively modest exit of K^+ that does occur across the blood-brain-barrier may partly occur *via* astrocytes, considering that these cells have a high intracellular K^+ content, that the astrocytic membrane has a large K^+ conductance, and that the major part of the periphery of the microvessel is covered by astrocytic processes.

In the mammalian CNS the spatial buffer appears to be of minor importance for K^+ clearance. A major indication in this direction is that the length constant, the distance along a cell process to the site where the voltage amplitude has decayed to 37% of its initial value, is shorter than the length of most astrocytic processes, which may abrogate cell-to-cell propagation (Walz, 2000, 2004). Nevertheless, inhibition of inward rectifying K^+ channels on astrocytes in hippocampal slices by cesium ions increases both the amplitude and the duration of stimulus-induced increases in extracellular K^+ concentration (Janigro et al., 1997). A large inhibitory effect of barium (Ba^{2+}) salts in hippocampal slices could also be taken as evidence supporting the conclusion that the spatial buffer is important for K^+ clearance (Jauch et al., 2002). However, the concentration of Ba^{2+} used (2 mM) was so high that it also inhibits active transport of K^+ (Walz et al, 1984a). A probably at most minor effect of the "spatial buffer" on extracellular K^+ clearance does not mean that it has no effect on short-distance intra-astrocytic K^+ movement, and K^+ currents through astrocytes may be responsible for the major part of slow potential shifts, which can be recorded epicranially in many species, including man, during anticipatory events (Laming, 2000). Based on such observations it has been suggested that neuronal-glial-neuronal K^+ signalling by the spatial buffer may have behavioural consequences; this is undoubtedly the case in non-mammalian species (Laming, 2004; Laming and Laming, 2004).

A different channel-mediated mechanism for removal of excess extracellular K^+ is accumulation of both Cl^- and K^+ ions through their respective channels, when extracellular K^+ is increased and extracellular Na^+ is decreased. This uptake depends upon high conductances for both K^+ and Cl^-, and it is driven by Donnan forces (a "modified Boyle-Conway mechanism" [Barres et al., 1990]). Such an uptake may occur in both neurons and astrocytes, although it might be more marked in the astrocytes, which have a higher K^+ conductance than neurons. However, in normal astrocytes the Cl- conductance is low (Walz, 2002). Reactive astrocytes have an increased Cl^- conductance and uptake of K^+ by a modified Boyle-Conway mechanism is therefore more important in reactive astrocytes, where it can fully substitute for the co-existing active uptake mechanisms (Walz, 2004).

Neuronal Return of K^+

Since CNS K^+ only slowly exchanges with the systemic circulation, all K^+ accumulated in astrocytes must eventually be re-accumulated in neurons in order to maintain CNS function. Little is known about the mechanisms involved. However, it has been established that K^+, which has been accumulated in astrocytes, is rapidly released after normalization of the extracellular K^+ concentration (Walz, 2000), perhaps in connection with changes in glial K^+ conductances, as has been suggested in peripheral nerves of invertebrates, where K^+ initially is accumulated in glial cells ensheathing the axon (Lieberman et al., 1994). Maintenance of extracellular K^+ concentrations close to (and below) their resting level facilitates neuronal, versus astrocytic, K^+ uptake on account of the higher affinity for extracellular K^+ of the neuronal Na^+, K^+-ATPase (Hajek et al., 1996). Even though K^+ uptake at an extracellular K^+ concentration of 6 mM is slower than in astrocytes, this might not be the case at 3 mM K^+.

The obligatory neuronal return of astrocytically accumulated K^+ raises the question why K^+ initially is accumulated in astrocytes (and/or redistributed through astrocytes). Although astrocytic involvement in K^+ homeostasis serves a signaling role in non-mammalian species (Laming, 2004), nothing is known about the possible behavioral importance of K^+ distribution and transport in the mammalian brain (except for the potential role of spatial buffer-mediated K^+ distribution in slow potential shifts). Yet, if the only purpose of astrocytic accumulation of K^+ were to remove excess K^+ from the extracellular space, then one would expect previous exposure to a reduced extracellular K^+ concentration to lead to a decrease in Na^+ K^+-ATPase activity in astrocytes. In fact, the opposite happens (Hajek et al., 1996), suggesting that astrocytic accumulation of K^+ *per se,* together with a possible subsequent transfer of accumulated K^+ via predermined pathways, is a metabolic priority (Hertz, 1965).

Na^+ Transport

Using fluorescent ratio imaging of a Na^+-sensitive fluorescent drug at different K^+ concentrations, in the presence and absence of a CO^2/bicarbonate (HCO_3^-) buffer, and during exposure to the cotransporter inhibitor bumetanide Rose and Ransom (1996) concluded that intracellular Na^+ in cultured hippocampal astrocytes is determined by Na^+,K^+-ATPase activity and cotransporter activity (an indirect confirmation of the importance of these carriers for K^+ uptake) as well as by inwardly directed Na^+/HCO_3 cotransport. The Na^+/ HCO_3^- cotransporter transports the two ionic species in a 1:2 ratio (McAlear and Bevensee, 2004) and accordingly carries one negative charge into the cell; it can therefore be expected to be activated by K^+-mediated depolarization.

In contrast to the complex mechanisms for Na^+ transport in hippocampal astrocytes only the Na^+, K^+-ATPase appears to be involved in Na^+ homeostasis in cultured neurons (Rose and Ransom, 1997).

Calcium Homeostasis

Free Cytosolic Calcium Concentration

The concentration of $[Ca^{2+}]_i$ serves as an essential intracellular messenger in virtually any type of cell. This concentration ($[Ca^{2+}]_i$).is regulated by Ca^{2+} entry or exit across the cell membrane and Ca^{2+} release and/or binding to intracellular structures, resulting in an average free intracellular $[Ca^{2+}]_i$ in mammalian cells of about 1 x 10^{-4} mM (100 nM). This value is 10,000 times lower than extracellular Ca^{2+} concentration (approximately 1 mM). Since $[Ca^{2+}]_i$ is so low, in can easily be dramatically increased, which facilitates its function as a signaling mechanism. The steep Ca^{2+} gradient across the cell membrane is maintained by control of channel-mediated Ca^{2+} entry into the cell (through transmitter-operated or voltage-dependent channels or in exchange with other ions), energy dependent, carrier-mediated transport of Ca^{2+} out of the cell, and controlled sequestration of Ca^{2+} in the cell interior by binding to or release from intracellular organelles (Scapagnini et al., 2004).

Quantitatively, much smaller amounts of Ca^{2+} are generally transported across cell membranes than the corresponding amounts of Na^+ and K^+ during neuronal activity; accordingly Ca^{2+} transport as such must have relatively little stimulatory effects on energy metabolism. Nevertheless, Ca^{2+} entry into synaptic vesicles during transmitter release, into Purkinje cell dendrites during Ca^{2+}- mediated action potential, and into the cytosol of both neurons and astrocytes during opening of Ca^{2+} channels, must be followed by energy requiring extrusion of Ca^{2+}. Indirectly, opening of astrocytic voltage sensitive astrocytic Ca^{2+} channels of the L-type by by elevated K^+ concentrations plays a major role in ion transport and energy metabolism because the entry of Ca^{2+} triggers stimulation of the cotransporter of Na^+, K^+ and Cl^- which is responsible for a major part of K^+ uptake into astrocytes (Su et al., 2000) and for the stimulation of oxidative metabolism (Hertz et al., 2002). Concentrations of K^+ that are achievable in the extracellular space also stimulate glycogenolysis in brain slices (Hof et al., 1988). In cultured astrocytes the stimulation of glycogenolysis is enhanced by benzodiazepines (Subbarao et al., 1995), acting on astrocytic $GABA_A$ receptors (Backus et al., 1988), which in astrocytes causes a partial depolarization due to a high intracellular concentration of Cl^- (Hansson and Rönnbäck, 2004) and thereby facilitate K^+-mediated depolarization and Ca^{2+} entry (Zhao et al., 1996).

Fascinating work during the last 10-15 years (Cornell-Bell et al., 1990, 2004; Dani and Smith, 1995; Newman and Zahs, 1998; Haydon and Parpura, 2000; Newman, 2003) has shown that glutamate release from neurons induces astrocytic increases in $[Ca^{2+}]_i$, which slowly (0.5-1 mm/min) spread from the stimulated centre as circular waves across long distances in the astrocytic syncytium and eventually modify the activity of distant neurons. Intercellular waves of increases in $[Ca^{2+}]_i$ have been found to propagate across the astrocytic syncytium after excitation of a single cell by glutamate, mechanical stimulation, or NMDA-induced stimulation of adjacent neurons, resulting in glutamate release, which stimulates astrocytic metabotropic glutamate receptors (Shuai et al., 2004; Nadkarni and Jung, 2004). The pathways along which the increase in $[Ca^{2+}]_i$ propagates across the astrocytic syncytium can be exceedingly complex (Harris-White et al., 1998), cover long distances and last during extended time periods (Figure 9). Since $[Ca^{2+}]_i$ is efficiently buffered in the cytosol, the propagation from cell to cell is not triggered by diffusion of Ca^{2+}, but by IP_3 transfer from cell to cell through gap junctions and subsequent IP_3-induced release of Ca^{2+}, and/or by stimulation of purinergic P2Y receptors by ATP released from the cells (Wang et al., 2000; Cornell-Bell et al., 2004). It is likely that other transmitters acting via the phosphatidyl inositide second messenger system or opening of voltage dependent Ca^{2+} channels might have a similar effect. Thus, acetylcholine shifts the zone of oscillating astrocytes farther away from the origin of mechanically induced Ca^{2+} waves, and serotonin increases the velocity of the waves (see Hertz et al., 2001). Glutamate and elevated extracellular K^+ concentrations increase astrocytic gap junction communication (Enkvist and McCarthy, 1994) whereas intracellular acidity closes gap junctions in astrocytes (Rozental et al., 2001). Many factors might accordingly modify Ca^{2+} wave propagation, including neuronal input (Rouach et al., 2004). Simultaneous involvement of different mechanisms, associated with different molecular cascades, may enable this system to integrate spatially, temporally and mechanistically different signals.

The astrocytic $[Ca^{2+}]_i$ waves are capable of exciting the neurons they encounter by vesicular release of glutamate (Parpura and Haydon, 2000; Newman, 2003b; Montana et al., 2004; Zhang et al., 2004; Fellin et al., 2004; Anlauf and Derouiche 2004) or inhibiting neuronal activity by release of ATP, which becomes converted extracellularly to adenosine (Newman, 2003a, b). The induced neuronal activity is almost abolished in the joint presence of an NMDA and an AMPA antagonist, which in the system used had no effect on astrocytic $[Ca^{2+}]_i$ (Parpura and Haydon, 2000). The entire circuit constitutes a neuronal-glial-neuronal signaling mechanism, which may be of considerable importance for temporal and spatial integration in the central nervous system. This signaling system might be involved in many phenomena, which are presently only poorly understood. It is remarkably sensitive to hypoxia, since15-30 min of superfusion with nitrogen-purged artificial cerebrospinal fluid results in permanent loss of Ca^{2+} signalling in organotypic hippocampal cultures (Harris-White et al., 1998).

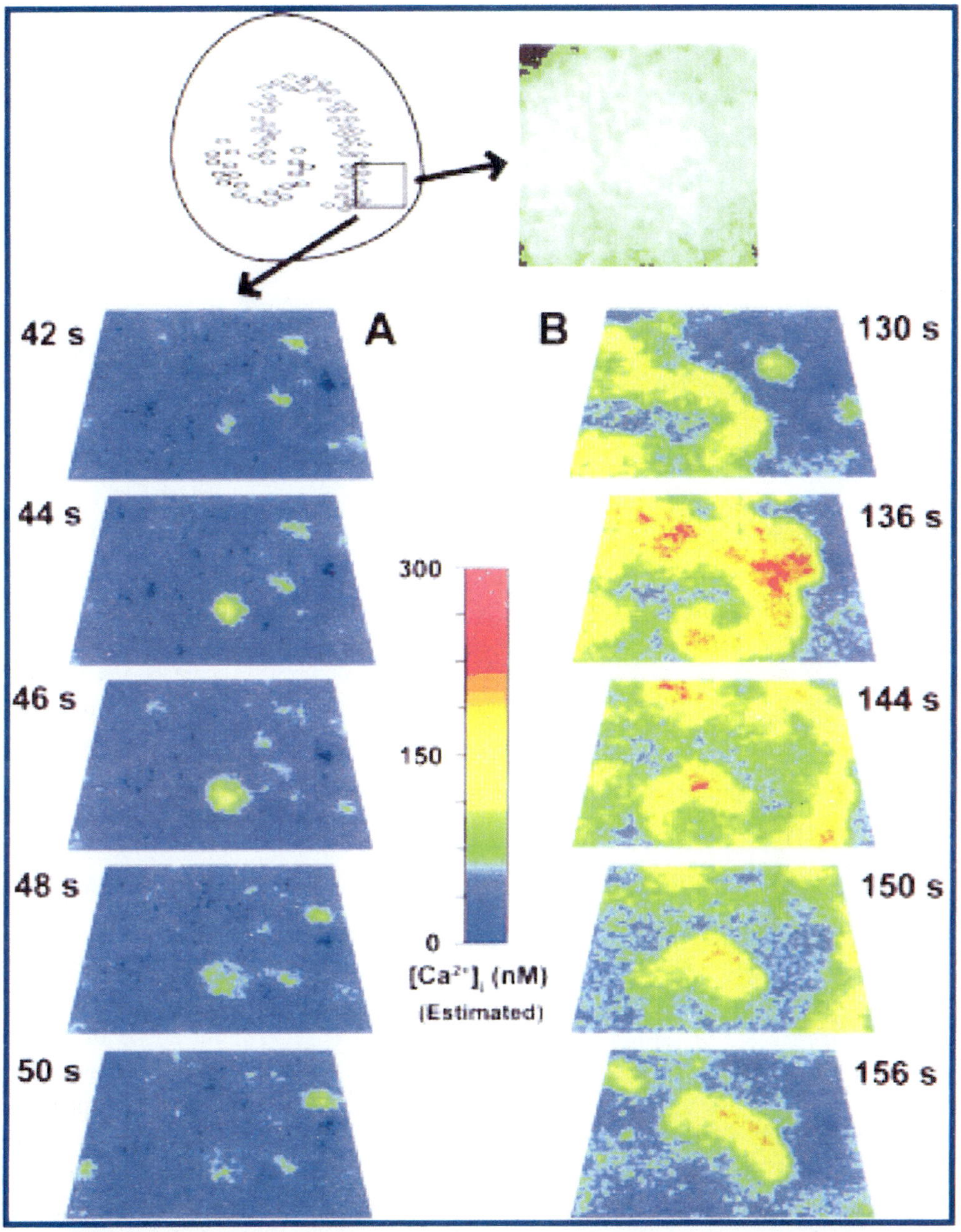

Figure 9. Spontaneous Ca^{2+} puffs (A) and the response to *N*-methyl-D-aspartate (NMDA) (B) in a mouse hippocampal slice culture. *Top left*: region of the slice explant that is imaged (box). *Top right*: baseline fura2 fluorescence in this region. *A*: spontaneous Ca^{2+} puffs that originate from multiple foci in the *stratum oriens*. *B*: same field 10 s following bath application of 20 μM NMDA (at 120 sec). A large spiral wave develops and propagates to involve most cells in the field. Scale: 550 μm horizontally × 400 μm vertically. From Harris-White et al., 1998.

Local cross talk at a glutamatergic synapse between neurons and astrocytes in the "tripartite synapse" can also maintain transmission in the stimulated synapse after cessation of the original stimulus. Mathematic modeling has demonstrated that at a high astrocytic density of metabotropic glutamate receptors electric stimulation of an adjacent neuron can be expected to cause astrocytic IP_3 to build up to a high level, causing a large release of Ca^{2+} from intracellular stores. Since $[Ca^{2+}]_i$ dynamics is much slower than neuronal dynamics, $[Ca^{2+}]_i$ spikes after the stimulation of the neurons is terminated; the increased $[Ca^{2+}]_i$ triggers astrocytic release of glutamate, which creates slow neuronal inward currents that keep the neuron firing after the termination of the original stimulus, further increasing IP_3 in the astrocytes (Figure 10). Thus the astrocytic feed-back keeps the neuron firing far beyond the electrical stimulation (Shuai et al., 2004; Nadkarni and Jung, 2004).

Intramitochondrial [Ca^{2+}]

In heart muscle, the increased $[Ca^{2+}]_i$ resulting from receptor activation causes an increase in intramitochondrial Ca^{2+} concentration, which regulates mitochondrial functions by activating specific dehydrogenases (Denton and Mc Cormack, 1993). In addition, utilization of glutamine as a metabolic substrate is increased. The functional advantage of a direct stimulation of enzyme activities, compared to the increase in metabolic activity evoked by utilization of ATP and the resulting increase in ADP, is that there is no concomitant decrease in the content of ATP, which, in contrast, may be elevated. Similar metabolic processes are stimulated in astrocytes by noradrenaline and some of its subtype-specific agonists (Subbarao and Hertz, 1991; Huang and Hertz, 1995; Chen and Hertz, 1999), which increase $[Ca^{2+}]_i$ (Hertz and Peng, 1992; Hertz et al., 2004b). There are excellent quantitative and subtype-specific correlations between the effects of noradrenaline on astrocytic $[Ca^{2+}]_i$ and on the activity of pyruvate dehydrogenase, the enzyme, which converts pyruvate to acetyl coenzyme A (acetyl CoA) (Chen and Hertz, 1999; Chen et al., 2000).

Acid-base Regulation

Extracellular pH is one of the factors regulating neuronal excitability, ion channel activity in both neurons and astrocytes and gap junction opening in astrocytes (Chesler and Kaila, 1992; Taira et al., 1993; Sykova et al., 1992; Ransom, 1992, 2000; Rozental et al., 2001). Changes in pH also affect the rate of energy metabolism (Erecinska et al., 1995; Brookes, 1997). Like K^+ and Ca^{2+}, hydrogen ions (H^+) are transported across cell membranes both passively through channels and actively by the aid of carriers. Bicarbonate (HCO_3^-) can be transported into cells together with Na^+ in a 2:1 ratio or it can be exchanged with Cl^- (Chesler and Kaila, 1992; McAlear and Bevensee, 2004). For pH homeostasis the presence of carbonic anhydrase (CA) is also essential. This enzyme catalyzes the bilateral exchange between carbonic acid (H_2CO_3) and CO_2 plus water. It has long been known to exist in glial cells (Giacobini, 1987), but it is also present in some neurons and, importantly, extracellularly (Chesler and Kaila, 1992).

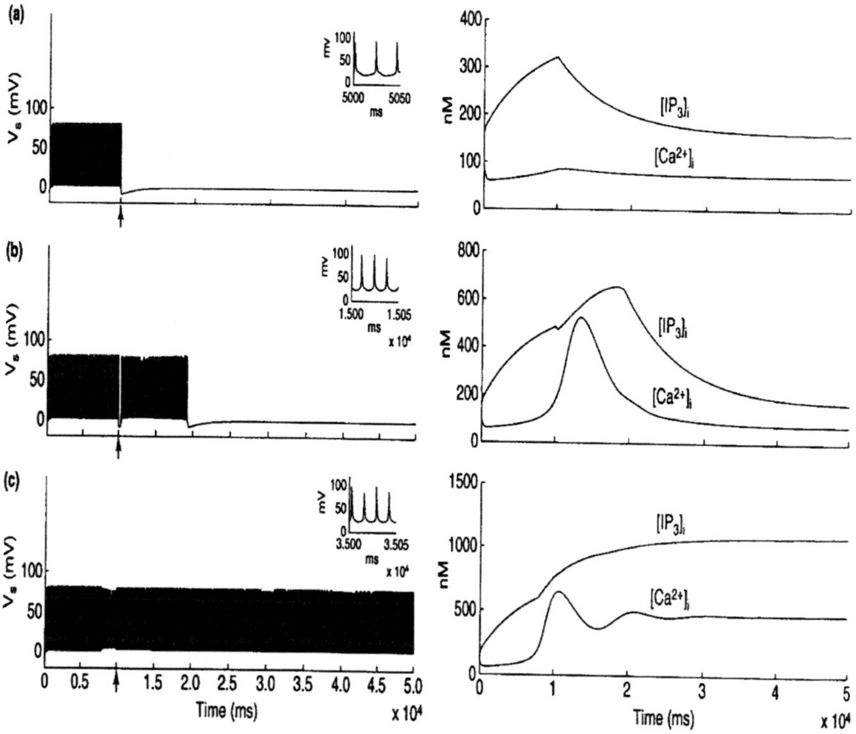

Figure 10. Mathematical modeling of effect of stimulation with DC current (0–10 s) of a neuron assumed to be surrounded by astrocytes with expression of metabotropic glutamate receptors, allowing inositol (1,4,5)-trisphosphate (IP_3) production rates which increase from a through b to c. Neuronal membrane potentials are shown on the left, and astrocytic concentrations of IP_3 and free cytosolic calcium concentrations ($[Ca^{2+}]_i$) as functions of time are shown on the right. Arrows at 1 x 10^4 msec indicate the end of the electric stimulation, which begins at zero time. Inserts show neuronal spiking during a 50 ms period on a faster timescale at different times during the stimulation (a) or the continued spiking evoked by astrocytic Ca^{2+} signaling (b and c). In (a), astrocytic IP_3 builds up and decays (owing to ongoing degradation) as the neuron stops firing; astrocytic $[Ca^{2+}]_i$ increases only slightly. In (b), IP_3 builds up faster and to a higher level (note the different scales on the ordinates). Because Ca^{2+} dynamics is much slower than neuronal dynamics, $[Ca^{2+}]_i$ spikes after stimulation of the neurons is terminated; the increased $[Ca^{2+}]_i$ triggers astrocytic release of glutamate, which creates slow neuronal inward currents that keep the neuron firing after stimulus termination, further increasing IP_3 in the astrocytes. Thus, astrocytic feedback keeps the neuron firing beyond the electrical stimulation. When IP_3 degradation eventually exceeds IP_3 production, spiking stops. In (c), the larger IP_3 production rate prevents decline of $[Ca^{2+}]_i$ below the level where it ceases to feedback to the neuron, and neuronal spiking continues. Reproduced from Hertz and Zielke, 2004, and modified from Nadkarni and Jung, 2004.

In neurons displaying GABAergic receptors stimulation can lead to extrusion through GABA activated chloride channels of HCO_3^- in exchange with Cl^-, causing extracellular alkalosis and intracellular acidosis (Chesler and Kaila, 1992) (Figure 11). A similar response has been observed in astrocytes expressing $GABA_A$ receptors (Kaila et al., 1991).

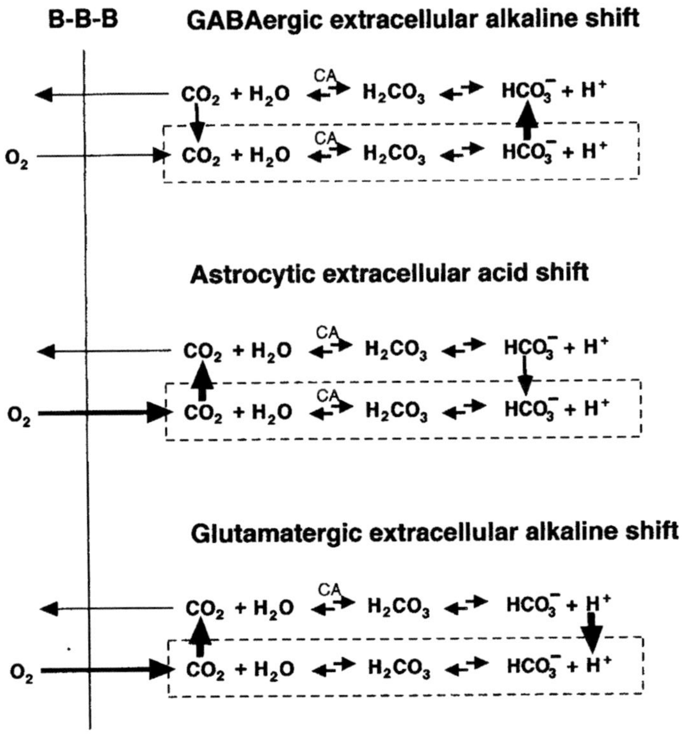

Figure 11. Transport processes between the intracellular space (within the boxes) and the extracellular space, which are involved in pH regulation in *stimulated neurons expressing either GABA receptors or glutamate receptors and in astrocytes*. In neurons with GABA receptors, HCO_3^- is extruded across the cell membrane in exchange with Cl^- and combines extracellularly with H^+ to form H_2CO_3, which in the presence of carbonic anhydrase (CA) activity dissociates to CO_2 and H_2O. CO_2 returns across the cell membrane by diffusion and associates, in the presence of intracellular CA activity, with H_2O to form H_2CO_3, which dissociates to H^+ and HCO_3^-, the latter of which is, again, transported across the cell membrane. The total process, which requires the presence of both intracellular and extracellular CA activity, leads to extracellular alkalinization/intracellular acidification. In neurons expressing glutamatergic receptors, CO_2 does probably not associate with water intracellularly, perhaps due to absence of intracellular carbonic anhydrase activity, but exits across the cell membrane and is converted extracellularly by the same processes as in the GABAergic neurons to HCO_3^- and H^+, the latter of which enters the cell, leading to extracellular alkalinization/intracellular acidification. Glucose metabolism is increased in neurons with glutamate receptors during stimulation, leading to enhanced CO_2 production; this increase in CO_2 may contribute to stimulus-induced intra- and extracellular shifts in the indicated directions. Oxygen (O_2) enters, and CO_2 leaves the CNS by diffusion across the blood-brain-barrier (B-B-B). In astrocytes uptake of HCO_3^- by the $Na^+/2\ HCO_3^-$ cotransporter is increased during K^+-mediated depolarization and it combines intracellularly with H^+ to form H_2CO_3, which in the presence of intracellular carbonic anhydrase (CA) activity dissociates to CO_2 and H_2O. Metabolic production of CO_2 is also increased and CO_2 from oxidative metabolism and from accumulated HCO_3^- can exit across the cell membrane and in the presence of extracellular CA form H_2CO_3, which dissociates to HCO_3^- and H^+, the former of which can again be transported across the cell membrane. Modified from Chesler and Kaila (1992).

Extracellular HCO_3^- can combine with H^+ to form carbonic acid (H_2CO_3), which in the presence of extracellular CA activity dissociates to water and CO_2. If the extracellular CO_2 tension exceeds the intracellular CO_2 tension (see below) CO_2 can enter the cells to form H_2CO_3 (contingent on the presence of intracellular CA), from which additional HCO_3^- can be formed and extruded, further enhancing extracellular alkalosis.

A stimulus-dependent extracellular alkalosis has also been demonstrated in neurons exposed to glutamate (Figure 11). In at least some neurons expressing glutamate receptors the mechanism for this alkaline shift is an influx of H^+, i.e., different process than in neurons expressing GABA receptors (Chesler and Kaila, 1992; Chesler, 2003), and no intracellular carbonic anhydrase activity is required. A pronounced increase in CO_2 production in cultured glutamatergic cerebellar granule neurons during K^+-mediated depolarization (Peng et al., 1994) might also lead to an exit of CO_2 across the cell membrane, contributing to an increase in extracellular CO_2 tension, as has been seen during stimulation of hippocampal slices (Chesler and Kaila, 1992). If extracellular CA is present H_2CO_3 can be formed and dissociate into HCO_3^- and H^+, which may again be accumulated.

In spite of the fact that stimulation of both GABAergic and glutamatergic neurons leads to extracellular alkalinization, the activity-induced alkalinization in intact tissue is in most cases transient (Sykova, 1992) and followed by extracellular acidification. This appears to be an astrocytic response to neuronal excitation, probably at least partly evoked by the increase in extracellular K^+ concentration (Chesler 2003). Among the many mechanisms by which astrocytes can acidify their surroundings is an uptake up HCO_3^- by the 2 HCO_3^-/Na^+ cotransporter (McAlear and Bevensee, 2004). K^+-induced depolarization can be expected to lead to an increased activity of this uptake, because the transporter mediates net entry of negative charge, which will be facilitated by reduction of the negative membrane potential. The elevation in K^+ concentrations also reverses the increased intracellular acidity evoked by stimulation of astrocytic $GABA_A$ receptors (Kaila et al., 1991). After uptake of HCO_3^- it combines with intracellular H^+ to form H_2CO_3, and CA activity in both the astrocyte and the extracellular fluid catalyzes the bidirectional reactions between CO_2 plus H_2O and H_2CO_3 and facilitates intracellular formation of CO_2 and extracellular formation of more HCO_3^-, which again can enter the cell (Figure 11. Increased energy metabolism in astrocytes exposed to elevated K^+ concentrations (see later) may also lead to increased release of both CO_2 and lactate, the latter cotransported with H^+ (reviewed by Hertz and Dienel, 2005).

Control of extracellular pH by astrocytes may play a major role in regulation of CNS activity, since extracellular acidosis decreases the sensitivity of the NMDA receptor to glutamate and thus neuronal excitability (Ransom, 1992; 2000; Taira et al., 1993; Chesler 2003).

Glucose Metabolism

CNS Energy Metabolism is Glucose Metabolism

Under normal, "resting" conditions the intact, mature brain *as a whole and at steady state* almost exclusively utilizes *complete* oxidative metabolism of glucose to CO_2 to fulfill its energetic requirements, as evidenced by virtually equivalent amounts of oxygen being extracted and CO_2 being produced, a glucose extraction which is almost consistent with the oxygen utilization during complete glucose oxidation (six molecules of oxygen per molecule of glucose or an oxygen/glucose ratio of 6.0), and lack of net extraction of other potential substrates for energy metabolism (Sokoloff, 1992). Under such conditions, but only under such conditions, *oxidative* metabolism of glucose can be measured as rate of utilization of labeled deoxyglucose, a glucose derivative, which readily forms deoxyglucose-6-phosphate, an analog of glucose-6-phosphate (Figure 12), but is not further metabolized. It is virtually unable to exit from the cell and therefore trapped intracellularly.

Much information has been obtained about energy metabolism in brain under normal and pathological conditions using labeled deoxyglucose in non-invasive "positron emission spectroscopy (PET)" scans, e.g., that mental activity is associated with an increased *local* rate of energy metabolism, and that the increased metabolism during brain activation (but not necessarily under resting conditions – see Akabayashi and Kato, 1993) mainly occurs in the neuropil (Sokoloff, 1992). This was the first metabolic "imaging technique" and it has now to a large extent been replaced by other methods in both research and the clinic. Among these, ^{13}C NMR (nuclear magnetic resonance) spectroscopy is a powerful methodology, which presently is yielding essential information about metabolic fluxes and metabolic pathways in the brain *in situ* (e.g., Curz and Cerdan, 1999; Gruetter et al., 2001; Shulman et al., 2004). Ordinary PET scanning does not have the resolution to distinguish between energy metabolism in different cell types, but very recently Nehlig et al. (2004) developed an ingenious approach, recording and identifiying individual *tracks* of electrons emitted during disintegrations of $[^{14}C]$2-deoxyglucose, which suggested that ~45% of the phosphorylated deoxyglucose was localized in astrocytes and 55% in neurons. NMR is able to demonstrate that pathways known to operate only in astrocytes are involved in synthesis of transmitter glutamate and GABA in the living brain (see below). Moreover acetate can replace the glucose metabolite pyruvate in the formation of acetyl coenzyme A (acetyl CoA), (Figure 12), but in contrast to pyruvate it is accumulated and oxidized in astrocytes but not in neurons and can therefore be used to specifically determine oxidative metabolism in astrocytes (Muir et al., 1986; Fonnum et al., 1997; Hassel et al., 1997; Waniewski and Martin, 1998; Blüml et al., 2002; Lebon et al., 2002). Both NMR studies and studies of ^{14}C-labeled acetate metabolism in brain have shown pronounced oxidative metabolism in astrocytes in the brain in vivo (15-20% of total pyruvate dehydrogenation in glial cells and another ~15% of pyruvate carboxylated to oxaloacetate). These data suggest approximately similar rate of oxidative metabolism per volume in neurons and glia, which is consistent with results from decades of studies using cultured astrocytes (reviewed by Hertz and Schousboe, 1975; Hertz, 1978; Hertz et al., 1988). The combined pyruvate dehydrogenation and carboxylation rates

(30-35% of total pyruvate utilization) are reasonably close to the estimate by Nehlig et al. (2004) that 45 % of glucose utilixation occurs in astrocytes.

Glycolysis and Glycogenolysis

Glucose metabolism is initiated by glycolysis, a cytosolic process, leading to the formation of pyruvate (2 molecules of pyruvate for each molecule of glucose), which can also be generated from glycogen, an astrocyte-specific energy store (Figure 12). In order to initiate glycolysis one molecule of oxidatively derived ATP is required to convert glucose to glucose- 6-phosphate (BeltrandelRio and Wilson, 1992; Cesar and Wilson, 1998). The further glycolytic degradation includes one oxidative process (formation of biphosphoglycerate from glyceraldehyde-phosphate) leading to conversion of NAD^+ to NADH + H^+ (not shown in Figure 12). In the presence of oxygen, cytosolic NADH is re-oxidized by transfer of a "reducing equivalent" to the mitochondria in a "malate-aspartate shuttle" and its oxidation in the TCA cycle (for details, see for example Hertz and Dienel, 2002). In its absence it is re-converted to NAD^+ by reduction of pyruvate to lactate (anaerobic glycolysis). However, even under oxygenated conditions an equilibrium is established between pyruvate and lactate, determined by the cytosolic redox potential, which leads to a higher concentration of lactate (normally about 1 mM) than of pyruvate. Pyruvate can also be transaminated to alanine (not shown in the Figure), an amino acid, which can supply the amino group for synthesis of transmitter glutamate (Westergaard et al., 1993; Erecinska et al., 1994), and pyruvate may be used for energy-requiring resynthesis of glycogen.

The net amount of ATP (2 molecules per molecule glucose) produced during glycolysis is much less than the energy yield during further oxidative metabolism of pyruvate. Nevertheless it has been claimed that glycolytically derived energy should be necessary for glutamate uptake in astrocytes when brain function is activated and that the generated lactate is transferred to neurons for oxidation (Pellerin and Magistretti, 1994, 2003; Magistretti et al., 1999). However, this belief is probably based upon a tissue culture artifact, and there is no solid evidence either that glutamate uptake requires aerobic glycolysis or that lactate is transported in net amounts from one cell type to another (Chih and Roberts, 2003; Hertz, 2004).

There is, however, very good evidence that functional activity in the CNS is correlated with turnover of the astrocyte-specific glycogen, even when glucose is freely available. The functional importance of stimulation of glycogenolysis is further supported by the observation that glycogen turnover, i.e., both glycogenolysis and resynthesis of glycogen, an energy-requiring process, are enhanced by brain activity (Swanson, 1992; Swanson et al., 1992; Dienel et al., 2002; Dienel and Cruz, 2003, 2004). Thus, glycogenolysis and subsequent re-synthesis of glycogen may possibly account for a sizeable fraction of energy metabolism in brain during and after enhanced activity. Recent experiments have shown that glycogen is present in the resting brain at a higher concentration than previously realized, i.e., ~10 μmol/glucose equivalent per g wet wt.or 100 nmol/mg protein (Cruz and Dienel, 2002; Dienel and Cruz, 2003), making glycogenolysis very important for energy production.

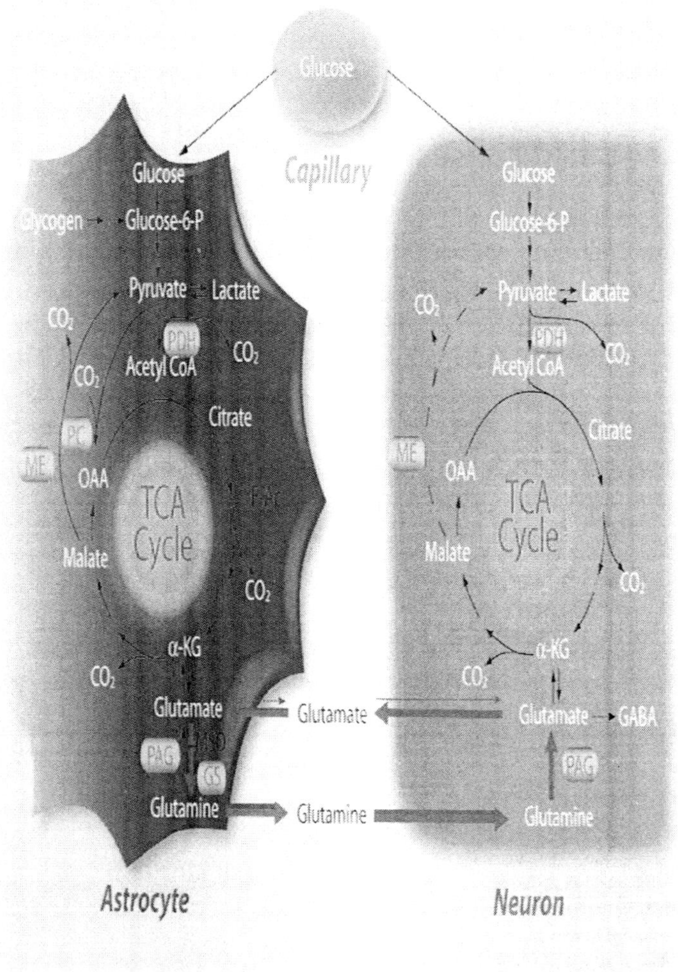

Figure 12. Compartmentalization of glucose metabolism between astrocytes and neurons (not all intermediates are shown). One molecule of glucose is glycolytically converted into two molecules of pyruvate in both neurons and astrocytes. In astrocytes, pyruvate is also formed from glycogen by glycogenolysis. In both cell types pyruvate production is via intermediates such as glucose-6-phosphate (glucose-6-P). Under anaerobic conditions, pyruvate is quantitatively converted to lactate, and in the brain some lactate is formed even under aerobic conditions. The main metabolic fate of pyruvate is entry into the tricarboxylic acid (TCA) cycle. In neurons, pyruvate enters the TCA cycle exclusively via acetyl coenzyme A (acetyl CoA), formed by oxidative decarboxylation, catalyzed by the pyruvate dehydrogenase complex (PDH). Acetyl CoA is metabolized in the TCA cycle after condensation with oxaloacetate (OAA) to form citrate, which is subsequently converted to OAA during one turn of the cycle, generating a substantial amount of ATP without net formation of any TCA cycle intermediate. In astrocytes, pyruvate can enter the TCA cycle both via acetyl CoA and via carboxylation, mediated by pyruvate carboxylase (PC). The latter reaction generates a new molecule of oxaloacetate, which condenses with acetyl CoA to produce a new molecule of citrate, from which α-ketoglutarate, the immediate precursor of glutamate (and thus also a precursor of GABA, which is formed in GABAergic neurons by decarboxylation of glutamate), is formed. α-Ketoglutarate as such might travel from astrocytes to neurons (not shown in the Figure), but it might also be converted in astrocytes via glutamate to glutamine, which is transported in the glutamate-glutamine cycle (lower rectangle) to neurons, where it is hydrolyzed to form transmitter glutamate by phosphate-activated glutaminase (PAG) activity. Released transmitter glutamate is predominantly accumulated in astrocytes, where approximately two-thirds are converted by glutamine synthetase (GS) into glutamine, which after return to neurons in the glutamate-glutamine cycle can be reused as a precursor for transmitter glutamate; the remaining one-third is reconverted to α-ketoglutarate, metabolized in the TCA cycle to malate and decarboxylated primarily by malic enzyme (ME) to pyruvate, predominantly although probably not exclusively in astrocytes (indicated by broken arrows in neurons). Another possible pathway is via oxaloacetate to phosphoenolpyruvate and from there to glycogen (not shown in the Figure). Inhibition of astrocyte-specific processes by methionine sulfoximine (MSO; a relatively specific inhibitor of glutamine synthetase) or by fluoroacetate (F-Ac), a specific inhibitor of the astrocytic TCA cycle) is indicated by breaks in the arrows. From Hertz and Zielke, 2004.

It is often believed that glycogen storage serves the purpose of providing the brain with glucose equivalents for use during failure of glucose delivery, but the increased turnover of glycogen during brain activity indicates a much more universal and dynamic role. Part of the energy invested in the formation of glycogen from glucose is recovered when glycogen is broken down, since the initial conversion to glucose-6-phosphate, in contrast to formation of glucose-6-phosphate from glucose, requires no ATP. Therefore glycogenolysis can freely occur in the absence of oxidatively derived ATP, and production of pyruvate from glycogen produces 3 ATP compared to the net production of 2 ATP when glucose is the substrate. Moreover glycogen particles are small enough, with diameters ranging between 10 and 40 nm (Cataldo and Broadwell, 1986; Wender et al., 2000) to reside within even the most peripheral astrocytic processes. Also, immunoreactivity for the glycogen degrading enzyme glycogen phosphorylase has shown that the enzyme is localized mainly in astrocytes, with diffuse staining throughout the cytoplasm and processes ensheathing capillaries, and in the fine processes and lamellae adjacent to synaptic structures (Richter et al., 1996). Glycogen is therefore able to serve at a rechargeable energy store in a spatio-temporal manner, i.e., provide energy in structures without mitochondria, to do so when the demand for energy is high (e.g., during uptake of K^+), and to subsequently be "re-charged" when oxidatively generated ATP can be delivered by creatine phosphate arriving from the mitochondria (see below). It is also possible that glycogenolysis can spread both intra- and intercellularly through gap junctions (Dienel and Cruz, 2003), since gap junctional communication has been found to coordinate vasopressin-induced glycogenolysis in rat hepatocytes (Eugenin et al., 1998). Similar to other substrates (Giaume et al., 1997) lactate is able to spread between astrocytes through gap junctions (Cruz et al., 1999). In contrast, it is unknown if any glycogen-derived lactate normally is broken down in neuronal constituents. Initial fueling of K^+ uptake by energy produced during formation of pyruvate/lactate from glycogen (and perhaps also from glucose) and subsequent dependence on the larger energy yield from oxidative metabolism (including oxidative metabolism of pyruvate formed from glycogen) are consistent with the observation that the initial clearance of K^+ from the extracellular space after neuronal stimulation is delayed by iodoacetate, an inhibitor of pyruvate formation from glucose and glycogen, whereas the later stages are inhibited by oxygen deprivation (Figure 13).

Pyruvate Oxidation

Production of Energy

Most of the pyruvate formed during glycolysis from glucose or from glycogen is under normal conditions further metabolized in the TCA cycle. For complete oxidative degradation, it initially enters the mitochondria and undergoes a pyruvate dehydrogenase- (PDH)-mediated oxidative decarboxylation (yielding 3 molecules of ATP per molecule of pyruvate or 6 molecules of ATP per molecule of glucose) to form acetyl CoA, which subsequently is condensed in the TCA cycle with oxaloacetate (OAA) to form citrate (Figure 12). Citrate is

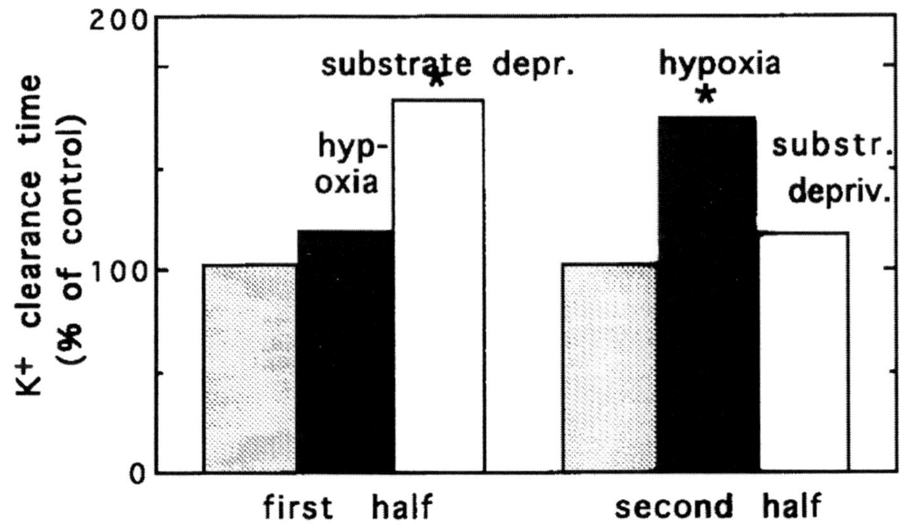

Figure 13. Effect of hypoxia (black columns) and of iodoacetate-evoked inhibition of glycolysis (white columns) on the initial ("first half") and final ("second half") phase of K^+ clearance from the extracellular space of rat cerebral cortex after electrical stimulation of the cerebral surface. The initial phase is defined as the half-time for reduction of the *evoked increase* in extracellular K^+ from its maximum value above the resting K^+ level (see Figure 6) to 50 % of this value, and the final phase as the half-time for further reduction to baseline level; control values, indicated as 100 %, are shown by gray columns. Asterisks indicate values, which are significantly different from control values. Modified from Rosenthal and Sick, 1992.

reconverted to oxaloacetate by one turn of the cycle, resulting in the production of CO_2, water, and a considerable amount of ATP (12 molecules ATP for each turn of the cycle or 24 molecules per molecule glucose) as well as the supply of another molecule oxaloacetate ready for condensation with another molecule of acetyl CoA, but *not* in *net* synthesis of any TCA cycle intermediate. The complete oxidation of 2 molecules of pyruvate, formed from one molecule of glucose leads to the synthesis of 30 molecules of ATP (24 plus 6 formed by pyruvate dehydrogenation), and the oxidation in the mitochondria of the "reducing equivalents" transferred in the malate-asparate cycle leads to the generation of another 6 molecules of ATP. Thus oxidative metabolism of one molecule of glucose produces a total of 36 molecules of ATP or 18 times more than the 2 molecules of ATP formed during glycolysis.

Eleven out of the 12 molecules of ATP generated during the turn of the TCA cycle originates from oxidative phosphorylation and the twelveth is formed by substrate phosphorylation in the mitochondria. Energy-requiring reactions are either fuelled directly by ATP or by creatine phosphate formed from creatine and ATP, catalyzed by mitochondrial creatine kinase. A major function of phoshocreatine is to act as a "spatial buffer", rapidly transferring metabolic energy to locations distant from the site of generation of ATP and serve as a substrate for regeneration of ATP from ADP and creatine phosphate (Hemmer and Wallimann, 1993). This function is catalyzed by cytosolic isoforms of the creatine kinase, and it may be especially important in the CNS with its complex histological structure and multitude of long and narrow processes, many of which do not contain mitochondria. Among the cytosolic forms, a muscle form (BM-CK) is characteristic for muscle and found at only very few locations in the CNS, notably in Purkinje cell dendrites, known to convey Ca^{2+} mediated action potentials. A brain form (BB-CK) is present in the CNS as well as in many other tissues and is enriched in astrocytes, including Bergmann glia (Hemmer and Wallimann, 1993, 1994; Tachikawa et al., 2004). In cultured astrocytes the contents of ATP reach levels of 35-50 nmol/mg protein and that of creatine phosphate 50 nmol/mg protein (Matz and Hertz, 1989; Yager et al., 1994). Since the Na^+,K^+-ATPase accumulates 2 K^+ for each molecule of ATP consumed, K^+ uptake alone (~400 nmol/min per mg protein or 7 nmol/sec per mg protein at 6 mM K^+ and more at highly elevated K^+ concentrations) would in the absence of regeneration of energy-rich phosphates consume all ATP and creatine phosphate in a quarter of one min. The presence of another 100 nmol/mg protein of glycogen (possibly more because glycogen is enriched in the peripheral processes) significantly increases available energy supply even though glycogenolysis per glucose equivalent provides less than 10% of the amount of ATP provided by oxidative metabolism.

In muscle fibers the diffusion of creatine phosphate is almost unrestricted, compared to free diffusion in watery substrates and it reaches 10-50 μm from the source within 1 sec (de Graaf et al., 2000). In the much narrover astrocytic processes and especially in the filopodia, diffusion will be considerably slower, but probably still of the correct order of magnitude to keep up with energy consumption even when it is increased. However, during stimulation of astrocytic energy demand, triggered e.g., by a higher extracellular K^+ concentration, the mitochondria will not "know about" the increased use of energy-rich phosphates until they are reached by elevated phosphate, reduced creatine phosphate or increased creatine levels, all of which are able to stimulate glucose degradation (see, e.g., Hertz and Dienel, 2002). Two responses may be crucial during this interval to meet energy demands, i) local glycogenolysis in the filopodia, and ii) direct stimulation of mitochondrial activity by increased mitochondrial Ca^{2+}, secondary to K^+-mediated opening of the voltage sensitive L channel for Ca^{2+}. Glycogenolysis is astrocyte-specific, and evidence suggesting an increase in astrocytic oxidative phosphorylation by transmitter-evoked elevations in intramitochondrial Ca^{2+} has been obtained in astrocytes (Hertz and Peng, 1992; Peuchen et al., 1996; Chen and Hertz, 1999; Hertz et al., 2004, whereas at present no such effects have been established in neurons.

Production of TCA Cycle Intermediates and their Metabolites

In addition to being converted to acetyl CoA, pyruvate can become condensed with one molecule of CO_2 to form oxaloacetate (by pyruvate carboxylation), a process which in the intact brain accounts for ~15% of the total metabolism of glucose (Lapidot and Gopher, 1994; Aureli et al., 1997; Gruetter et al., 2001). This process, which is catalyzed by pyruvate carboxylase (PC), has been shown both histochemically and by cell culture studies to be

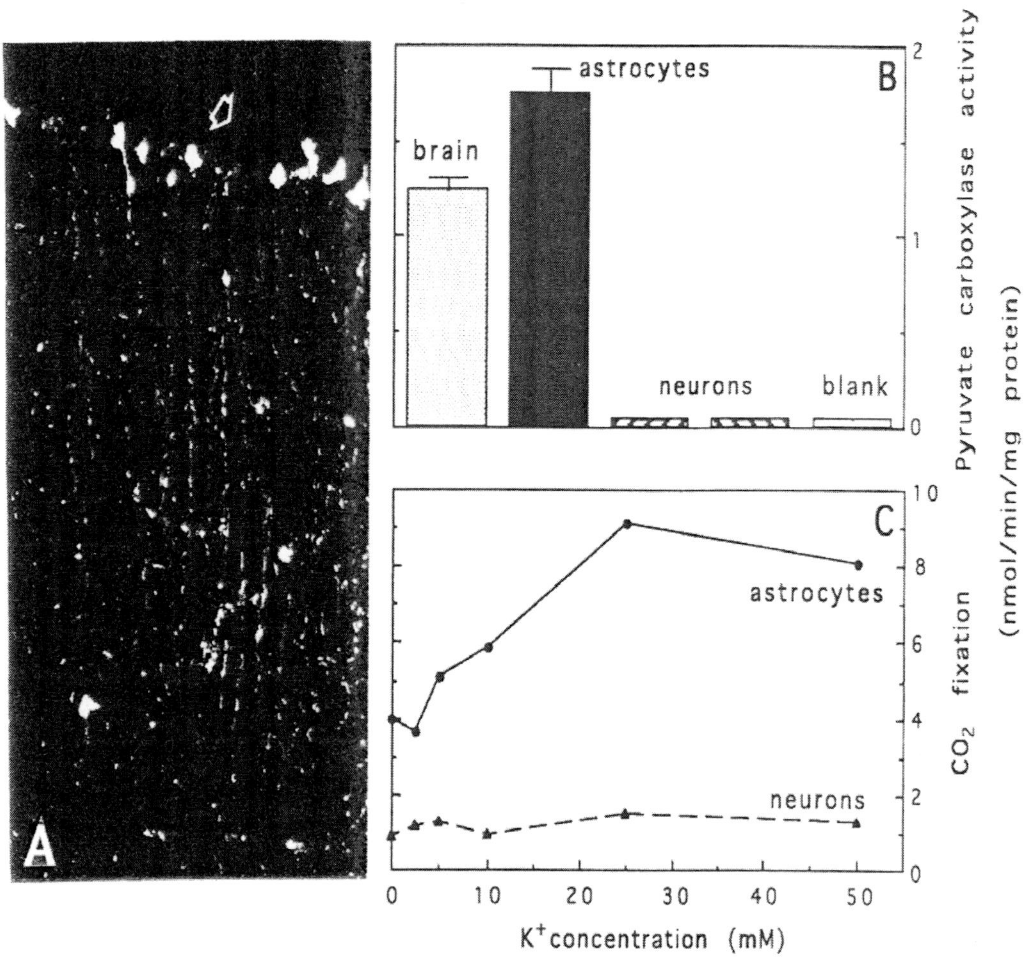

Figure 14. A: Fluorescence micrograph obtained after incubation of mouse cerebellar tissue with goat anti-pyruvate carboxylase antibody followed by a fluorescent anti-goat antibody. Pyruvate carboxylase expression is seen in Bergmann fibers and their cell bodies (Golgi epithelial cells) (from Shank et al., 1985). B: Pyruvate carboxylase activity in primary cultures of mouse astrocytes, cerebral cortical neurons or cerebellar granule cell neurons and in adult mouse brain (from Yu et al., 1983). C: Rates of CO_2 fixation in primary cultures of rat astrocytes and striatal neurons at different K^+ concentrations (from Kaufman and Driscoll, 1992).

confined to glial cells, specifically astrocytes (Figure 14), and it is absent in neurons (Yu et al., 1983; Shank et al., 1985; Kaufman and Driscoll, 1992). Pyruvate carboxylation does not generate energy (rather, it consumes one molecule of ATP), but it leads to net synthesis of a TCA cycle intermediate (Figure 12) and is essential not only to replace tear and wear in the TCA cycle but also for net formation of derivatives of TCA cycle intermediates (Hertz et al., 1999), notably the amino acid transmitters glutamate and γ-aminobutyric acid (GABA), which are formed from the TCA cycle intermediate α-ketoglutarate (α-KG).

From Figure 12 it can be seen that net formation of one molecule of α-ketoglutarate requires not only pyruvate carboxylation, forming one molecule of oxaloacetate, but also condensation of one molecule of oxaloacetate with one molecule of acetyl CoA (formed from another molecule of pyruvate) and turnover in the TCA cycle from the condensation product, citrate, to α-ketoglutarate. Since pyruvate carboxylase activity is absent in neurons, and citrate is not transferred between the two cell types, this turnover must occur in the astrocytic, not the neuronal, TCA cycle. It can therefore be selectively inhibited by the inhibitor fluoroacetate (F-Ac in Figure 12), which is astrocyte-specific, due to a muh more active uptake in astrocytes than in neurons. This is the explanation, why this astrocyte specific toxin has been found to inhibit electrophysiological indications of glutamatergic transmission in brain (reviewed by Hertz and Zielke, 2004).

Turnover of Glutamate, Glutamine and GABA

For *de novo* synthesis from glucose of the neurotransmitter glutamate and its metabolite, the neurotransmitter GABA, α-ketoglutarate formed in astrocytes must be converted to glutamate, probably by a transamination and a transfer must occur from astrocytes to neurons (Figure 12). Transamination is catalyzed by a transaminase, such as aspartate aminotransferase (AAT), and involves an additional amino acid, which in the process is converted to its corresponding ketoacid. A likely candidate for the supply of the amino group is alanine, which is formed in astrocytes from pyruvate (Westergaard et al., 1993; Erecinska et al., 1994). In the transamination reaction, alanine is converted to pyruvate, which then can then be used as a metabolic substrate. Alternativety, amino groups might be supplied from essential amino acids, which are accumulated in the CNS from the systemic circulation, e.g., one of the branched-chain amino acids (Hutson et al., 2001). It is unknown whether α-ketoglutarate is first transferred and then converted to glutamate in neurons or whether glutamate is generated from α-ketoglutarate in astrocytes and then transferred to neurons *via* the glutamate-glutamine cycle described below. Only the latter possibility is illustrated in Figure 12.

Glutamate, as the major excitatory transmitter, and GABA, as the major inhibitory transmitter, are both released in large amounts from neurons during brain function. After the release, a small fraction of the extracellular glutamate and a relatively large percentage of released GABA are reaccumulated into glutamatergic and GABAergic neurons for continued use as transmitters. However, most glutamate (Danbolt, 2001) and a considerable fraction of GABA (Schousboe et al., 2004) are accumulated into adjacent astrocytes, and cultured astrocytes exposed to a high glutamate concentration can accumulate glutamate at a maximum rate of ~50 nmol/min per mg protein or 2.5 μmol/g wet wt. (Hertz et al., 1978), i.e., an almost 10 times lower uptake rate than that for K^+ at 6 mM.

The glutamate uptake in astrocytes necessitates either renewed *de novo* synthesis from glucose in the manner described above or re-cycling of astrocytically accumulated transmitter back to neurons in the glutamate-glutamine cycle. The re-cycling of glutamate from astrocytes to neurons occurs *via* formation of glutamine, a process, which due to the absence of glutamine synthetase (GS) activity in neurons (Norenberg and Martinez-Hernandez, 1979;

Derouiche 2004) exclusively occurs in astrocytes. Glutamine can be released from astrocytes and accumulated into glutamatergic or GABAergic neurons (Broër and Brookes, 2001) and after conversion to glutamate (catalyzed by phosphate activated glutaminase [PAG]) form transmitter glutamate (Figure 12) and GABA, which is formed in GABAergic neurons by decarboxylation of glutamate. However, about one third of accumulated glutamate and some GABA are oxidatively degraded in astrocytes, and glutamine is also partly utilized as a metabolic substrate both in astrocytes and in neurons (Hertz et al., 1999; Gruetter et al., 2001; Hertz and Hertz, 2003; Hertz, 2004a).

Although oxidative metabolism of glutamate has also been described in synaptosomes (McKenna et al., 2000) it is likely that it mainly occurs in astrocytes (Waagepetersen et al., 2002), where glutamate can be oxidatively deaminated to α-ketoglutarate (Yu et al., 1982), which is metabolized via succinate to malate. Malate can leave the TCA cycle by two different routes, i) it can be decarboxylated by malic enzyme (ME) to pyruvate/lactate, which is re-introduced into the TCA cycle, *via* acetyl CoA, for complete oxidative degradation (Figure 12), or ii) it can be oxidized in the TCA cycle to oxaloacetate, which is then decarboxylated to phosphoenolpyruvate (not shown in the Figure). Phosphoenolpyruvate can either be converted to pyruvate or used for synthesis of glycogen through the same intermediates as glycolytic breakdown. GABA is introduced into the TCA via a different process and converted to α-ketoglutarate in the cycle. Utilization of glutamate and GABA as metabolic substrates does not violate the principle that the intact CNS utilizes oxidation of glucose for its metabolic needs, because both glutamate and GABA have been formed behind the blood-brain-barrier from glucose. Formation and subsequent oxidative degradation of glutamate, including the oxidation of the generated pyruvate provides almost the same amount of energy as the complete oxidative degradation of one molecule of glucose (Hertz, 2004a). Astrocytes in primary cultures are able to carry out the complete process (Hertz and Hertz, 2003), but it is unknown if the generated pyruvate also in the brain in vivo is metabolized by astrocytes. If this is the case they would account for ~30% of total glucose metabolism in brain.

One may ask what advantage the CNS can gain from utilization of glutamate, GABA and glutamine, formed from glucose, as metabolic substrates. The answer to this question probably is that regulation of the extracellular concentrations of glutamate and GABA has very high priority. There is evidence that increased glutamatergic function is associated with enhanced formation of glutamate from glucose (Hertz et al., 2003). Subsequently glutamatergic neurotransmission is terminated by an intense, active uptake of glutamate, mainly in astrocytes, where the accumulated glutamate can be either converted to glutamine and returned to neurons in the glutamate-glutamine cycle, or oxidatively degraded. The oxidative degradation is the major way to effectively dispose of excess glutamate, since it does not exit rapidly across the blood-brain-barrier. It is consistent with this concept that the fraction of astrocytically accumulated glutamate that is oxidatively degraded, increases with an increasing extracellular concentration of glutamate (McKenna et al., 1996). Thus, formation of glutamate from glucose in the CNS and its further metabolic degradation can be seen as successive parts of a metabolic pathway for glucose oxidation, which initially leads to production of the excitatory transmitter glutamate (and the inhibitory transmitter GABA), and subsequently to the oxidative degradation of glutamate (and to some extent also of GABA).

The possibility to form glucose and glycogen from glutamate enables the central nervous system to metabolize more glutamate and glutamine than needed at that specific time for energy production and store the surplus for later oxidative degradation.

Brain Activation

Great efforts have gone into the examination of metabolic events, which are increased during activation of brain function, e.g., by sensory stimulation, provocation of motor activity involving increased glutamatergic function, or simply by opening the animals cages. Increased metabolic activity is indicated by enhanced blood flow in the brain, increased cerebral glucose utilization, and increased consumption of oxygen in the brain. In many cases, a mismatch has been observed between large increases in cerebral blood flow and glucose utilization and a much smaller increase in oxygen consumption during the activation (Fox and Raichle, 1986; Barinaga, 1997), resulting in a ratio between oxygen consumption and glucose utilization of well below 6.0. If attention also is paid to the additional glycogenolysis the mismatch is even more pronounced (Dienel and Criz, 2003, 2004). In some instances a higher rate of oxygen consumption than corresponding to the utilization of glucose has been observed *after* the stimulation (Madsen et al., 1999). These observations suggest that during the activation a glucose metabolite accumulates, which is completely oxidized only after a considerable delay. The lactate concentration in the tissue increases from ~1 to ~2 mM (Kuhr et al., 1988; Prichard et al., 1991), and during a 5-min stimulation in the rat approximately one half of the mismatch between glucose and oxygen utilization could be explained by lactate accumulation (Dienel et al., 2002). Other glucose metabolites, which are not fully oxidized, perhaps primarily glutamate and glutamine, may also contribute to the higher rate of glucose utilization than corresponding to oxygen consumption rate (Hertz, 2004a). Moreover, some lactate may move away from the activated region through astrocytic gap junctions (Cruz et al., 1999).

When lactate production from glycogen is also considered the rate of lactate production may be higher in astrocytes than in neurons, and excess lactate concentrations must be released from the cells in order not to disturb the redox equilibrium to such an extent that aerobic glycolysis cannot continue (Dienel and Hertz, 2001). Accordingly the extracellular lactate concentration also increases (Kuhr et al., 1988; Hu and Wilson, 1997), but a flow of lactate from astrocytes to neurons as proposed by Magistretti et al. (1999) has not been established (Hertz, 2004b). Under a few conditions lactate is released to blood, e.g., during spreading depression (Cruz et al., 1999), when the lactate content in brain becomes approximately tripled (Scheller et al., 1992) and the glycogen level decreases (Krivanek, 1958; Lauritzen et al., 1990). During spreading depression a wave of transient EEG depression spreads across the entire cortex, accompanied by a huge increase of K^+ in the extracellular fluid (to ~50 mM), and subsequent rapid cellular re-accumulation of K^+ after the passage of the wave (Somjen, 1979).

Stimuli of Energy Metabolism

The precise character of the events which cause a stimulation of glucose metabolism and the cellular and subcellular localizations where the stimulation occurs are not well understood. Neuronal excitation leads to several physiological processes that increase energy metabolism: i) during neuronal excitation K^+ leaves the cell and Na^+ enters, and extrusion of the increased intraneuronal Na^+ content requires energy and is coupled with neuronal K^+ uptake; ii) a large part of the initial removal of K^+ from the extracellular fluid occurs in astrocytes; iii) Ca^{2+} is taken up into synaptic vesicles in association with transmitter release, into Purkinje cells during Ca^{2+} mediated action potentials, and into the cytosol of both neurons and astrocytes during depolarization, and subsequent re-extrusion of Ca^{2+} is energy requiring; iv) following release of glutamate and GABA transmitter activity is terminated by energy-requiring cellular re-uptake of the amino acid transmitters into both neurons and astrocytes, by energy-requiring glutamine synthesis in astrocytes and by energy-requiring uptake of glutamine in neurons. It is presently not possible to distinguish with any kind of accuracy between the importance of each of these energy-requiring processes during CNS stimulation, and their relative importance may differ according to the type of stimulation. Moreover, the phenomena occurring during many types of CNS stimulation are obviously interrelated, with glutamatergic activity leading to postsynaptic release of K^+ and influx of Na^+ and both pre-and postsynaptic increase in $[Ca^{2+}]_i$. A robust correlation has been described between glutamate-glutamine cycle activity and rate of stimulation of oxidative metabolism in animals during different levels of anesthesia and convulsive activity (Shulman et al, 2004). This does not indicate that the energy-requiring processes associated with the cycle as such (astrocytic uptake of glutamate and formation of glutamine; neuronal uptake of glutamine) account for the main part of the stimulation, but reflects that glutamate is the major excitatory transmitter, and that *the sum of processes elicited by glutamate release, including ion fluxes associated with action potential generation and subsequent restoration of ion distribution*, accounts for a major fraction of the increased energy metabolism. It is likely that post-stimulation extrusion of Na^+ and uptake of K^+ across cell membranes in most cases will account for most of the increased energy demand.

Restoration of Na^+ and K^+ gradients in neurons is triggered by the increased intracellular concentration of Na^+ which stimulates Na^+, K^+-ATPase activity, Na^+ extrusion, neuronal K^+ uptake and energy metabolism in neurons. In addition K^+ ions are taken up by astrocytes, triggered by the elevated extracellular K^+ concentration and stimulating astrocytic energy metabolism. Although it is known that the resulting stimulation of energy metabolism mainly occurs in the neuropil (Sokoloff, 1992), little or no information is available about its precise cellular and subcellular localization. Much of the information which has been obtained about metabolic manifestations of restoration of K^+ and Na^+ gradients at the cellular level has been obtained using simpler systems such as isolated synaptosomes and primary cultures.

Experiments using cultures of the glutamatergic cerebellar granule cell neurons may provide some clues. Histologically these cells are similar to their counterparts in the brain, with a well developed neuropil, displaying synaptic profiles and dendrites (Figure 15A.). An increase of the extracellular K^+ concentration to 50 mM leads to a distinct increase in oxidative metabolism of glucose, probably at least partly subsequent to a depolarization-induced glutamate release (Peng et al., 1994). One hundred μM ouabain decreases the stimulation of oxidative metabolism by at least 75% (Peng et al., 1996), indicating that most of the excitation-induced increase in energy demand reflects stimulation of Na^+, K^+-ATPase activity. Cerebellar granule cells can also be grown in such a manner that they show massive degeneration of dendrites (Figure 15B), but normal presynaptic structures (Peng et al., 1991). In such neurons elevated K^+ evokes less stimulation of glucose oxidation, and the remaining stimulation is virtually insensitive to ouabain (Peng and Hertz, 1993). The resulting conclusion that stimulation of Na^+, K^+-ATPase activity and thus of energy metabolism in normal neurons mainly may be a dendritic phenomenon is consistent with the histochemical demonstration that key metabolic enzymes are enriched in dendrites (Kageyama and Wong-Riley, 1982). It is also in agreement with the observation that the "mitochondrial" form of the creatine kinase in the cerebellum is primarily localized in the glomeruli (Hemmer and Wallimann, 1993), i.e., the synaptic connections between the glutamatergic mossy fibers and the cerebellar granule cells (Figure 1), which contain many cerebellar granule cell dendrites. That enhanced postsynaptic activity is a major determinant for increased energy demand during brain activation has also been concluded from in vivo experiments using brain imaging (Lauritzen, 2001; Gjedde et al., 2002).

Exposure to a concentration of K^+ similar to the maximum increase in the brain after neuronal excitation, i.e., 12 mM, has no effect on energy metabolism in cultured neurons. However, it stimulates deoxyglucose phosphorylation in primary cultures of astrocytes in an ouabain sensitive manner (Peng et al., 1994; Peng and Hertz, 1993; Hertz et al., 1998; Honneger et al., 1999). Exposure to higher K^+ concentrations (e.g., 30-50 mM) causes no further increase in deoxyglucose phosphorylation in astrocytes than 12 mM K^+, reflecting maximum stimulation of astrocytic Na^+, K^+-ATPase activity at this K^+ concentration. It does, however, stimulate breakdown of glycogen in cultured astrocytes. This is because K^+-induced glycogenolysis is secondary to Ca^{2+}entry and therefore has similar concentration dependence as the K^+ effect on $[Ca^{2+}]_i$. The same may apply to a transient, ouabain-resistant stimulation of glucose oxidation in astrocytes exposed to high concentrations of extracellular K^+, which activates the cotransporter of Na^+, K^+ and Cl^- (Hertz et al., 2004). In the brain *in vivo* there will accordingly be a triple effect of excitatory neuronal activity at the dendritic level: i) a Na^+-induced stimulation of dendritic Na^+, K^+-ATPase activity and energy metabolism; an ii) an increase in astrocytic Na^+, K^+-ATPase activity and energy metabolism stimulated by the elevated extracellular K^+ concentration; and iii) an increase of cotransporter activity, which stimulates oxidative metabolism due to intracellular accumulation of Na^+, and of glycogenolysis, both of which are secondary to a K^+-mediated increase of Ca^{2+} uptake through voltage-sensitive L-channels. The stimulation of energy metabolism in both dendrites and adjacent astrocytes are likely to fuel an intense active clearance of excess extracellular K^+, without which further impulse transmission might be blocked after transmission of a few impulses, due to accumulation of extracellular K^+ (Walz and Hertz, 1983). Such a dendritic-astrocytic interaction in dendrites carrying excitatory, glutamatergic receptors is consistent with the ultrastructure of Purkinje cell dendrites shown in Figure 2, which except for the very ends are covered by astrocytic processes. It is also consistent with re-accumulation of K^+ in astrocytes that the monumental re-accumulation of K^+ required after a wave of spreading

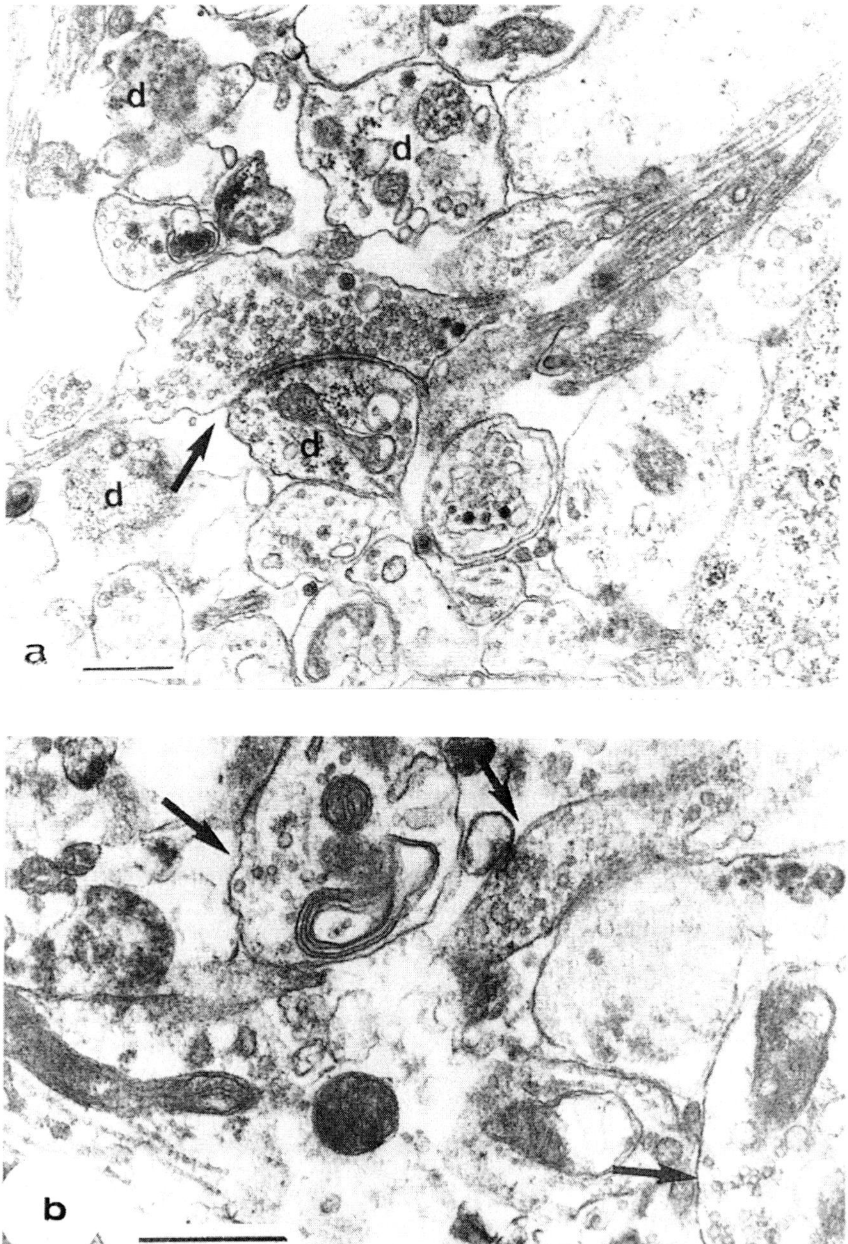

Figure 15. Electron micrographs of neuropil in mouse cerebellar granule cell neurons in primary cultures. A: Conventional culture with intact pre- and postsynaptic structures. A number of dendritic processes (d) can be recognized by the presence of ribosomes, and the arrow indicates an axo-dendritic synapse. B: Corresponding culture grown in such a manner that the dendrites are degenerated, whereas the presynaptic elements, shown by arrows, are intact. Bar equals 0.5 µm. From Peng et al., 1991.

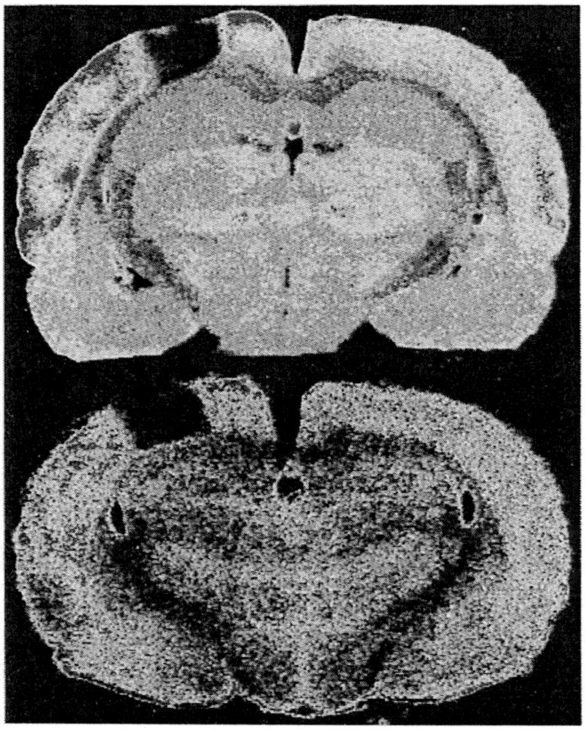

Figure 16. Metabolic imaging of unilateral spreading cortical depression evoked by localized application of KCl on the dura of the left cerebral cortex of the conscious rat. Red indicates high metabolic rate, with progressively lower rates represented by yellow, green, blue, and black. An intravenous pulse of [^{14}C] tracer (deoxyglucose or acetate) was injected at 20 min after the topical application of KCl to the dura. The labeling period was five min for both tracers, and autoradiographs were prepared from serial coronal sections. Spreading depression caused heterogeneous increases in labeling of the left compared to the untreated right cerebral cortex with both [^{14}C]deoxyglucose and [1-^{14}C]acetate. The dark area in the left cortex is the KCl application site. From Hertz and Dienel, 2002.

depression is accompanied not only by an increase in lactate content and a decreased glycogen level (see above), but also by a large increase in both deoxyglucose phosphorylation (which probably occurs in both neurons and astrocytes) and accumulation of labeled acetate (Figure 16), an indication of stimulation of oxidative metabolism specifically in astrocytes (Dienel et al., 2001). The constellation of K^+-mediated increases in both glycolysis and oxidative metabolism, including oxidative metabolism in astrocytes, during spreading depression is in agreement with the concept that the initial phase of K^+ uptake depends on glycolysis, including glycogenolysis, and the later phase is fueled by oxidative metabolism. Other types of brain activation, e.g., acoustic and visual stimulation, are also known to lead to an increase in acetate metabolism (Dienel et al., 1999; Cruz et al., 2005; Hertz and Dienel, 2002).

Uptake of glutamate and GABA in cotransport with Na^+ along its concentration gradient and subsequent extrusion of Na^+ also stimulates Na^+, K^+-ATPase activity, in the case of glutamate much more so in astrocytes (which accumulate most glutamate than in neurons.

This is reflected by a stimulation of oxygen consumption in cultured astrocytes (Eriksson et al., 1995). Since the actual ionic fluxes are several times lower than those of Na^+ extrusion and K^+ uptake following action potentials the energy demands are probably lower. However, glutamine synthesis in astrocytes and glutamine uptake in neurons consume additional ATP. So does pyruvate carboxylation, needed for de novo synthesis of glutamate from glucose. This process is also stimulated by excess extracellular K^+ (Kaufman and Driscoll, 1992). The mechanism is unknown, but it seems neither to be by stimulation of Na^+, K^+-ATPase activity (since the stimulation is not inhibited by ouabain) or by depolarization-induced Ca^{2+} entry (since it is not inhibited in the absence of Ca^{2+}). Because stimulation of pyruvate carboxylase activity promotes net synthesis of TCA cycle constituents it may be an important step in the adjustment of neuronal-astrocytic interactions to CNS stimulation.

Restoration of Ca^{2+} gradients is energy requiring in both neurons and astrocytes. The observation that ouabain inhibits most of the K^+-induced stimulation in normal cerebellar granule cells but has little, if any, effect on the smaller stimulation in neurons with dendritic degeneration may suggest that the remaining stimulation in these cells is at least partly a metabolic manifestation of Ca^{2+} extrusion. This concept is supported by the finding that a K^+-induced stimulation of oxygen uptake in synaptosomes is smaller than in normal cultured cerebellar granule cells, and not inhibited by ouabain, but abolished by Ca^{2+} depletion (Erecinska et al., 1991). In astrocytes the probably relatively small amounts required for reversal of increases in $[Ca^{2+}]_i$ evoked by K^+-mediated depolarization or by transmitters, such as noradrenaline, should not be confused with the quite substantial increases in both oxidative metabolism and glycogen breakdown elicited by K^+-mediated depolarization and ensuing entry of Ca^{2+} through voltage-activated channels or with receptor-mediated release of Ca^{2+} from intracellular stores.

Many types of CNS stimulation lead to release of noradrenaline and perhaps other monoamines or neuropeptides. Noradrenaline (both β-adrenergic and α-adrenergic agonists), vasoactive intestinal protein VIP, a neuropeptide and serotonin are all capable of inducing glycogenolysis in cultured astrocytes (Subbarao and Hertz, 1990; Hertz and Peng, 1992; Magistretti et al., 1993). Noradrenaline increases many additional aspects of energy metabolism, including glycolysis, pyruvate dehydrogenation, TCA cycle activity and oxidative metabolism of glutamine (Subbarao and Hertz, 1991; Hertz and Peng, 1992; Huang and Hertz, 1995; Hertz, 2004a). These effects may explain why intracranially administered norepinephrine causes an increase in oxidative metabolic activity in the brain *in situ* (McKenzie et al., 1976), and noradrenergic antagonists reduce deoxyglucose utilization in many brain nuclei (Savaki et al., 1982). Norepinephrine also enhances glutamate uptake and Na^+, K^+-ATPase activity in astrocytes (Hansson and Rönnbäck, 1988; Hajek et al., 1996; Alexander et al., 1997), i.e., promotes energy requiring mechanisms involved in the removal of neuroactive compounds from the interstitial space. In addition noradrenaline-activated glycogenolysis may also serve as a means to provide precursor for glutamate formation (Hertz et al., 2003).

Conclusion

The objective of the present chapter has been to provide the reader with an appreciation of the enormous histological complexity of the central nervous system (CNS) at both the cellular and subcellular level and of the extent of metabolic and functional interactions between its different components. During increased brain activity resting distribution of Na^+ and K^+ are disturbed, and restoration of pre-stimulus values leads to a major stimulation of energy metabolism, not only in neurons, but also in closely apposed astocyte processes. The initial phase of the restoration is largely fueled by glycolytically derived energy (exemplified by the inhibition of the initial phase of K^+ clearance by the glycolytic inhibitor iodoacetate), perhaps mainly obtained by glycogenolysis, but subsequently oxidative metabolism, generating much larger amounts of ATP takes over and fuels not only energy-requiring processes, including K^+ clearance from the extracellular fluid, but also resynthesis of glycogen. These processes may be specially pronounced in dendrites carrying excitatory activity and their associated astrocytic processes. Other activity-stimulated energy-requiring processes include astrocytic glutamate uptake and glutamine synthesis as well as neuronal uptake of glutamine, which together are likely to account for a somewhat smaller, although still sizeable increase in energy metabolism. Restoration of resting Ca^{2+} and H^+ distribution require relatively low rates of ion transport and may be correlated with smaller energy expenditure, but intramitochondrial Ca^{2+} plays a major role in activation of oxidative metabolism of glucose and of glycogenolysis. Extracellular K^+ is also an important signal in the regulation of metabolic activities, partly operating by a direct effect on the astrocytic Na^+,K^+-ATPase, partly via an increased influx of Ca^{2+}.

Besides the prominent role of glucose in energy metabolism it is an essential precursor for glutamate and GABA, which neurons are unable to synthesize because they are deficient in the enzyme pyruvate carboxylase, an enzyme that has a high activity in astrocytes. The dependence of neurons on astrocytes for their supply of transmitter enables astrocytes to modulate glutamatergic activities. Glutamatergic neurons do not even reaccumulate most of the transmitter they have released, but it is taken up in astrocytes and either metabolized or returned to neurons for further use. Recently an even more direct role of astrocytes in glutamatergic signalling has been identified by the demonstration that glutamate release from stimulated neurons results in increases in $[Ca^{2+}]_i$ in astrocytes and in propagation of Ca^{2+} waves across an astrocytic syncytium, responses that on one hand can maintain glutamatergic synaptic activity after the incoming stimulus has ceased, and on the other hand can exert either a stimulatory or an inhibitory effect on neurons which are not synaptically connected with the initially activated neuron(s).

Major emphasis has been placed on astrocytes in this review, partly because they are the glial cells about which most information is presently available. This should not be taken as an indication that other non-neuronal cells may not be equally important for the function of the nervous sytem. This includes not only oligodendrocytes (Szuchet and Seeger, 2004) but also endothelial cells of the microvessels (Prevot et al., 2004), ependyma cells and tanycytes (Szele and Szuchet, 2004), choroid plexus epithelial cells (Weaver et al., 2004), meningeal and perivascular cells (Mercier and Hatton, 2004) and, not the least, microglial cells, which are of key importance under pathological conditions (Rock et al., 2004), to a large extent via an interaction with astrocytes. We presently possess reasonably good working knowledge of some mechanisms by which neurons and astrocytes interact, but our understanding is fragmentary of interactions that may exist between different non-neuronal cell types as well as between these cells and neurons.

Acknowledgements

My son, Tomas Hertz, MD, is cordially thanked for help during the preparation of this review.

References

Akabayashi, A; Kato, T. Glucose utilization rates in single neurons and neuropil determined by injecting nontracer amounts of 2-deoxyglucose. *J. Neuroche* 1993, 60, 931-935.

Alexander, GM; Grothusen, JR; Gordon, SW; Schwartzman, RJ. Intracerebral microdialysis study of glutamate reuptake in awake, behaving rats. *Brain Res* 1997, 766, 1-10.

Anlauf, E; Derouiche, A. Astrocytic exocytosis vesicles and glutamate: A high-resolution immunofluorescence study. *Glia* 2004, Aug 6 prepublished

Aureli, T; Di Cocco, ME; Calvani, M; Conti, F. The entry of [1-^{13}C]glucose into biochemical pathways reveals a complex compartmentation and metabolite trafficking between glia and neurons: a study by ^{13}C-NMR spectroscopy. *Brain Res* 1997, 765, 218-227.

Balestrino, M; Young, J; Aitken, P. Block of (Na^+,K^+)ATPase with ouabain induces spreading depression-like depolarization in hippocampal slices. *Brain Res* 1999, 838, 37-44.

Barinaga, M. What makes brain neurons run? *Science* 1997, 276, 196-198.

Barres, BA; Chun, LLY; Corey, D; P. Ion channels in vertebrate glia. *Ann. Rev. Neurosci* 1990, 13, 441-474.

Bass, NH; Hess, HH; Pope, A; Thalheimer, C. Quantitative cytoarchitectonic distribution of neurons, glia and DNA in rat cerebral cortex. *J Comp Neurol* 1971, 143, 481-490.

BeltrandelRio, H; Wilson, JE. Interaction of mitochondrially bound rat brain hexokinase with intramitochondrial compartments of ATP generated by oxidative phosphorylation and creatine kinase. *Arch Biochem Biophys* 1992, 299, 116-124.

Blüml, S; Moreno-Torres, A; Shic, F; Nguy, C.H; Ross, BD. Tricarboxylic acid cycle of glia in the in vivo human brain. NMR Biomed 2002, 15, 1-5.

Bringmann, A; Francke, M; Reichenbach, A. Muller cells in retinopathies. In: Hertz, L, editor. *Non-Neuronal Cells of the Nervous System: Function and Dysfunction.* Elsevier; Amsterdam; 2004, pp.1117-1132.

Broër, S; Brookes, N. Transfer of glutamine between astrocytes and neurons. *J Neurochem* 2001, 77, 705-719.

Brookes, N. Intracellular pH as a regulatory signal in astrocyte metabolism. *Glia* 1997. 21, 64-73.

Bushong, EA; Martone, ME; Jones, YZ; Ellisman, MH. Protoplasmic astrocytes in CA1 stratum radiatum occupy separate anatomical domains. *J Neurosci* 2002, 22, 183-192.

Butt, AM; Duncan, A; Berry, M. Astrocyte associations with nodes of Ranvier: ultrastructural analysis of HRP-filled astrocytes in the mouse optic nerve. *J Neurocytol* 1994, 23, 486-499.

Cataldo, AM; Broadwell, RD. Cytochemical staining of the endoplasmic reticulum and glycogen in mouse anterior pituitary cells. *J Histochem Cytochem* 1984, 32, 1285-1294.

Cesar, M de C; Wilson JE. Further studies on the coupling of mitochondrially bound hexokinase to intramitochondrially compartmented ATP, generated by oxidative phosphorylation. *Arch Biochem Biophys* 1998, 350, 109-117.

Chen, Y; Hertz, L. Noradrenaline effects on pyruvate decarboxylation - correlation with calcium signaling. *J Neurosci Res* 1999, 58, 599-606.

Chen, Y; McNeill, JR; Hajek, I; Hertz, L. Effect of vasopressin on brain swelling at the cellular level - do astrocytes exhibit a furosemide-vasopressin-sensitive mechanism for volume regulation ? *Can J Physiol Pharmacol* 1992, 70, S367-S373.

Chen, Y; Zhao, Z; Code, WE; Hertz, L. Correlation between dexmedetomidine-induced biphasic increases in free cytosolic calcium concentration and in energy metabolism in astrocytes. *Anesthesia & Analgesia* 2000, 91, 353-357.

Chesler, M; Kaila, K. Modulation of pH by neuronal activity. *Trends Neurosci* 1992, 15, 396-402.

Chih, C-P; Roberts, EL Jr. Energy substrates for neurons during neural activity: a critical review of the astrocyte-neuron lactate shuttle hypothesis. *J Cereb Blood Flow Metab* 2003, 23, 1263-1281.

Cornell-Bell, AH; Finkbeiner, SM; Cooper, MS; Smith, SJ. Glutamate induces calcium waves in cultured astrocytes: long-range glial signalling. *Science* 1990, 247, 470-473.

Cornell-Bell, AH; Jung, P; Trinkaus-Randall, V. Decoding calcium wave signaling. In: Hertz, L, editor. *Non-Neuronal Cells of the Nervous System: Function and Dysfunction.* Elsevier; Amsterdam; 2004, pp.661-687.

Cruz, F; Cerdán, S. Quantitative ^{13}C NMR studies of metabolic compartmentation in the adult mammalian brain. *NMR Biomed* 1999, 12, 451-462.

Cruz, NF; Dienel, GA. High glycogen levels in brains of rats with minimal environmental stimuli: implications for metabolic contributions of working astrocytes. *J Cereb Blood Flow Metab* 2002, 22, 1476-1489.

Cruz, NF; Adachi, K; Dienel, GA. Rapid efflux of lactate from cerebral cortex during K^+-induced spreading cortical depression. *J Cereb Blood Flow Metab* 1999. 19, 380-392.

Cruz, NF; Lasater, A; Zielke, HR; Dienel, GA. Activation of astrocytes in brains of conscious rats during acoustic stimulation: acetate utilization in working brain. J Neurochem 2005, in press.

Cserr, HF; Bundgaard M. The neuronal microenvironment; a comparative view. *Ann NY Acad Sci* 1986, 48, 1-6.

D'Ambrosio, R; Gordon, DS; Winn, HR. Differential role of KIR channel and Na(+)/K(+)-pump in the regulation of extracellular K(+) in rat hippocampus. *J Neurophysiol* 2002, 87, 87-102.

Danbolt, N. Glutamate uptake. *Prog Neurobiol* 2001, 65, 1-105.

Dani, JW; Smith, SJ. The triggering of astrocytic calcium waves by NMDA-induced neuronal activation. *Ciba Found Symp* 1995, 188, 195-205.

de Graaf, RA; van Kranenburg, A; Nicolay, K. In vivo (31)P-NMR diffusion spectroscopy of ATP and phosphocreatine in rat skeletal muscle. *Biophys J* 2000, 78, 1657-1164.

Derouiche A. The perisynaptic astrocyte process as a glial compartment – immunolabeling for glutamine synthetase and other glial markers. In: Hertz, L, editor. *Non-Neuronal Cells of the Nervous System: Function and Dysfunction.* Elsevier; Amsterdam; 2004, pp.147-163.

Derouiche, A; Frotscher M. Peripheral astrocyte processes: monitoring by selective immunostaining for the actin-binding ERM proteins. *Glia* 2001, 36, 330-341.

Derouiche, A; Anlauf, E; Aumann, G; Muhlstadt, B; Lavialle, M. Anatomical aspects of glia-synapse interaction: the perisynaptic glial sheath consists of a specialized astrocyte compartment. *J Physiol Paris* 2002, 96, 177-182.

Dienel, GA; Cruz, NF. Neighborly interactions of metabolically-activated astrocytes in vivo. *Neurochem Int* 2003, 43, 339-354.

Dienel, GA; Cruz, NF. Nutrition during brain activation: does cell-to-cell lactate shuttling contribute significantly to sweet and sour food for thought? *Neurochem Int* 2004, 45, 321-351.

Dienel, GA; Hertz, L. Glucose and lactate metabolism during brain activation. *J Neurosci Res* 2001, 66, 824-838.

Dienel, GA; Liu, K; Popp, D; Cruz, NF. Enhanced acetate and glucose utilization during graded photic stimulation. Neuronal-glial interactions in vivo. *Ann NY Acad Sci* 1999, 893, 279-281.

Dienel, GA; Liu, K; Cruz, NF. Local uptake of ^{14}C-labeled acetate and butyrate in rat brain in vivo during spreading cortical depression. *J Neurosci Res* 2001, 66, 812-820.

Dienel, GA; Wang, RY; Cruz, NF. Generalized sensory stimulation of conscious rats increases labeling of oxidative pathways of glucose metabolism when the brain glucose-oxygen uptake ratio rises. *J Cereb Blood Flow Metab* 2002, 22, 1490-1502.

Enkvist, MOK; McCarthy K.D. Astroglial gap junction communication is increased by treatment with either glutamate or high K^+ concentration. *J Neurochem* 1994, 62, 489-495.

Erecinska, M; Nelson, D; Chance, B. Depolarization-induced changes in cellular energy production. *Proc Nat Acad Sci US* 1991, 88, 7600-7604.

Erecinska, M; Nelson, D; Nissim, I; Daikhin, Y; Yudkoff, M. Cerebral alanine transport and alanine aminotransferase reaction: alanine as a source of neuronal glutamate. *J Neurochem* 62, 1994, 1953-1964.

Erecinska, M; Deas, J; Silver, IA. The effect of pH on glycolysis and phosphofructokinase activity in cultured cells and synaptosomes. *J Neurochem* 1995, 65, 2765-2772.

Eriksson G; Peterson A; Iverfeldt K; Walum E. Sodium-dependent glutamate uptake as an activator of oxidative metabolism in primary astrocyte cultures from newborn rat. *Glia* 1995, 15, 152-156.

Eugenin, EA; Gonzalez, H; Saez, CG; Saez, JC. Gap junctional communication coordinates vasopressin-induced glycogenolysis in rat hepatocytes. *Am J Physiol* 1998, 274, G1109-G1116.

Fellin, T; Pascual, O; Gobbo, S; Pozzan, T; Haydon, PG; Carmignoto, G. Neuronal synchrony mediated by astrocytic glutamate through activation of extrasynaptic NMDA receptors. *Neuron* 2004, 43, 729-743.

Fonnum, F; Johnsen, A; Hassel, B. Use of fluorocitrate and fluoroacetate in the study of brain metabolism. *Glia* 1997, 21, 106-113.

Fox, PT Raichle, ME. Focal physiological uncoupling of cerebral blood flow and oxidative metabolism during somatosensory stimulation in human subjects. *Proc Nat Acad Sci US* 1986, 83, 1140-1144.

Giacobini, E. Carbonic anhydrase: the first marker of glial development. *Curr Top Dev Biol* 1987, 21, 207-215.

Giaume C; Tabernero A; Medina JM. Metabolic trafficking through astrocytic gap junctions. *Glia* 1997, 21, 114-123.

Gjedde, A; Marrett, S; Vafaee, M. Oxidative and nonoxidative metabolism of excited neurons and astrocytes. *J Cereb Blood Flow Metab* 2002, 22, 1-14.

Grisar, T; Frere, JM; Franck, G. Effect of K^+ ions on kinetic properties of the (Na^+; K^+)-ATPase (EC 3.6.1.3) of bulk isolated glial cells; perikarya and synaptosomes from rabbit brain cortex. *Brain Res* 1979, 165, 87-103.

Grosche, J; Matyash, V; Möller, T; Verkhratsky, A; Reichenbach, A; Kettenmann, H. Microdomains for neuron-glia interaction: parallel fiber signaling to Bergmann glial cells. *Nat Neurosci* 1999, 2, 139-143.

Gruetter, R; Seaquist, ER; Ugurbil, K. A mathematical model of compartmentalized neurotransmitter metabolism in human brain. *Am J Physiol Endocrinol Metab* 2001, 282, E100-112.

Hajek, I; Subbarao, KV; Hertz, L. Stimulation of Na^+,K^+-ATPase activity in astrocytes and neurons by K^+ and/or noradrenaline. *Neurochem Int* 1996, 28, 335-342.

Hansson, E; Rönnbäck, L. Regulation of glutamate and GABA transport by adrenoceptors in primary astroglial cell cultures. *Life Sci* 1988, 44, 27-34.

Hansson, E; Rönnbäck, L. Astrocytic receptors and second messenger systems. In: Hertz, L, editor. *Non-Neuronal Cells of the Nervous System: Function and Dysfunction.* Elsevier; Amsterdam; 2004, pp. 475-501.

Hargittai, PT; Lieberman, EM. Axon-glia interactions in the crayfish: glial cell oxygen consumption is tightly coupled to axon metabolism. *Glia* 1991, 4, 417-423.

Harris-White, ME; Zanott, SA; Frautschy, SA; Charles, AC. Spiral intercellular calcium waves in hippocampal slice cultures. *J Neurophysiol* 1998, 79, 1945-1952.

Hassel, B; Bachelard, H; Jones, P; Fonnum, F; Sonnewald, U. Trafficking of amino acids between neurons and glia in vivo. Effects of inhibition of glial metabolism by fluoroacetate. *J Cerebr Blood Flow Metab* 1997, 17, 1230-1238.

Hatton, GI. Function-related plasticity in the hypothalamus. *Annu Rev Neurosci* 1997, 20, 375-397.

Häusser, M; Spruston, N; Stuart, GJ. Diversity and dynamics of dendritic signaling. *Science* 2000, 290, 739-744.

Hemmer, W; Wallimann, T. Functional aspects of creatine kinase in brain. *Dev Neurosci* 1993, 15, 249-260.

Herreras, O; Somjen, G G. Analysis of potential shifts associated with recurrent spreading depression and prolonged unstable spreading depression induced by microdialysis of elevated K^+ in hippocampus of anesthetized rats. *Brain Res* 1993, 610, 283-294.

Hertz, L. Possible role of neuroglia: A potassium-mediated neuronal-neuroglial-neuronal impulse transmission system. *Nature* 1965, 206, 1091-1094.

Hertz, L. Neuroglial localization of potassium and sodium effects on respiration in brain. *J Neurochem* 1966, 13, 1373-1387.

Hertz, L. Energy metabolism of glial cells. In Schoffeniels, E; Franck, G; Hertz, L; Tower, DB. editors. *Dynamic Properties of Glial Cells*. Oxford, Pergamon Press 1978, pp. 121-132.

Hertz, L. Potassium transport in astrocytes and neurons in primary cultures. *Ann NY Acad Sci* 1986, 481. 318-333.

Hertz, L. Potassium effects on transport of amino acids, inorganic ions and water: Ontogenetic and quantitative differences. In Levi, G; Battistin, L: Lajtha, A, editors *Transport Phenomena in the Nervous System, Physiological and Pathological Aspects* New York; Plenum Press, 1990, pp. 371-383.

Hertz, L. Autonomic control of neuronal-astrocytic interactions regulating metabolic activities and ion fluxes in the CNS. *Brain Res Bull* 1992, 29, 303-338.

Hertz, L. Intercellular Metabolic Compartmentation in the Brain: Past, Present and Future, *Neurochem Internat* 2004a, 45, 285-296.

Hertz, L. The astrocyte-neuron lactate shuttle: a challenge of a challenge. *J Cereb Blood Flow Metab* 2004b, 24, 1241-1348.

Hertz, L; Dienel, GA. Energy metabolism in the brain. *Int Rev Neurobiol* 2002, 51, 1-102.

Hertz, L; Dienel, GA. Lactate transport and transporters: general principles and functional roles in brain cells, *J Neurosci Res* 2005, 79, 11-18.

Hertz E; Hertz L. Polarographic measurement of oxygen uptake by astrocytes in primary cultures using the tissue culture flask as the respirometer chamber. *In Vitro* 1979, 15, 429-436.

Hertz, L; Hertz, E. Determination of rate of glutamate-sustained oxygen consumption in primary cultures of astrocytes as a means to estimate cataplerotic TCA cycle flux. *Neurochem Internat* 2003, 43, 355-361.

Hertz, L; Peng, L. Effects of monoamine transmitters on neurons and astrocytes: correlation between energy metabolism and intracellular messengers. *Progr Brain Res* 1992, 94, 83-301.

Hertz, L; Schousboe, A. Ion and energy metabolism of the brain at the cellular level. *Internat Rev Neurobiol* 1975, 18, 141-211.

Hertz, L; Zielke, HR. Regulation of glutamatergic activities: astrocytes as the stars of the show. *Trends Neurosci* 2004, 27, 735-743.

Hertz, L; Dittmann, L; Mandel, P. K^+ induced stimulation of oxygen uptake in cultured cerebral glial cells. *Brain Res* 1973, 60, 517-520.

Hertz, L; Schousboe, A; Boechler, N; Mukerji, S; Fedoroff, S. Kinetic characteristics of the glutamate uptake into normal astrocytes in cultures. *Neurochem. Res* 1978, 3, 1-14.

Hertz, L; Drejer, J; Schousboe, A. Energy metabolism in glutamatergic neurons, GABAergic neurons and astrocytes in primary cultures. *Neurochem. Res*. 1988, 13, 605-610.

Hertz, L; Bender, AS; Woodbury, D; White, SA. Potassium induced calcium uptake in astrocytes and its potent inhibition by a calcium channel blocker, *J Neurosci Res* 1989, 22, 209-215.

Hertz, L; Gibbs, ME; O'Dowd, BS; Sedman, GL; Robinson, SR; Peng, L; Huang, R; Hertz, E; Hajek, I; Sykova, E; Ng, KT. Astrocyte-neuron interaction during one-trial aversive learning in the neonate chick. *Neurosci Biobehavioral Rev* 1996, 20, 537-551.

Hertz, L; Peng, L; Lai, JCK. Functional studies in cultured astrocytes. *Methods - A Companion to Methods in Enzymology* 1998, 16, 293-310.

Hertz, L; Dringen, R; Schousboe, A; Robinson, SR. Astrocytes: glutamate producers for neurons. *J Neurosci Res* 1999, 57, 417-428.

Hertz, L; Hansson, E; Rönnbäck, L. Signaling and gene expression in the neuron-glia unit during brain function and dysfunction: Holger Hydén in memoriam. *Neurochem Internat* 2001, 39, 227-252.

Hertz, L; Chen Y; Spatz, M. Involvement of non-neuronal brain cells in AVP-mediated regulation of water space at the cellular, organ and whole-body level. *J Neurosci Res* 2002, 62, 480-490.

Hertz, L; O'Dowd, BS; Ng, KT; Gibbs, ME. Reciprocal changes in forebrain contents of glycogen and of glutamate/glutamine during early memory consolidation in the day-old chick. *Brain Res* 2003, 994, 226-233.

Hertz, L; Peng, L; Kjeldsen, CC; O'Dowd, BS; Dienel, GA. Ion, transmitter and drug effects on energy metabolism in astrocytes. In: Hertz, L, editor. *Non-Neuronal Cells of the Nervous System: Function and Dysfunction.* Elsevier; Amsterdam; 2004a, pp. 435-460.

Hertz, L; Chen, Y; Gibbs, ME; Zang, P; Peng, L. Astrocytic adrenoceptors: a major drug target in neurological and psychiatric disorders? *Current Drug Targets – CNS and Neurological Disorders* 2004b, 3, 239-268.

Hof, PR; Pascale, E; Magistretti, PJ. K^+ at concentrations reached in the extracellular space during neuronal activity promotes a Ca^{2+}-dependent glycogen hydrolysis in mouse cerebral cortex. *J Neurosci* 1988, 8, 1922-1928.

Honneger, P; Pardo, B. Separate neuronal and glial Na^+,K^+-ATPase isoforms regulate glucose utilization in response to membrane depolarization and elevated extracellular potassium. *J Cereb Blood Flow Metab* 1999, 19, 1051-1059.

Hösli, E; Hösli L. Colocalization of neurotransmitter receptors on astrocytes in explant cultures of rat CNS. *Neurochem Int* 2000 36, 301-311.

Hu, Y; Wilson, GS. A temporary local energy pool coupled to neuronal activity: fluctuations of extracellular lactate levels in rat brain monitored with rapid-response enzyme-based sensor. *J. Neurochem* 1997, 69, 1484-90.

Huang, R; Hertz, L. Noradrenaline induced stimulation of glutamine metabolism in primary cultures of astrocytes. *J Neurosci Res* 1995, 41, 677-683.

Hutson, SM; Lieth, E; LaNoue, KF. Function of leucine in excitatory neurotransmitter metabolism in the central nervous system. *J Nutr* 2001, 131, 846S-850S.

Isaacson, JS. Glutamate spillover mediates excitatory transmission in the rat olfactory bulb. *Neuron* 1999, 23, 377-384.

Janigro, D; Gasparino, S; D'Ambrosio, R; McKhann, G 2nd; Di Francesco, D. Reduction of K^+ uptake in glia prevents long-term depression maintenance and causes epileptiform activity. *J Neurosci* 1997, 17, 2813-2824.

Jauch, R; Windmuller, O; Lehmann, TN; Heinemann, U; Gabriel, S. Effects of barium, furosemide, ouabaine and 4,4'-diisothiocyanatostilbene-2,2'-disulfonic acid (DIDS) on ionophoretically-induced changes in extracellular potassium concentration in hippocampal slices from rats and from patients with epilepsy. *Brain Res* 2002, 925, 18-27.

Johnston, D; Magee, JC; Colbert, CM; Cristie BR. Active properties of neuronal dendrites. *Annu Rev Neurosci* 1996, 19, 165-86.

Kageyama, GH; Wong-Riley, MTT. Histochemical localization of cytochrome oxidase in the hippocampus: Correlation with specific neuronal types and afferent pathways. *Neuroscience* 1982, 7, 2337-2361.

Kaila, K; Panula, P; Karhunen, T; Heinonen E. Fall in intracellular pH mediated by $GABA_A$ receptors in cultured rat astrocytes. *Neurosci Lett* 1991, 126, 9-12.

Kaufman, EE; Driscoll, BF. Carbon dioxide fixation in neuronal and astroglial cells in culture. *J Neurochem* 1992, 58, 258-262.

Kimelberg, HK; Frangakis, MV. Furosemide- and bumetanide-sensitive ion transport and volume control in primary astrocyte cultures from rat brain. *Brain Res* 1985, 361, 125-134.

Kimelberg, HK; Cai, Z; Schools, G; Zhou, M. Acutely isolated astrocytes as models to probe astrocytic functions. *Neurochem Int* 2000, 36, 359-367.

Krivanek, J. Changes of brain glycogen in the spreading EEG-depression of Leao .J Neurochem 1958, 2, 337-343.

Kuffler, SW; Nicholls, JG; Orkand, RK. Physiological properties of glial cells in the central nervous system of amphibia, *J Neurophysiol* 1966, 29, 768-787.

Kuhr, WG; van den Berg, CJ; Korf, J. In vivo identification and quantitative evaluation of carrier-mediated transport of lactate at the cellular levels in the striatum of conscious, freely moving rats. *J Cereb Blood Flow Metab* 1988, 8, 848-846.

Laming, PR. Potassium signalling in the brain: its role in behaviour. *Neurochem Int* 2000, 36, 271-290.

Laming, P. Potassium and glia-derived slow potential shifts in relation to behaviour. In: Hertz, L, editor. *Non-Neuronal Cells of the Nervous System: Function and Dysfunction.* Elsevier; Amsterdam; 2004, pp.611-633.

Laming, GE; Laming, PR. Tectal responses to potassium loads and subsequent visual stimuli in the toad, Bufo bufo. *Comp Biochem Physiol A Mol Integr Physiol* 2004, 137, 665-674.

Lapidot, A; Gopher, A. Cerebral metabolic compartmentation. Estimation of glucose flux via pyruvate carboxylase/pyruvate dehydrogenase by ^{13}C NMR isotopomer analysis of D-[U-^{13}C]glucose metabolites. *J Biol Chem* 1994, 269, 27198-27208.

Largo, C; Cuevas, P; Somjen, GG; Martin del Rio, R; Herreras, O. The effect of depressing glial function in rat brain in situ on ions homeostasis, synaptic transmission, and neuron survival. *J Neurosci* 1996, 16, 1219-1229.

Lauritzen, M. Relationship of spikes, synaptic activity, and local changes of cerebral blood flow. *J Cereb Blood Flow Metab* 2001, 21, 1367-1383.

Lauritzen, M; Hansen, AJ; Kronborg D; Wieloch, T. Cortical spreading depression is associated with arachidonic acid accumulation and preservation of energy charge. J Cereb Blood Flow Metab 1990, 10, 115-122.

Lebon, V; Petersen, KF Cline, GW; Shen, J; Mason, GF; Dufour, S; Behar, KL; Shulman, GI; Rothman, DL.. Astroglial contribution to brain energy metabolism in humans revealed by ^{13}C nuclear magnetic resonance spectroscopy: elucidation of the dominant pathway for neurotransmitter glutamate repletion and measurement of astrocytic oxidative metabolism. J Neurosci 2002, 22, 1523-31.

Lian, XY; Stringer JL. Astrocytes contribute to regulation of extracellular calcium and potassium in the rat cerebral cortex during spreading depression. *Brain Res* 2004, 1012, 177-184.

Lieberman, EM; Hargittai, PT; Grossfeld, RM. Electrophysiological and metabolic interactions between axons and glia in crayfish and squid. *Progr Neurobiol* 1994, 44, 333-376.

Llinas, R; Sugimori, M. Electrophysiological properties of in vitro Purkinje cell dendrites in mammalian cerebellar slices. *J Physiol*, 1980 305,197-213.

MacKenzie, ET; McCulloch, J; Harper, AM. Influence of endogenous norepinephrine on cerebral blood flow and metabolism. *Am J Physiol* 1976, 231, 489-494.

MacVicar, BA; Feighan, D; Brown, A; Ransom, B. Intrinsic optical signals in the rat optic nerve: role for K(+) uptake via NKCC1 and swelling of astrocytes. *Glia* 2002, 37, 114-123.

Madsen, PL; Cruz, NF; Sokoloff, L; Dienel, GA. Cerebral oxygen/glucose ratio is low during sensory stimulation and rises above normal during recovery: excess glucose consumption during stimulation is not accounted for by lactate efflux from or accumulation in brain tissue. *J Cereb Blood Flow Metab* 1999, 19, 393-400.

Magistretti, PJ; Sorg, O; Yu, N; Martin, JL; Pellerin, L. Neurotransmitters regulate energy metabolism in astrocytes: implications for the metabolic trafficking between neural cells. *Dev Neurosci* 1993, 15, 306-312.

Magistretti, PJ; Pellerin, L; Rothman, DL; Shulman, RG. Energy on demand. *Science* 1999, 283, 496-497.

Martina, M; Yao, GL; Bean, BP. Properties and functional role of voltage-dependent potassium channels in dendrites of rat cerebellar Purkinje neurons. *J Neurosci* 2003, 23, 5698-5707.

Matz, H; Hertz, L. Adenosine metabolism in neurons and astrocytes in primary cultures, *J Neurosci Res* 1989, 24, 260-267.

McAlear, SD; Bevensee, MO. PH regulation in non-neuronal brain cells and interstitial fluid. In: Hertz, L, editor. *Non-Neuronal Cells of the Nervous System: Function and Dysfunction.* Elsevier; Amsterdam; 2004, pp.707-745.

McKenna, MC; Sonnewald, U; Huang, X; Stevenson, JH; Zielke, HR. Exogenous glutamate concentration regulates the metabolic fate of glutamate in astrocytes. *J Neurochem* 1996, 66, 386-393.

McKenna, MC; Stevenson, JH; Huang, X; Hopkins, IB. Differential distribution of the enzymes glutamate dehydrogenase and aspartate aminotransferase in cortical synaptic mitochondria contributes to metabolic compartmentation in cortical synaptic terminals. *Neurochem Int* 2000, 37, 229-241.

Mercier, F; Hatton, GI. Meninges and perivasculature as mediators of CNS plasticity. In: Hertz, L, editor. *Non-Neuronal Cells of the Nervous System: Function and Dysfunction.* Elsevier; Amsterdam; 2004, pp.215-253.

Miyata, Y; Otsuka M. Quantitative histochemistry of gamma-aminobutyric acid in cat spinal cord with special reference to presynaptic inhibition. *J Neurochem*, 1975, 25, 239-244.

Montana, V; Ni, Y; Sunjara, V; Hua, X; Parpura, V. Vesicular glutamate transporter-dependent glutamate release from astrocytes. *J Neurosci* 2004, 24, 2633-2642.

Muir, D; Berl, S; Clarke, DD. Acetate and fluoroacetate as possible markers for glial metabolism in vivo. *Brain Res* 1986, 380, 336-340.

Nadkarni, S; Jung, P. Dressed neurons: modeling neural-glial interactions. *Phys Biol* 2004, 1, 35-41.

Nehlig, A; Wittendorp-Rechenmann, E; Lam CD. Selective uptake of [^{14}C]2-deoxyglucose by neurons and astrocytes: high-resolution microautoradiographic imaging by cellular ^{14}C-trajectography combined with immunohistochemistry. *J Cereb Blood Flow Metab* 2004, 24, 1004-1014.

Newman, EA. Glial cell inhibition of neurons by release of ATP. *J Neurosci* 2003, 23, 1659-1666.

Newman, EA. New roles for astrocytes: regulation of synaptic transmission. *Trends Neurosci* 2003b, 26, 536-542.

Newman, EA; Zahs, KR. Modulation of neuronal activity by glial cells in the retina. *J Neurosci* 1998, 18, 4022-4028.

Newman, EA; Frambach, DA; Odette, LL. Control of extracellular potassium levels by retinal glial cell K^+ siphoning. *Science* 1984, 225, 1174-1175.

Nilsson, M; Hansson, E; Rönnbäck, L. Heterogeneity among astroglial cells with respect to 5HT-evoked cytosolic Ca^{2+} responses. A microspectrofluorimetric study on single cells in primary culture. *Life Sci* 49, 1339-1350.

Norenberg, MD; Martinez-Hernandez, A. Fine structural location of glutamine synthetase in astrocytes of rat brain. *Brain Res* 1979, 161, 303-310.

O'Kane, RL; Martinez-Lopez, I; DeJoseph, MR; Vina, JR; Hawkins, RA. Na^+-dependent glutamate transporters (EAAT1, EAAT2, and EAAT3) of the blood-brain barrier. A mechanism for glutamate removal. *J Biol Chem* 1999, 274, 31891-31895.

Palay, SL Chan-Palay, V. Cerebellar Cortex, Cytology and Organization, 1974. New York, Springer-Verlag.

Parpura, V; Haydon, PG. Physiological astrocytic calcium levels stimulate glutamate release to modulate adjacent neurons. *Proc Nat Acad Sci US* 2000, 97, 8629-8634.

Pellerin, L; Magistretti, PJ. Glutamate uptake into astrocytes stimulates aerobic glycolysis: A mechanism coupling neuronal activity to glucose utilization. *Proc Nat Acad Sci US* 1994, 91, 10625-10629.

Pellerin, L; Magistretti, PJ. Food for thought: challenging the dogmas. *J Cereb Blood Flow Metab* 2003, 23, 1282-1286.

Peng, L; Hertz, L. Potassium induced stimulation of oxidative metabolism of glucose in cultures of intact cerebellar granule cells but not in corresponding cells with dendritic degeneration. *Brain Res* 1993, 629, 331-334.

Peng, L; Juurlink, BHJ; Hertz, L. Differences in transmitter release, morphology, and ischemia-induced cell injury between cerebellar granule cell cultures developing in the presence and in the absence of a depolarizing potassium concentration. *Dev Brain Res* 1991, 63, 1-12.

Peng, L; Zhang, X; Hertz, L. High extracellular potassium concentrations stimulate oxidative metabolism in a glutamatergic neuronal culture and glycolysis in cultured astrocytes but have no stimulatory effect in a GABAergic neuronal culture. *Brain Res* 1994, 663, 168-172.

Peng, L; Juurlink, BHJ; Hertz L. Pharmacological and developmental evidence that the potassium-induced stimulation of deoxyglucose uptake in astrocytes is a metabolic manifestation of increased Na^+-K^+-ATPase activity. *Dev Neurosci* 1996, 18, 353-359.

Peng, L; Swanson, RA; Hertz, L. Effects of L-glutamate, D-aspartate and monensin on glycolytic and oxidative glucose metabolism in mouse astrocyte cultures. *Neurochem Internat* 2001, 38, 437-443.

Peters, A; Palay, SL; Webster, H deF. The Fine Structure of the Nervous System. Neurons and their Supporting Cells. 3rd edn. 1991. Oxford, Oxford University Press.

Peuchen, S; Duchen, MR; Clark, JB. Energy metabolism of adult astrocytes in vitro. *Neuroscience* 71, 1996, 855-870.

Piet, R; Vargova, L; Sykova, E; Poulain, DA; Oliet, SH. Physiological contribution of the astrocytic environment of neurons to intersynaptic crosstalk. *Proc Nat Acad Sci US* 2004, 101, 2151-2155.

Pope, A. Neuroglia: Quantitative aspects. In: Schoffeniels, E; Franck, G; Hertz, L; Tower, D.B, editors. *Dynamic Properties of Glia Cells.* 1978. Oxford; Pergamon Press, pp. 13-20.

Pouille, F; Cavelier, P; Desplantez, T; Beekenkamp, H; Craig, PJ; Beattie, RE; Volsen, SG; Bossu JL. Dendro-somatic distribution of calcium-mediated electrogenesis in Purkinje cells from rat cerebellar slice cultures. *J Physiol*, 2000, 527, 265-282.

Prevot, V; De Seranno, S; Estrella, C. Glial-neuronal-endothelial interactions and the neuroendocrine control of GnRH secretion. In: Hertz, L, editor. *Non-Neuronal Cells of the Nervous System: Function and Dysfunction.* Elsevier; Amsterdam; 2004, pp.199-214.

Prichard, J; Rothman, D; Novotny, E; Petroff, O; Kuwabara, T; Avison, M; Howseman, A; Hanstock, C; Shulman, R. Lactate rise detected by 1H NMR in human visual cortex during physiologic stimulation. *Proc Nat Acad Sci US* 1991, 88, 5829-5831.

Raffin, CN; Rosenthal, M; Busto, R; Sick, TJ. Glycolysis, oxidative metabolism, and brain potassium ion clearance. *J Cereb Blood Flow Metab* 1992, 12, 34-42.

Ransom, BR. Glial modulation of neural excitability mediated by extracellular pH: A hypothesis. *Prog Brain Res*, 1992, 94, 37-46.

Ransom, BR. Glial modulation of neural excitability mediated by extracellular pH: a hypothesis revisited. *Prog Brain Res* 2000, 125, 217-228.

Ransom, CB; Ransom, BR; Sontheimer, H. Activity-dependent extracellular K^+ accumulation in rat optic nerve: the role of glial and axonal Na^+ pumps. *J Physiol* 2000, 522, 427-442.

Rash, JE; Duffy, HS; Dudek, FE; Bilhartz, BL; Whalen, LR; Yasumura, T. Grid-mapped freeze-fracture analysis of gap junctions in gray and white matter of adult rat central nervous system, with evidence for a "panglial syncytium" that is not coupled to neurons. *J Comp Neurol* 1997, 388, 265-292.

Reichenbach, A; Stolzenburg, J-U; Eberhardt, W; Chao, TI; Dettmar, D; Hertz, L. What do retinal Muller (glial) cells do for their neuronal "small siblings"? *J Chem Neuroanat* 1993, 6, 201-213.

Richter, K; Hamprecht, B; Scheich, H. Ultrastructural localization of glycogen phosphorylase predominantly in astrocytes of the gerbil brain. *Glia* 1996, 17, 263-273.

Robinson, SR; Dreher, Z. Muller cells in adult rabbit retinae: Morphology, distribution and implications for function and development. *J Comp Neurol* 1990, 292, 178-192.

Rock, RB; Gekker, G; Hu, S; Sheng, WS; Cheeran, M; Lokensgard, JR; Peterson, PK. Role of microglia in central nervous system infections. *Clin Microbiol Rev* 2004, 17, 942-964.

Rose, CR; Ransom, BR. Intracellular sodium homeostasis in rat hippocampal astrocytes. *J Physiol* 1996, 491, 291-305.

Rose, CR; Ransom, BR. Regulation of intracellular sodium in cultured rat hippocampal neurones. *J Physiol* 1997, 499, 573-587.

Rouach, N; Koulakoff, A; Giaume, C. Neurons set the tone of gap junctional communication in astrocytic networks. *Neurochem Int* 2004, 45, 265-272.

Rozental, R; Srinivas, M; Spray, DC. How to close a gap junction channel. In: Bruzzone, R; Giaume, C, editors. *Methods in Molecular Biology, vol. 154. Connexin Methods and Protocols*. Totowa NJ, Humana Press; 2001; pp. 447-476.

Rosenthal, M; Sick, TJ. Glycolytic and oxidative metabolic contributions to potassium ion transport in rat cerebral cortex. *Can J Physiol Pharmacol* 1992, 70, S165-S169.

Salm, AK; Ayoub, AE; Lally, BE. Structural plasticity of nonneuronal cells in the hypothalamo-neurohypophyseal system: in the right place at the right time. In: Hertz, L, editor. *Non-Neuronal Cells of the Nervous System: Function and Dysfunction.* Elsevier; Amsterdam; 2004, pp.181-198.

Savaki, HE; Kadekaro, M; McCulloch, J; Sokoloff, L. The central noradrenergic system in the rat: metabolic mapping with alpha-adrenergic blocking agents. *Brain Res* 1982, 234, 65-79.

Scapagnini, G; Nelson, TJ; Alkon, DL. Regulation of Ca^{2+} stores in glial cells. In: Hertz, L, editor. *Non-Neuronal Cells of the Nervous System: Function and Dysfunction.* Elsevier; Amsterdam; 2004, pp.635-660.

Scemes, E; Spray DC. The astrocytic syncytium. In: Hertz, L, editor. *Non-Neuronal Cells of the Nervous System: Function and Dysfunction.* Elsevier; Amsterdam; 2004, pp.165-179.

Scheller, D; Kolb, J; Tegtmeier, F; Lehmenkühler A. Extracellular changes of inorganic phosphate are different during spreading depression and global cerebral ischemia of rats. *Neurosci Lett* 1992, 141, 269-272.

Schousboe, A; Sarup, A; Bak, LK; Waagepetersen, HS; Larsson, OM. Role of astrocytic transport processes in glutamatergic and GABAergic neurotransmission. Neurochem Int 2004, 45, 521-527.

Shank, RP; Bennett, GS; Freytag, SO; Campbell, GL. Pyruvate carboxylase: An astrocyte-specific enzyme implicated in the replenishment of amino acid neurotransmitter pools. *Brain Res* 329, 364-367.

Shuai, J-W; Nadkarni, S; Jung, P; Cornell-Bell, A; Trinkaus-Randall, V. Decoding calcium wave signaling. In: Hertz, L, editor. *Non-Neuronal Cells of the Nervous System: Function and Dysfunction.* Elsevier; Amsterdam; 2004, pp.689-706.

Shulman, RG; Rothman, DL; Behar, KL; Hyder F. Energetic basis of brain activity: implications for neuroimaging. *Trends Neurosci* 2004, 27, 489-495.

Skou, JC. The influence of some cations on an adenosine triphosphatase from peripheral nerve. *Biochim Biophys Acta* 1957, 23, 394-401.

Skou, JC. Nobel Lecture. The identification of the sodium pump. *Biosci Rep* 1998, 18, 155-169.

Smith, QR. Transport of glutamate and other amino acids at the blood-brain barrier. *J Nutr* 2000, 130, 1016S-1022S.

Sokoloff, L. The brain as a chemical machine. *Prog Brain Res* 1992, 94, 19-33.

Somjen, GG. Extracellular potassium in the mammalian central nervous system, *Ann Rev Physiol* 1979, 41, 159-177.

Sontheimer, H. Voltage-dependent ion channels in glial cells. *Glia* 1994, 11, 156-172.

Su, G; Haworth, RA; Dempsey, RJ; Sun, D. Regulation of Na^+-K^+-Cl^- cotransporter in primary astrocytes by dibutyryl cAMP and high $[K^+]_o$. *Am J Physiol Cell Physiol* 2000, 279, 1710-1721.

Subbarao, KV; Hertz, L. Effects of adrenergic agonists on glycogenolysis in primary cultures of astrocytes. *Brain Res* 1990, 536, 220-226.

Subbarao, KV; Hertz, L. Stimulation of energy metabolism in astrocytes by adrenergic agonists. *J Neurosci Res* 1991, 28, 399-405.

Subbarao, KV; Stolzenburg, J-U; Hertz, L. Pharmacological characteristics of potassium-induced glycogenolysis in astrocytes. *Neurosci Lett* 1995, 196, 45-48.

Swanson, RA. Physiologic coupling of glial glycogen metabolism to neuronal activity in brain. *Can J Physiol Pharmacol* 1992, 70, S138-S144.

Swanson, RA; Morton, MM; Sagar, SM; Sharp, FR. Sensory stimulation induces local cerebral glycogenolysis: Demonstration by autoradiography. *Neuroscience* 1992, 51, 451-461.

Sykova, E. Ionic and Volume Changes in the Microenvironment of Nerve and Receptor Cells. Progress in Sensory Physiology, Vol. 13. Springer-Verlag, New York 1992.

Sykova, E; Chvatal, A. Extracellular ionic and volume changes: The role in glia-neuron interaction. *J Chem Neuroanat* 1993 6, 247-260.

Sykova, E; Jendelova, P; Svoboda, J; Chvatal, A. Extracellular K^+, pH, and volume changes in spinal cord of adult rats and during postnatal development. *Can J Physiol Pharmacol* 1992, 70, S301-S309.

Szele, FG; Szuchet, S. Cells lining the ventricular system: evolving concepts underlying developmental events in the embryo and adult. In: Hertz, L, editor. *Non-Neuronal Cells of the Nervous System: Function and Dysfunction.* Elsevier; Amsterdam; 2004, pp.127-146.

Szuchet, S; Seeger, MA. Oligodendrocyte phenotypical and morphological heterogeneity: a reexamination of old concepts in view of new findings. In: Hertz, L, editor. *Non-Neuronal Cells of the Nervous System: Function and Dysfunction.* Elsevier; Amsterdam; 2004, pp.53-73.

Tachikawa, M; Fukaya, M; Terasaki, T; Ohtsuki, S; Watanabe, M. Distinct cellular expressions of creatine synthetic enzyme GAMT and creatine kinases uCK-Mi and CK-B suggest a novel neuron-glial relationship for brain energy homeostasis. *Eur J Neurosci* 2004, 20, 144-160.

Taira, T; Smirnov, S; Voipio; Kaila, K. Intrinsic proton modulation of excitatory transmission in rat hippocampal slices. *NeuroReport* 1993, 4, 93-96.

Tas, PWl; Massa, PT; Kress, HG; Koschel, K. Characterization of an Na^+, K^+, Cl^- cotransport in primary cultures of rat astrocytes. *Biochim Biophys Acta* 1987, 903, 411-416.

Van Harreveld, A; Steiner, J. Extracellular space in frozen and ethanol substituted central nervous tissue. *Anat Rec* 1970, 166, 117-129.

Vannucci, SJ; Maher, F; Simpson IA. Glucose transporter proteins in brain: delivery of glucose to neurons and glia. *Glia* 1997, 21, 2-21.

Vetter, P; Roth, A; Häusser, M. Propagation of action potentials in dendrites depends on dendritic morphology. *J Neurophysiol* 2001, 85, 926-37.

Volterra, A; Magistretti, PJ; Haydon, PG, editors. The Tripartite Synapse, New York; Oxford University Press 2002.

Waagepetersen, HS; Qu, H; Hertz, L; Sonnewald, U; Schousboe, A. Demonstration of pyruvate recycling in primary cultures of neocortical astrocytes but not in neurons. *Neurochem Res* 2002, 27, 1431-1437.

Wallraff A; Odermatt B; Willecke K; Steinhauser C. Distinct types of astroglial cells in the hippocampus differ in gap junction coupling. *Glia* 2004, 48, 36-43.

Walz, W. Role of astrocytes in the clearance of excess extracellular potassium. *Neurochem Int* 2000, 36, 291-300.

Walz, W. Chloride/anion channels in glial cell membranes. *Glia* 2002, 40, 1-10.

Walz, W. Potassium homeostasis in the brain at the organ and cell level. In: Hertz, L, editor. *Non-Neuronal Cells of the Nervous System: Function and Dysfunction.* Elsevier; Amsterdam; 2004, pp. 595-609.

Walz, W; Hertz, L. Intense furosemide-sensitive potassium accumulation into astrocytes in the presence of pathologically high extracellular potassium levels. *J Cereb Blood Flow Metab* 1984, 4, 301-304.

Walz, W; Hertz, L. Functional interactions between neurons and astrocytes II. Potassium homeostasis at the cellular level. *Prog Neurobiol* 1993, 20, 133-183.

Walz, W; Hinks. A transmembrane sodium cycle in astrocytes. *Brain Res* 1986, 368, 326-332.

Walz, W; Wuttke, W. Independent mechanisms of potassium clearance by astrocytes in gliotic tissue. *J Neurosci Res* 1999 56, 595-603.

Walz, W; Shargool, M; Hertz, L. Barium-induced inhibition of K^+ transport mechanisms in cortical astrocytes - its possible contribution to the large Ba^{2+}-evoked extracellular K^+ signal in brain. *Neuroscience* 1984a,13, 945-949.

Walz, W; Wuttke, W; Hertz, L. Astrocytes in primary cultures: Membrane potential characteristics reveal exclusive potassium conductance and potassium accumulator properties. *Brain Res* 1984b, 292, 367-376.

Wang, Z; Haydon, PG; Yeung, ES. Direct observation of calcium-independent intercellular ATP signaling in astrocytes. *Anal Chem* 2000, 72, 2001-2007.

Waniewski, RA; Martin, DL. Preferential utilization of acetate by astrocytes is attributable to transport. *J Neurosci* 1998, 18, 5225-5233.

Weaver, CE; McMillan, PN; Duncan, JA; Stopa, EG; Johanson, CE. Hydrocephalus disorders; their biophysical and neuroendocrine impact on the choroid plexus epithelium. In: Hertz, L, editor. *Non-Neuronal Cells of the Nervous System: Function and Dysfunction.* Elsevier; Amsterdam; 2004, pp.269-293.

Wender, R; Brown, AM; Fern, R; Swanson, RA; Farrell, K; Ransom, BR. Astrocytic glycogen influences axon function and survival during glucose deprivation in central white matter. *J Neurosci* 2000, 20, 6804-6810.

Westergaard, N; Varming, T; Peng, L; Sonnewald, U; Hertz, L; Schousboe, A. Uptake, release and metabolism of alanine in neurons and astrocytes in primary cultures. *J Neurosci Res* 1993, 35, 540-545.

Williams, V; Grossman, RG; Edmunds, SM. Volume and surface area estimates of astrocytes in the sensorimotor cortex of the cat. *Neuroscience* 1980, 5, 1151-1159.

Wofchuk, ST; Rodnight, R. Age-dependent changes in the regulation by external calcium ions of the phosphorylation of glial fibrillary acidic protein in slices of rat hippocampus. *Dev Brain Res* 1995, 85, 181-186.

Wolff, J; Chao, TI. Cytoarchitectonics of non-neuronal cells in the central nervous system. In: Hertz, L, editor, *Non-Neuronal Cells of the Nervous System: Function and Dysfunction.* Amsterdam. Elsevier. 2004, pp 1-52.

Xiong, ZO; Stringer, JL. Astrocytic regulation of the recovery of extracellular potassium after seizures in vivo. *Eur J Neurosci* 1999, 11, 1677-1684.

Xiong, ZQ; Stringer, JL. Sodium pump activity, not glial spatial buffering, clears potassium after epileptiform activity induced in the dentate gyrus. *J Neurophysiol* 2000, 83, 1443-1451.

Yager, J; Kala, G; Hertz, L; Juurlink, BHJ. Correlation between content of high-energy phosphates and hypoxic-ischemic damage in immature and mature astrocytes. *Dev Brain Res* 1994, 82, 62-68.

Yu, ACH; Schousboe, A; Hertz, L. Metabolic fate of ^{14}C-labeled glutamate in astrocytes in primary cultures. *J Neurochem* 1982, 39, 954-960.

Yu, ACH; Drejer, J; Hertz, L; Schousboe, A. Pyruvate carboxylase activity in primary cultures of astrocytes and neurons. *J Neurochem* 1983, 41, 1484-1487.

Zhang, Q; Fukuda, M; Van Bockstaele, E; Pascual, O; Haydon, PG. Synaptotagmin IV regulates glial glutamate release. *Proc Nat Acad Sci US* 2004, 101, 9441-9446.

Zhao, Z; Hertz, L; Code, WE. Effects of benzodiazepines on potassium induced increase in free cytosolic calcium concentration in astrocytes and neurons in primary cultures. *Can J Physiol Pharmacol* 1996, 74, 273-277.

Zhou, M; Kimelberg HK. Freshly isolated astrocytes from rat hippocampus show two distinct current patterns and different $[K^+]_o$ uptake capabilities. *J Neurophysiol* 2000, 84, 2746-2757.

In: Focus on Neurochemistry Research
Editor: Robert M. Coleman, pp.61-99
ISBN 1-59454-417-4

Chapter II

Ethanol Induced Adaptive Changes in Structure and Function of NMDA Receptors

József Nagy,* Sándor Kolok, András Boros, Péter Dezső
Gedeon Richter Ltd., Pharmacological and Drug Safety Research
Laboratory of Molecular and Cellular Biology, Budapest, Hungary

Abstract

Ethanol is a small molecule acting on several neurotransmitter systems in the brain. Accumulating evidences suggest that the primary excitatory – i.e. the glutamatergic – neurotransmitter system is a particularly important site of ethanol's action. Several studies showed that ethanol is a potent and selective inhibitor of the N-methyl-D-aspartate (NMDA) receptors and prolonged ethanol exposition leads to a compensatory "up-regulation" of these receptors resulting in enhanced NMDA receptor mediated functions after removal of ethanol. These alterations are supposed to contribute to the development of ethanol tolerance, dependence as well as the acute and delayed signs of ethanol withdrawal. Recently, alterations in subunit composition of NMDA receptors were reported after long-term ethanol exposure. mRNA and/or protein levels of NR2A and NR2B types of subunits were found elevated both by *in vivo* and *in vitro* experiments. In primary cultures of rat hippocampal or cortical neurons, especially the NR2B subunit expression and levels of certain NR1 splice variants were increased after 3 days of intermittent ethanol treatment. According to the high calcium permeability, the increased agonist sensitivity and the relatively slow closing kinetics of NMDA ion channels composed of NR1 and NR2B subunits, these changes may underlie the enhanced NMDA receptor activation observed after long-term ethanol exposure. Consequently, novel type of NMDA receptor antagonists - the NR2B subunit selective

* Fax: 36-1-2605000; e-mail: jozsef.nagy@richter.hu

agents potently inhibited the neurotoxic effect of ethanol withdrawal seen in primary cultures of rat cortical neurons pre-treated with ethanol. Consistent with these observations, enhanced NR2B type of NMDA receptor subunit expression is likely involved in development of alcohol dependence and withdrawal symptoms, therefore this subunit may serve as a possible drug target in pharmacological interventions against alcoholism.

Introduction

Alcohol use disorders represent a substantial public health problem all over the world. The World Health Organization (WHO) estimated 2 billion alcohol users in 2000, compared with 1.3 billion smokers and 185 million users of other psychoactive drugs. According to the data released by WHO in 2003, the global prevalence of alcohol use disorders was 1.7%, and these disorders accounted for 1.4% of the total world disease burden. Data released by The National Institute on Alcohol Abuse and Alcoholism (NIAAA) in August 2004 showed that whereas in 1991–1992 the total prevalence of 12-month alcohol abuse and dependence – according to the "Diagnostic and Statistical Manual of Mental Disorders, Fourth Edition" (DSM-IV) – was 7.41% in the USA representing 13.8 million adult Americans, this prevalence rose to 8.46% representing 17.6 million Americans in 2001–2002. As it was published by the editorial board of an international general medical journal, The Lancet, in March 2004, alcohol use is the fourth largest cause of disability in men and the tenth largest in women in developed regions of the world. Given the harmful effects of alcohol use disorder on the afflicted individuals and society as a whole, alcohol use disorders continue to represent a substantial public health problem (Grant 2004).

Investigation of the mechanisms underlying the actions of ethanol in the central nervous system (CNS) has been ongoing for more than a century. Early research on acute alcohol actions focused on the physical properties of alcohols and related anesthetics and established the widely appreciated Meyer-Overton relation that describes the direct relationship between the hydrophobicity of an alcohol and its potency for producing intoxication (Meyer 1901; Overton 1896). According to this "lipid theory" of alcohols' action, ethanol is thought to produce its behavior effects by actions on the membrane lipids bringing about changes in the function of the membrane proteins. Consistent with this hypothesis, the intoxicating potency of aliphatic n-alcohols with up to five carbon atoms is correlated with both their lipid solubility and their membrane lipid disordering potency. However, the potency of aliphatic n-alcohols for producing intoxication reaches a maximum at six to eight carbon atoms and then decreases (Peoples and Weight 1995; Godden 2001; Akinshola 2001). The molecular bases of this "cut-off" effect cannot be explained by the "lipid theory", since both lipid solubility and membrane lipid disordering potency increase exponentially with the increase in the carbon chain length. The hypothesis of ethanol's non-specific membrane effects has also been challenged by data demonstrating definite subunit compositions and certain amino-acid residues of sensitive receptors critical for allosteric modulation by ethanol (Allgaier 2002; Mihic 1997; Kumari and Ticku 2000; Ueno 2000).

Effect of Ethanol on Neurotransmission

Although, the exact mechanism by which ethanol exerts its effect is still a matter of debate, studies over the last decade showed that ethanol affecting several neurotransmitter systems differentially modifies neurotransmission processes in the CNS. Besides altering the release of neurotransmitters like dopamine (DA) – resulting in increased DA levels in the "central reward" pathway –, ethanol alters the function of a number of neurotransmitter receptors (e.g.: γ-amino butyric acid A ($GABA_A$), glycine, glutamate, norepinephrine, DA, serotonin, acetylcholine) as well as transporters (adenosine, norepinephrine, DA, serotonin). Especially the ligand-gated ion channels including the brain's major amino acid neurotransmitter systems – the inhibitory GABA and the excitatory glutamate – were shown highly sensitive to the acute effect of ethanol at relevant (<100 mM) concentrations (Lovinger 1990, Li 1994, Franks and Lieb 1994). In the past years, there has been increasing evidence that acute ethanol facilitates GABAergic transmission by enhancing chloride conductance through the $GABA_A$ receptors, and inhibits glutamatergic function by decreasing cationic conductance especially through the N-methyl-D-aspartate (NMDA) activated subclass of glutamate receptors. Conversely, the development of tolerance and dependence associated with chronic ethanol consumption leads to reduced GABAergic and increased glutamatergic functions.

Structure and Function of NMDA Receptors

NMDA receptors (NMDARs) belong to the family of ionotropic glutamate receptors. These receptors are highly expressed in the CNS. The involvement of these receptors in diverse processes like excitatory synaptic transmission, synaptic plasticity, and excitotoxicity rests upon their unique features. Their special characteristics are the high permeability to Ca^{2+} ions, the relatively slow activation / deactivation kinetics and their voltage-sensitive blockage by extracellular Mg^{2+} ions. The Mg^{2+} block results in a unique mechanism of channel activation. Glutamate, the native agonist of the NMDARs, can open the ion-channel only if the plasma membrane became depolarized and the Mg^{2+} blockage was displaced. Thus, NMDARs act as coincidence perceptive elements, which become active only when electrical and chemical signals are present concurrently.

Besides glutamate, NMDARs are sensitive to several other endogenous modulators. For full receptor activation, the co-agonist glycine (Meguro et al., 1992) or D-serine (Mothet 2000) must also bind to the receptor. Endogenous polyamines, spermine and spermidine also facilitate (Lerma 1992; Rock and Macdonald 1992), whereas extracellular Zn^{2+} ions (Christine and Choi 1990) and protons (Tang 1990; Traynelis and Cull-Candy 1990) suppress NMDAR activation. In addition, NMDARs interact with various intracellular scaffolding, anchoring, and signaling molecules associated with the postsynaptic density (Figure 1). The sensitivity of NMDARs to different ligands, its permeation, and block by divalent ions, kinetic properties, and interaction with intracellular proteins highly depend on their subunit composition (Danysz and Parsons 1998; Husi and Grant 2001).

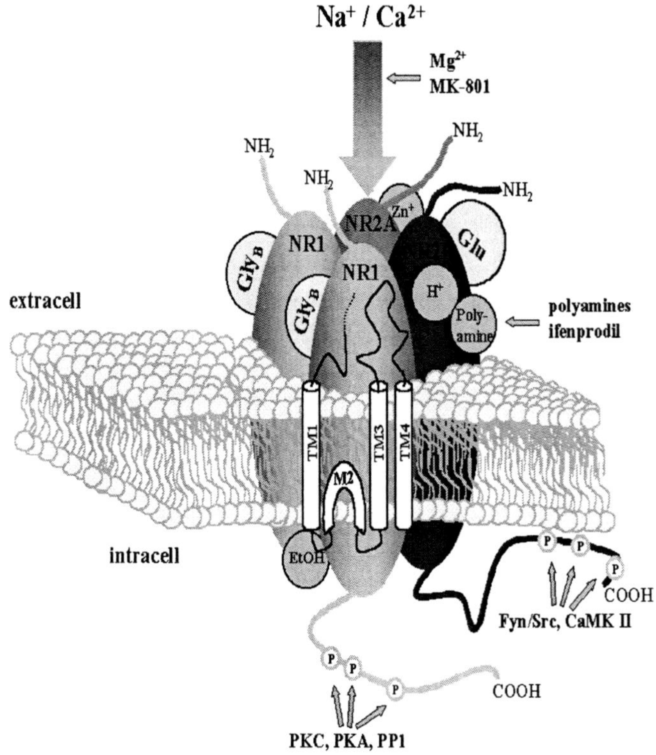

Figure 1. Schematic model of the NMDA receptor showing the heteromeric assemblies of four subunits in NMDA receptors. Each subunit has four hydrophobic regions, although only three form membrane-spanning domains (TM1, TM3, and TM4). M2 makes a hairpin bend within the membrane and forms the channel pore. Functional NMDA receptor complexes are formed by combinations of NR1 and NR2 subunits, which contain the glycine and glutamate recognition sites, respectively. Supposed binding sites of agonists, some antagonists, and other modulators are shown.
From: József Nagy (2004) NR2B type of NMDA receptor subunits: a possible target in treatment for alcohol dependence. *Current Drug Targets – CNS & Neurological Disorders 3,* 169-179.

NMDARs, like other ion-channel receptors, are multimeric transmembrane proteins, made up of different types of subunits. Several distinct NMDAR subunits have been identified in neurons (Figure 2). The ubiquitously expressed NR1 subunit occurs as eight distinct isoforms because of three independent sites (N1, C1 and C2 cassettes) of alternative splicing. Families of four distinct NR2 (A, B, C and D) and two NR3 (A, B) subunits are also identified (Moriyoshi 1991; Hollmann 1999; Das 1998). Although, the precise subunit composition and stoichiometry of native NMDARs are difficult to determine and largely unknown (Premkumar and Auerbach 1997; Laube 1998), it is believed that NMDARs exist as

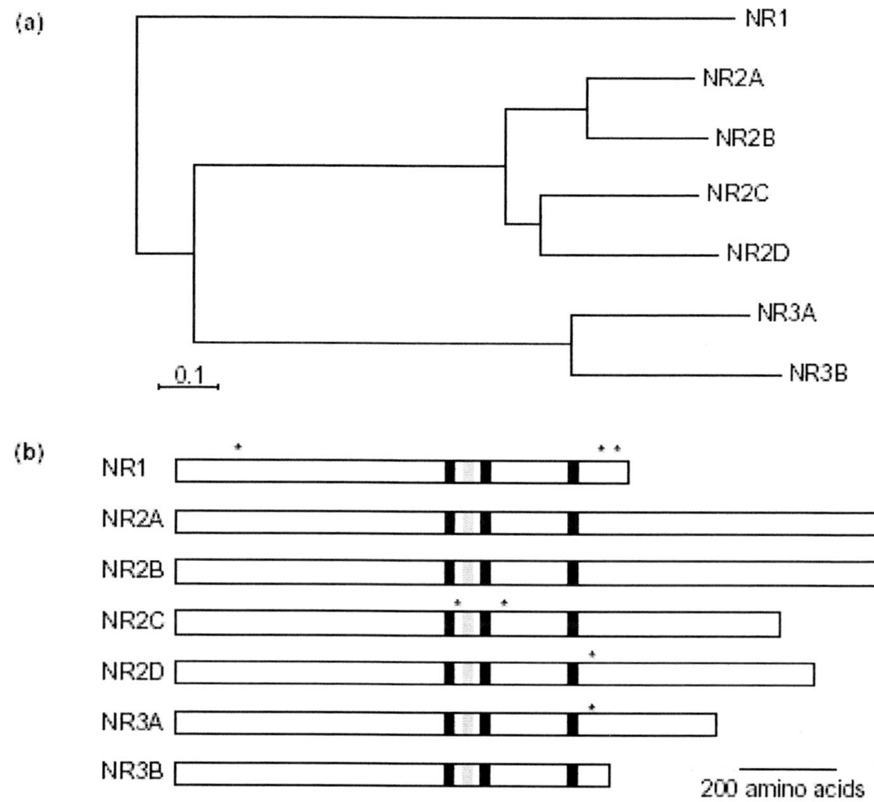

Figure 2. NMDAR subunit diversity. (a) Dendrogram of complete amino-acid sequences for rat NMDAR subunits (except NR3B; human). Sequences were aligned using ClustalX and the tree was generated with NJPlot (http://pbil.univ-lyon1.fr/software/njplot.html) ClustalX (ftp://ftp-igbmc.u-strasbg.fr/pub/ClustalX). (b) Representation of NMDAR subunit polypeptides. Black boxes indicate transmembrane domains, and grey boxes show the transmembrane TM2 re-entrant loop. Asterisks denote regions at which alternative splicing takes place. This is best characterized for the NR1 subunit, which has three regions of alternative splicing: the amino-terminal N1 cassette (exon 5); and the carboxy-terminal C1 (exon 21) and C2 (exon 22) cassettes. Splicing at these sites can generate eight distinct isoforms of NR1 subunit (NR1-1a, -1b, -2a, -2b, -3a, -3b, -4a and -4b). Splicing of the NR2C subunit leads to truncated polypeptides ending after TM1 or TM3. The NR2D subunit can be spliced in the carboxyl terminus, producing a 33-amino-acid insert. Likewise, NR3A splicing leads to a 20-aminoacid insert in the carboxy-terminal domain. NR2B, NR2C and NR2D also have splice sites in their 5′-untranslated regions but no splice variants have been reported for NR2A.
From: Stuart Cull-Candy, Stephen Brickley and Mark Farrant (2001) NMDA receptor subunits: diversity, development and disease. *Current Opinion in Neurobiology 11,* 327–335.

tetrameric complexes consisting of at least one NR1 and one NR2 subunits (Moriyoshi 1991; Monyer 1992, 1994). The subunits are most probably arranged as dimer of dimers with an NR1-NR1-NR2-NR2 orientation in the channel (Schorge 2003). Each subunit has four hydrophobic regions, although only three of them form membrane-spanning domains (TM1,

TM3, and TM4). The fourth one (M2) makes a hairpin bend within the membrane and participate in the formation of the ion-channel pore (Danysz and Parsons 1998).

Two important composition dependent properties of the NMDARs include their single-channel conductance (Brimecombe 1997) and their block by extracellular Mg^{2+} ions. Diheteromeric NMDARs composed of NR1/NR2A or NR1/NR2B subunits generate 'high-conductance' channel openings with a high sensitivity to Mg^{2+} blockade. On the contrary, receptors containing NR2C or NR2D subunits give rise to 'low-conductance' channels with a lower sensitivity to extracellular Mg^{2+} ions. The NR3 subunits are thought to be regulatory in nature since they do not form functional channels alone but co-assemble with NR1/NR2 complexes (Das 1998). NMDARs containing NR3 subunits form 'low-conductance' channels; insertion of an NR3 subunit into NR1/NR2A complexes show a roughly fivefold reduction in relative Ca^{2+} permeability compared to those composed of merely NR1/NR2A subunits (Perez-Otano 2001).

One of the NR2 subunits, the NR2B type appears to be specifically important in respect of several functional properties of the NMDAR. It participates in regulation of the Mg^{2+} block, since replacement of Trp607 of this subunit to Asp or Ala greatly attenuates while substitution of this amino acid side chain for Leu completely abolishes the Mg^{2+} block. According to these observations, Williams et al. (1998) proposed a model in which the M2 loop of the NR2B subunit containing the Trp607 is part of the binding site for Mg^{2+} ions. The presence of an NR2B subunit in the receptor complex affects the binding of the co-agonist glycine as well. Whereas the binding site for glycine is thought to be located on the NR1 subunit (Danysz and Parsons 1998), a high-affinity site for CGP61594, a glycine site antagonist, was shown exclusively displayed by NR1/NR2B receptors as compared with receptors containing other types of NR2 subunits (Honer 1998). One of the unique characteristics of the NR2B subunit is that it binds polyamines. Polyamines appear to facilitate NMDAR function, probably by reducing the inhibitory impact of protons (Gallagher 1997). The NR2B subunit is a highly relevant candidate for mediating the modulatory effects of polyamines, such as spermine and spermidine. In addition, similarly to polyamines, ifenprodil (Avenet 1996), the prototype of the recently discovered NR2B subunit selective NMDAR antagonists (SSNAs) and related compounds – e.g. eliprodil, CP-101,606, Ro25-6981, RG-1103 (Avenet 1997; Williams 1993, 1995; Kashiwagi 1996; Gallagher 1996; Borza 2003) – selectively antagonize NMDA responses in those receptors that contain NR2B subunit.

Ethanol is a Potent Inhibitor of NMDA Receptors

Glutamatergic system is considered an important factor in the mediation of the addictive effect of alcohol (Di Chiara 1998). Among glutamate receptors, AMPA and especially NMDA type of receptors represent the highest affinity targets for ethanol in the CNS (Lovinger 1989; Hoffman 1990; Grant 1995). Biochemical, electrophysiological and behavioral evidences show that ethanol is a potent and selective inhibitor of these receptors (Dildy 1989; Hoffman 1989; Lovinger 1989; Lovinger 1993; Costa 2000) (Table 1A).

Table 1A. Effect of Acute Ethanol Treatment on NMDA Ion Channels

Effects	Ethanol conc.	Preparation	References
Inhibition of NMDA-evoked currents	2.5 – 100 mM	Hippocampal neurons, sensory neurons, Recombinant NR1, ε1/ζ1 and ε2/ζ1, as well as NR1/NR2A, NR1/NR2B and NR1/NR2A/NR2B subunits expressed in Xenopus oocytes	Lovinger 1989, White 1990, Koltchine 1993, Masood 1994, Mirshahi and Woodward 1995
Inhibition of NMDA-mediated signal transmission	1 – 50 mM	rat hippocampal slice preparation	Lovinger 1990
Inhibition of NMDA-induced cytoplasmic Ca^{2+}-increase	10 – 100 mM	Newborn rat neurons, Cerebellar granule cells	Dildy and Leslie 1989, Hoffman 1989a
Inhibition of NMDA- induced cGMP production	10 – 50 mM	Cerebellar granule cells	Hoffman 1989a, 1989b
Inhibition of NMDA excitotoxicity	100 mM	Rat cortical culture	Lustig 1992
Increase the possibility of open state of NMDA channels	1.74 – 174 mM	Rat hippocampal culture	Lima-Landman and Albuquerque 1989

Several studies involving recombinant receptors have demonstrated that receptors containing different types of NR2 subunits have differential sensitivity to the inhibitory effect of ethanol (Popp 1998; Smothers 2001). The ability of ethanol to depress NMDA evoked currents paralleled with the neuroprotective action of ifenprodil in rat cultured cortical neurons (Lovinger 1995). Similar results were obtained when NMDA-induced release of (^{3}H)-norepinephrine was measured in slices from rat cerebral cortex (Fink and Göthert 1996). Since ifenprodil is known as an NR2B subunit selective NMDAR antagonist (Williams 1993), it was assumed that ethanol acts on the NR2B subunit. Indeed, studies performed on recombinant NMDARs showed that heteromers containing either NR2A or NR2B subunits are preferentially sensitive to ethanol inhibition vs. heteromers containing NR2C or NR2D subunits (Kuner 1993; Masood 1994; Buller 1995; Chu 1995; Mirshahi 1995; Wirkner 1999). Moreover, NMDARs with NR1/NR2B subunit combination were more susceptible to the effect of ethanol compared to those composed of NR1/NR2A subunits (Blevins 1995, 1997; Anders 1999; Smothers 2001).

Earlier it was thought that ethanol binds to a hydrophobic pocket distinct from other modulatory binding sites of the NMDAR (Peoples and Weight 1995; Peoples 1997). Recently it was suggested that this pocket is associated with the third transmembrane domain (TM3) of the NR1 subunit. While mutation of Phe639 to Ala in this region of the NR1 subunit expressed in either oocytes or HEK-293 cells significantly decreased the inhibitory effect of ethanol, the substitution of this residue for Trp resulted in receptors that were slightly more sensitive to ethanol inhibition than the wild-type receptors. These observations suggest that the 639 position of the NR1 subunit is an important determinant of ethanol sensitivity (Ronald 2001; Allgaier 2002).

Action of ethanol on NMDARs may also be mediated by changes in the phosphorylation status of the receptor subunits. According to Alvestad et al. (2003) tyrosine phosphorylation of NR2A and NR2B subunits was significantly reduced following *in situ* exposure of hippocampal slices to 100 mM ethanol. Specifically, phosphorylation of Tyr1472 of the NR2B subunit was reduced by 23.5 percent. Addition of a phosphotyrosine phosphatase inhibitor – bpV(phen) – in the recording medium prior to and during ethanol exposure significantly reduced the inhibitory effect of ethanol on NMDAR mediated excitatory postsynaptic potentials (Alvestad 2003; Ferrani-Kile 2003). These data suggest a possible mechanism by which ethanol may inhibit the NMDAR via activation of a tyrosine phosphatase and phosphatase-mediated dephosphorylation of NMDAR subunits may play a role in mediating the inhibitory effect of ethanol.

Effect of Chronic Ethanol Treatment on NMDA Receptors

Changes in NMDAR Functions

Data from studies on neuroadaptation following chronic ethanol exposure indicate a significant role of NMDARs in expression of alcohol withdrawal syndrome as well as in associated neuronal damage (Table 1B). Initial *in vivo* studies showed that seizures evoked by withdrawal of ethanol in alcohol dependent animals were attenuated by NMDAR antagonists and exacerbated by administration of NMDA at doses that are not convulsant in control animals. According to these data, it has been hypothesized that an enhancement of NMDAR mediated neurotransmission following the removal of ethanol blockade mediates ethanol withdrawal seizures (Grant 1990; Gonzalez 2001).

Indeed, several papers reported that chronic ethanol exposure leads to a selective enhancement of NMDAR function in cultured hippocampal and cortical neurons (Hoffman 1996; Smothers 1997; Chandler 1999; Nagy 2001; Floyd 2003). For instance, whereas in hippocampal (Whittington 1995; Thomas 1998) and cortical (Ibbotson 1997) slices prepared from ethanol-naive animals, synaptic potentials were rather AMPA-mediated, in slices from mice chronically treated with ethanol the electrical stimulation evoked synaptic potentials involved a large extent of NMDA-mediated components. In addition, when ethanol was superfused for one hour, in naïve slices a transient depression of NMDA-induced depolarization developed, which vanished at the end of the superfusion period and washout of ethanol caused a withdrawal-type increase in NMDA responses. Similarly, while the amount of non-viable cells in hippocampal brain slice explants was significantly reduced in presence of ethanol, cytotoxic effect of NMDA was significantly higher in ethanol-exposed samples after 24h withdrawal. These data also suggest that acute administration of ethanol has a minor neuroprotective effect but following long-term ethanol exposure and withdrawal neurons became more sensitive to NMDA (Thomas 2000; Harris 2003). Correspondingly, when cultures of rat cortical cells were treated with ethanol the morphology of neurons was not altered, whereas obvious signs of neuronal damage and increased release of lactate dehydrogenase (LDH) were observed after 24 hours of withdrawal (Figure 3A) (Nagy 2001). Interestingly, neurotoxic effect of ethanol withdrawal was observed only in those cultures which were pre-treated with ethanol repeatedly, once daily at least for three consecutive days (Figure 3B) (Nagy 2002). This kind of daily ethanol treatment produced an *in vitro* neuronal model of alcohol-dependence similar to that of Hu and Ticku (1997) who used chronic but intermittent ethanol treatment (CIE, 12h ethanol : 12h withdrawal) providing a feasible model for studying the mechanism underlying the CIE-induced kindling-like phenomenon observed in humans (Ballenger 1978; Becker 1993). This *in vitro* test system, when ethanol treatment is interrupted and the cycle of treatment and withdrawal is repeated several times, may be a useful model for studying the development of ethanol dependence and withdrawal symptoms observed in alcoholics who consume alcohol intermittently for an extended period.

Table 1B. Effect of chronic ethanol treatment on NMDA ion channels

	Effects	Duration of treatment	Ethanol conc.	Preparation	References
in vitro experiments	Enhancement of NMDA-induced cytoplasmic Ca^{2+}-increase	2 – 4 days	20 – 100 mM	Rat cortical culture, Cerebellar granule cells	Blevins 1995 Iorio 1992
	Enhancement of NMDA excitotoxicity	3 – 4 days	100 mM	Rat cortical culture, Cerebellar granule cells	Chandler 1993a, Iorio 1993 Nagy and Laszlo 2002
	Enhancement of NMDA-induced NO production	4 days	100 mM	Rat cortical culture	Chandler 1997
	Increase of NR2B mRNA levels	5 days	50 mM	Rat cortical culture	Hu 1996
	Increase of NR2B protein levels	3 days, intermittent	100 mM	Rat cortical culture Rat hippocampal culture	Nagy 2003
	Enhancement of inhibitory potential of NR2B subunit selective NMDA receptor antagonists	3 days, intermittent	100 mM	Rat cortical culture	Nagy 2003
in vivo experiments	Increase of MK-801 binding	7 days	7% ethanol in liquid diet	hippocampus, cortex, striatum of C57BL mice	Grant 1990, Gulya 1991
	Increase of NR1 immuno-labeling	12 weeks	5% ethanol in liquid diet	Rat hippocampus	Trevisan 1994
	Increase of NR2A and NR2B mRNA levels	5 days	20% ethanol intragastrically	Rat brain	Follesa and Ticku 1995
	Increase of ornithine decarboxylase enzyme activity	5 days	>2 mg/ml blood alcohol level	Rat hippocampus	Davidson 1998
	Enhancement of glutamate release	4 days + 12h withdrawal	1 – 3 g/kg	Rat nucleus accumbens and hippocampus	Dahchour 1996, 1999, 2000

After such kind of ethanol treatment, NMDA responses and sensitivity of NMDARs for ethanol were enhanced in primary cultures of rat cortical neurons (Figure 4) (Nagy 2002). Furthermore, the neurotoxic effect of withdrawal could be reduced by re-addition of ethanol or administration of NMDAR antagonists (Figure 5). On the other hand, the effect of the $GABA_A$ receptor agonist muscimol was insignificant. These observations support the conception that NMDARs may play a role in the development of *in vitro* ethanol dependence and in alcohol-withdrawal evoked neurotoxicity.

A. Control

B. Pre-treated with 50 mM ethanol for 3 days and withdrawn for 24h

C. Pre-treated with 50 mM ethanol for 3 days and treated with 50 mM ethanol for 24h

Figure 3A. In vitro model of ethanol dependence and withdrawal
Differential interference photomicrographs of control primary cultures of rat cortical neurons (A) and cultures previously adapted to 50 mM alcohol for 3 days (B, C). 24 hours before fixation cultures were exposed to control medium (A, B), or medium containing 50 mM ethanol (C). Red coloured neurons are immuno-labelled with anti-neurofilament antibody. Note the extensive granulation and disintegration of neuritic processes and cell bodies in culture under alcohol-withdrawal (B) and its protection by ethanol (C). (magnification: 200x)
From: Nagy J., Müller F., László L. (2001) Cytotoxic effect of alcohol-withdrawal on primary cultures of cortical neurons. *Drug Alcohol Depend. 61,* 155–162.

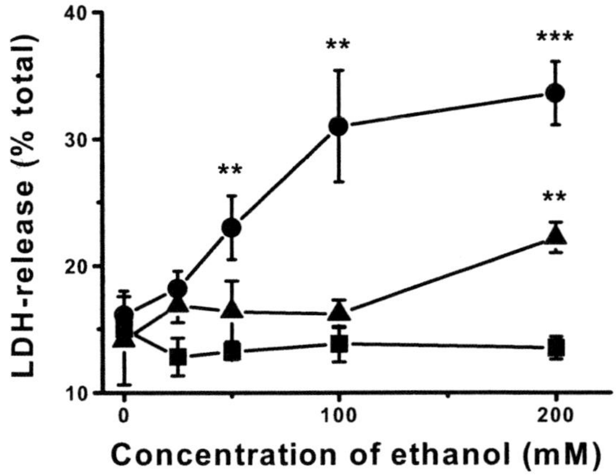

Figure 3B. Toxic effect of 24-hour ethanol-withdrawal and its inhibition by re-addition of ethanol in ethanol pre-treated primary cortical cultures.

LDH activity of the culture medium expressed as percentage of total activity was measured in cultures pre-treated with different concentrations of ethanol once for 24 (squares) or 72 (triangles) hours as well as daily for 3 successive days (circles).

(*: $p<0.05$, **: $p<0.01$, ***: $p<0.001$ as compared to the respective control not treated with ethanol)

From: Nagy J., László L. (2002) Increased sensitivity to NMDA is involved in alcohol-withdrawal induced cytotoxicity observed in primary cultures of cortical neurons chronically pre-treated with ethanol. *Neurochem. Int. 40,* 585–591.

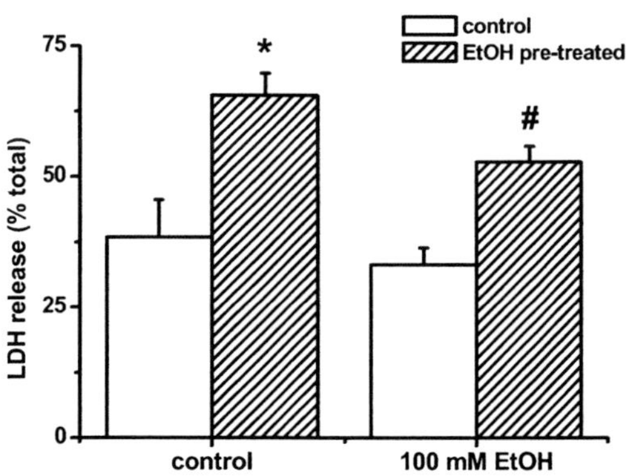

Figure 4A. Altered excitotoxic effect of NMDA after ethanol pre-treatment

Effect of acute and chronic ethanol treatment on NMDA induced cytotoxicity. Control cortical cultures (open columns) and cortical cultures pre-treated with 100mM ethanol repeatedly, once daily for 3 days (dashed columns) were exposed to 300 μM NMDA for 15 min in presence or absence of 100 mM ethanol. LDH-release, expressed as percentage of total LDH content, was measured 24h after NMDA wash out. (*, $P < 0.01$ compared to NMDA induced LDH-release in control cultures; #, $P < 0.01$ compared to LDH-release in absence of ethanol).

From: József Nagy (2004) Renaissance of NMDA receptor antagonists: Do they have a role in the pharmacotherapy for alcoholism? *IDrugs: The Investigational Drugs Journal 7(4),* 339-350.

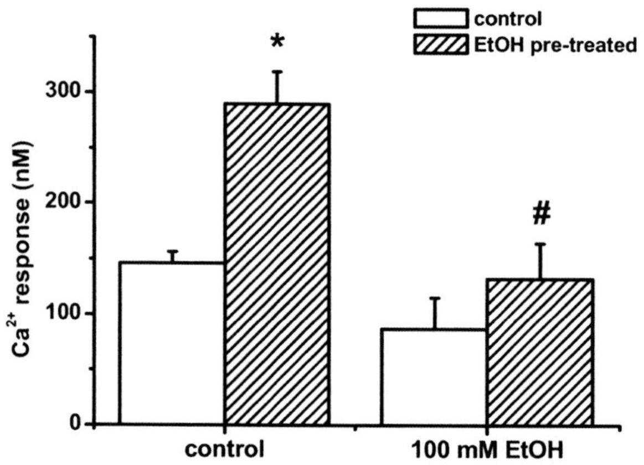

Figure 4B. Altered NMDA evoked calcium signals after ethanol pre-treatment.
Effect of chronic ethanol pre-treatment on NMDA induced changes in cytosolic calcium ion concentration. Control cortical cultures (open columns) and cortical cultures pre-treated with 100mM ethanol repeatedly, once daily for 3 days (dashed columns) were exposed to 10 μM NMDA in presence or absence of 100 mM ethanol. (*, $P < 0.01$ compared to NMDA induced elevation of [Ca2+]i in control cultures; #, $P < 0.05$ compared to elevation of [Ca2+]i in absence of ethanol).
From: József Nagy (2004) Renaissance of NMDA receptor antagonists: Do they have a role in the pharmacotherapy for alcoholism? *IDrugs: The Investigational Drugs Journal 7(4),* 339-350.

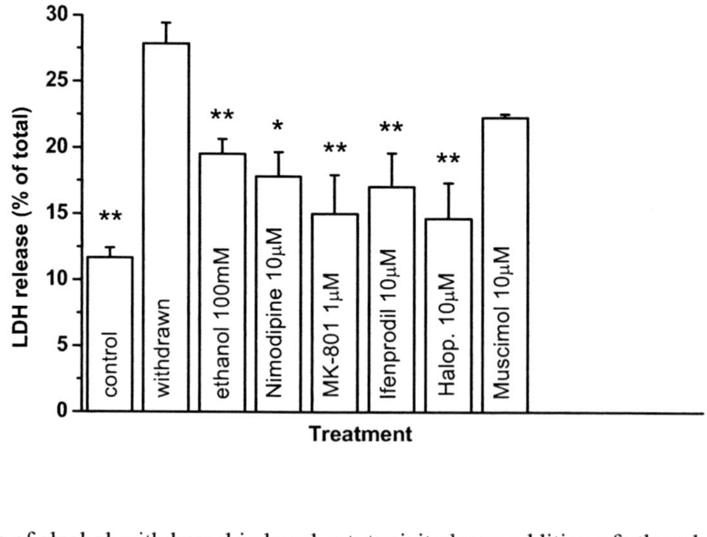

Figure 5. Inhibition of alcohol-withdrawal induced cytotoxicity by re-addition of ethanol and by treatment with different drugs. Columns represent LDH-release after 24 hours of the indicated treatment in primary cultures of rat cortical neurons pre-treated with 100 mM ethanol daily for 3 successive days. Control column shows LDH-release in cultures not adapted to ethanol.
(*: $p<0.05$, **: $p<0.01$ compared to the alcohol-adapted and withdrawn group)
From: József Nagy (2004) NR2B type of NMDA receptor subunits: a possible target in treatment for alcohol dependence. *Current Drug Targets – CNS & Neurological Disorders 3,* 169-179.

Changes in Structure of NMDARs

According to earlier reports, it was hypothesized that chronic ethanol treatment leads to an increase in the number of NMDARs leading to facilitated receptor function (Wirkner 1999). However, up-regulation of NMDAR expression has not been found in all studies after chronic ethanol exposition (Rudolph 1997; Ferreira 2001). According to the recently emerging view, the increased NMDAR function is presumably due to a differential up-regulation of the various NMDAR subunits. This conception is supported by several papers presenting evidence for altered NMDAR subunit composition after chronic ethanol treatment.

Notwithstanding, there is a disagreement in respect of the expression of the different NR1 splice variant forms and NR2 subunits. On the one hand, some authors reported no changes in subunit expression at all (Cebere 1999) and others found changes solely in the expression of the NR2A subunit due to long-term ethanol exposure (Darstein 2000). On the other hand, there are papers concluding that besides several other types of subunits and certain NR1 splice variants the expression of the NR2B subunit is increased. Lots of *in vitro* studies showed increased NR2B subunit mRNA levels, with no change in NR1 and/or NR2A subunit transcription in cultured cortical neurons following chronic ethanol administration (Morrow 1994; Hu 1996; Hardy 1999). On the contrary, the levels of NR2B as well as NR2A and NR1 subunit proteins were found increased in the cortex and hippocampus of rats or mice (Trevisan 1994; Follesa 1996a, 1996b; Kalluri 1998; Narita 2000; Henniger 2003). Furthermore, in cultured cerebellar granule cells, the 'developmental switch' of NR2B subunit for NR2A was found delayed resulting in higher NR2B and lower NR2A subunit levels (Snell 2001).

Similarly to these later observations, in primary cultures of rat cortical neurons treated with ethanol for three consecutive days, whereas the efficiency of the non-subunit selective NMDAR antagonists was not changed, the maximal inhibitory effect of ethanol as well as the NR2B subunit selective NMDAR antagonists tested was significantly increased (Figure 6, Table 2). Accordingly, whereas the immuno-labelling of the NR2A, NR2C, and NR2D subunits was not changed, increased NR2B immuno-labelling was observed in ethanol pre-treated compared to control cultures of rat cortical as well as hippocampal neurons. In addition, the expression of the C1 and C2' cassette containing splice variant forms of the NR1 subunit was also increased in these cultures (Figure 7) (Nagy 2003). Considering that i) the deactivation time of NMDARs composed of NR1/NR2B subunits is longer than those built up of NR1/NR2A subunits (Monyer 1992), ii) the deactivation rate is four times faster for receptors composed of NR1 subunit containing the N1 cassette than those lacking this cassette (Rumbaugh 2000; Cull-Candy 2001), and iii) NMDARs assembled of NR1 splice variants containing C1 and/or C2 cassettes may form functionally more active ion channels (Rameau 2000), the increased expression of the C1 and C2' cassette containing splice variant forms of the NR1 and/or the NR2B subunits may underlie the enhanced NMDAR function and subsequently the increased sensitivity of neurons to excitotoxic insults seen after ethanol withdrawal.

Table 2. Maximal Inhibition and Calculated IC_{50} Values of NMDAR Antagonists

Compound	Control		Ethanol pre-treated	
	I_{max} (%)	IC_{50}	I_{max} (%)	IC_{50}
MK-801	89 ± 4	104 ± 18 nM	92 ± 5	126 ± 18 nM
5,7-DCK	82 ± 2	23 ± 2 μM	85 ± 6	20 ± 4 μM
ethanol	34 ± 1	242 ± 6 mM	56 ± 6 *	177 ± 14 mM *
t-ifenprodil	63 ± 2	210 ± 29 nM	74 ± 6 *	167 ± 40 nM
CP-101606	48 ± 4	18 ± 3 nM	69 ± 6 *	18 ± 6 nM
CI-1041	51 ± 6	17 ± 5 nM	83 ± 9 **	12 ± 3 nM

Note: Maximal inhibition value (I_{max}) shows the mean ± S.E. of inhibition produced by the given antagonist at the maximal concentration used. IC_{50} data indicate the mean ± S.E. of the calculated IC_{50} values of individual measurements using Microcal™ Origin® 6.0 data analysis software.

From: Nagy, J., Kolok, S., Dezső, P., Boros, A., Szombathelyi, Z. (2003) Differential alterations in the expression of NMDA receptor subunits following chronic ethanol treatment in primary cultures of rat cortical and hippocampal neurons. ***Neurochem. Int. 42(1),*** 35-43.

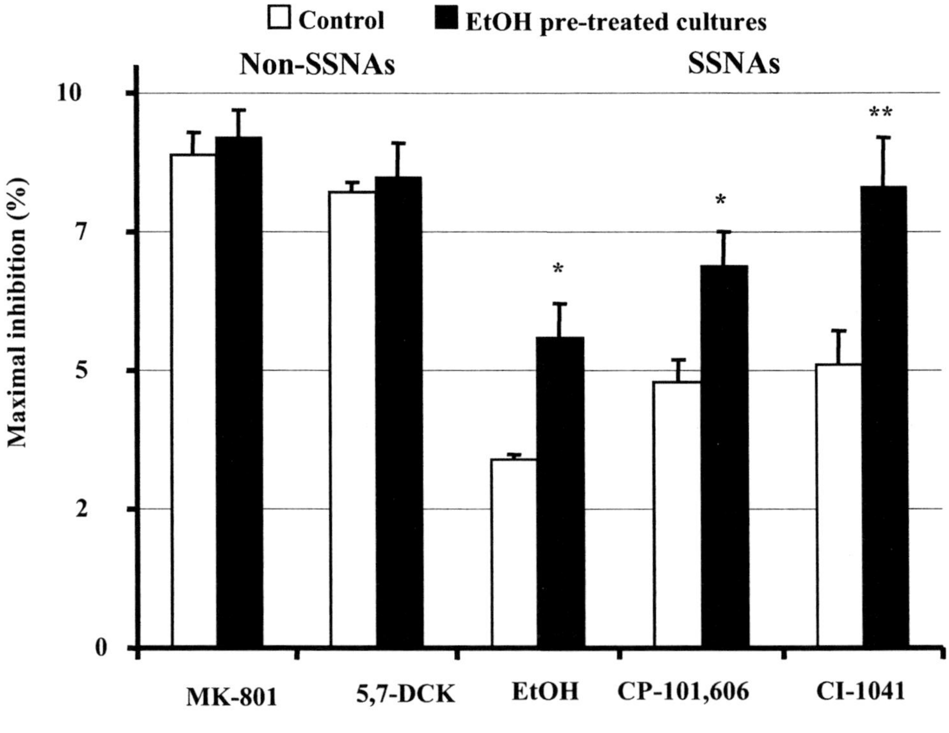

Figure 6. Effect of chronic ethanol pre-treatment on the inhibitory potential of different NMDA receptor antagonists

Maximal inhibitory effect of different NMDAR antagonists on NMDA induced cytosolic calcium elevation in control (white columns) and 100 mM ethanol pre-treated (black columns) primary rat cortical cultures. (*, $P < 0.05$; **, $P < 0.01$ compared to the adequate NMDA induced calcium increase in control cultures).

From: Nagy, J., Kolok, S., Dezső, P., Boros, A. (2003) Szombathelyi, Z. Differential alterations in the expression of NMDA receptor subunits following chronic ethanol treatment in primary cultures of rat cortical and hippocampal neurons. *Neurochem. Int., 42(1),* 35-43.

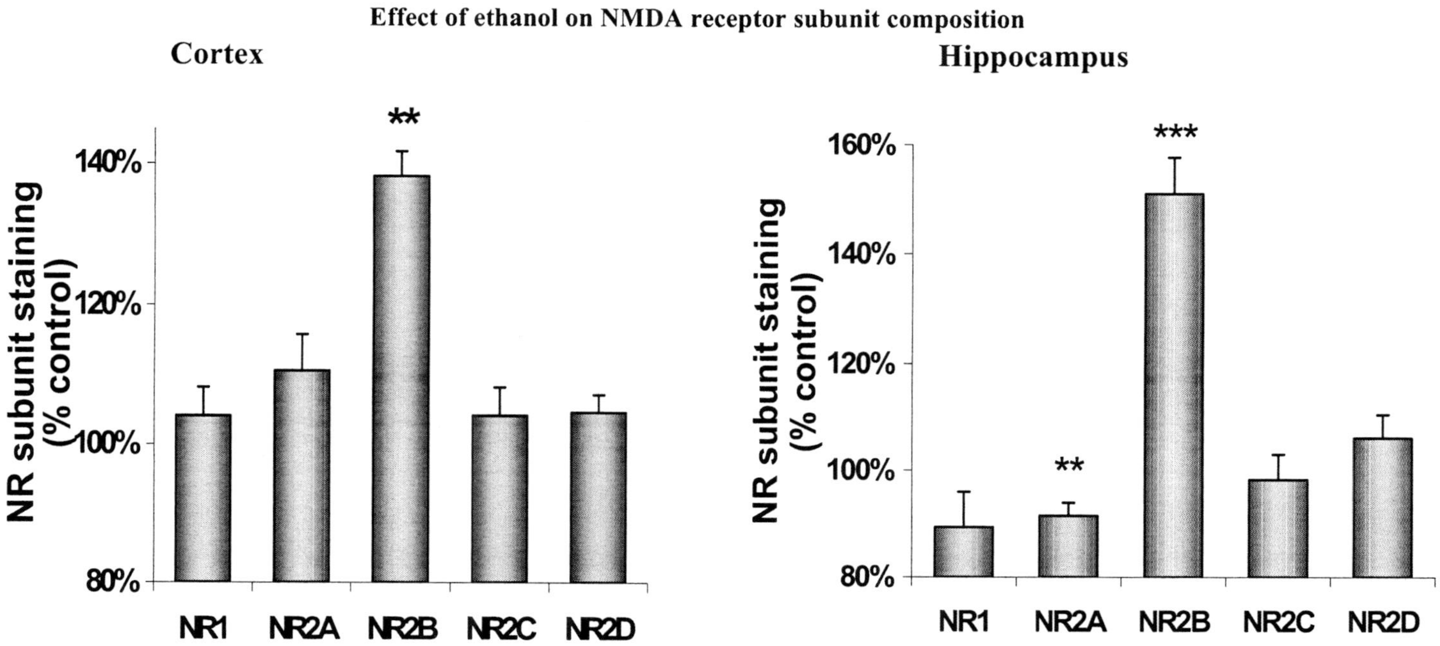

Figure 7. Effect of chronic ethanol pre-treatment on the expression of NMDA receptor subunits and NR1 splice cassettes

Expression of NMDA receptor subunit proteins was estimated by a flow-cytometry based immuno-cytochemical method in primary cultures of rat cortical and hippocampal neurones. Background related fluorescence values measured in 100 mM ethanol pre-treated cultures are expressed as percentage of those obtained in control cultures.

(*, $P < 0.05$; **, $P < 0.01$; ***, $P < 0.001$ compared to control).

From: Nagy, J., Kolok, S., Dezső, P., Boros, A. (2003) Szombathelyi, Z. Differential alterations in the expression of NMDA receptor subunits following chronic ethanol treatment in primary cultures of rat cortical and hippocampal neurons. *Neurochem. Int., 42(1),* 35-43.

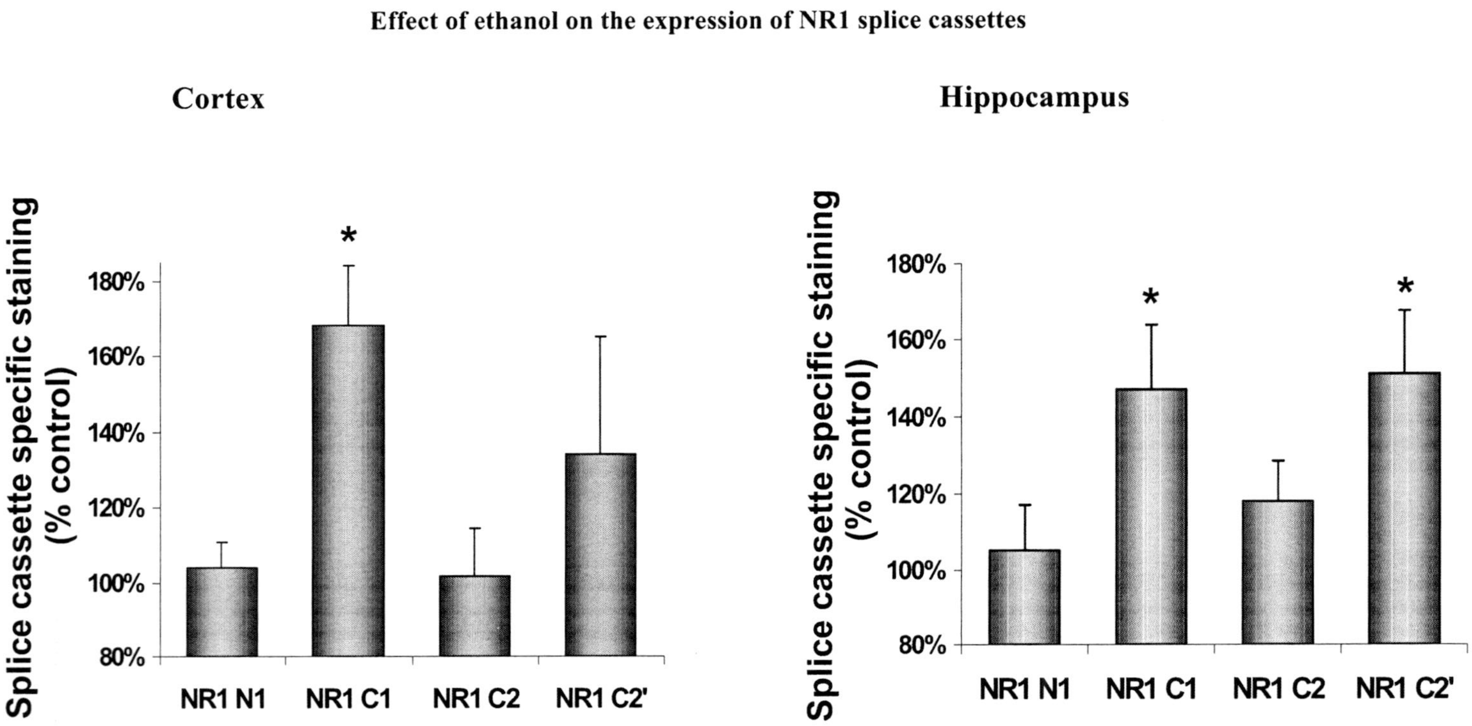

Figure 7. Effect of chronic ethanol pre-treatment on the expression of NMDA receptor subunits and NR1 splice cassettes

Expression of NMDA receptor subunit proteins was estimated by a flow-cytometry based immuno-cytochemical method in primary cultures of rat cortical and hippocampal neurones. Background related fluorescence values measured in 100 mM ethanol pre-treated cultures are expressed as percentage of those obtained in control cultures.

(*, $P < 0.05$; **, $P < 0.01$; ***, $P < 0.001$ compared to control).

From: Nagy, J., Kolok, S., Dezső, P., Boros, A. (2003) Szombathelyi, Z. Differential alterations in the expression of NMDA receptor subunits following chronic ethanol treatment in primary cultures of rat cortical and hippocampal neurons. *Neurochem. Int., 42(1),* 35-43. (Continued)

According to Carpenter-Hyland et al. (2004), the co-localization of NR1 clusters with the presynaptic marker protein synapsin was increased in rat hippocampal neurons exposed to 50 mM ethanol for 4 days. This was accompanied by significant increases in the size and density of these synapsin-associated clusters with no change observed in non synapsin-associated NR1 clusters. Similar effects were observed with NR2B clustering after chronic ethanol exposure. The increase in synaptic NMDA receptor clustering was prevented by addition of a protein kinase A inhibitor or by co-exposure to a low concentration of NMDA and was reversed when ethanol was removed from the cultures. On the contrary, no changes were observed in the synaptic content, cluster size, or density of AMPA receptors after ethanol exposure. Electrophysiological measurements on ethanol-treated neurons revealed a similar enhancement in synaptic NMDA currents with no change in AMPA-mediated events.

It is known that Src family of protein tyrosine kinases, specifically Src and Fyn, up-regulates NMDAR function (Salter and Kalia 2004) thereby reducing the sensitivity of NMDARs to ethanol. Previously it was shown that both c-Src and Fyn kinases potentiate the NMDA-activated currents in *in vitro* recombinant systems (Chen and Leonard 1996; Kohr and Seeburg 1996; Anders 1999; Salter and Kalia 2004) as well as in spinal neurons (Wang and Salter 1994). Furthermore, phosphorylation status of NR1 and NR2B subunits was found increased in consequence of ethanol exposure (Miyakawa 1997). Concerning the NR1 subunit, truncation ($NR1_{858stop}$) (Anders 2000) or phosphorylation of Ser897 of this subunit (Maldve 2002) decreased the ability of ethanol to inhibit NMDA receptor function. In addition, the reduced sensitivity of NMDARs to ethanol was linked up with the dopamine receptor-1 activation via dopamine and cAMP-regulated phosphoprotein-32 kD (DARPP-32) phosphorylation (Maldve 2002). According to Yaka et al. (2003a, 2003b), the scaffolding protein RACK1 that binds Fyn kinase to the NR2B subunit dissociates from the complex due to ethanol exposure, consequently facilitating Fyn-mediated phosphorylation of the NR2B subunit leading to enhanced channel activity counteracting the inhibitory actions of ethanol. In addition, transgenic mice over-expressing the Fyn tyrosine kinase and withdrawn from alcohol failed to show any increase of anxiety-like behavior or reduction of exploratory activity like it was observed in case of their wild-type littermates (Stork 2002). This apparent lack of alcohol withdrawal-induced behavioral effects was associated with increased Fyn kinase activity and tyrosine phosphorylation of several proteins including the NR2B subunit.

Taken together, the changes in subunit expression, phosphorylation states and synaptic clustering of NMDAR subunits due to ethanol exposure may lead to the enhancement of NMDA responses. These changes may also explain the occurrence of acute ethanol tolerance leading to reinforcement of ethanol consumption and may underlie the development of ethanol dependence.

Consequences of Increased NMDAR Function

Presumably, in consequence of the increased function of NMDARs, enhanced release of glutamate was observed after chronic ethanol exposure both in *in vitro* as well as *in vivo* experiments. Besides several other factors (e.g. adaptive changes in GABA receptors and calcium channels), the NMDARs are major contributors to the increased glutamate release during alcohol withdrawal since in the brain of ethanol-dependent rats, the extracellular concentration of glutamate shows a transient, NMDAR mediated increase after cessation of ethanol intake and these changes are time-locked to the behavioral signs of ethanol withdrawal (Rossetti 1995; Fadda 1998). This enhanced glutamate release may contribute to the further shift towards the excitatory dominance in the CNS after ethanol withdrawal. Furthermore, the up-regulation of the NMDARs can enhance the activity of the noradrenergic system (Engberg and Hajos 1992a, 1992b) as well, that may account for the vegetative instability seen in serious states of alcohol withdrawal, especially in delirium tremens (Tsai 1995, 1998).

Recently, it has been hypothesized that the same neuronal system including the mesolimbic dopaminergic pathway mediates the reinforcement for alcohol and other addictive drugs like opiates or cocaine (Koob 1992; Nestler 1993). Indeed, ethanol has been reported to stimulate dopamine (DA) release in the nucleus accumbens (Imperato 1986) and electrophysiological studies have demonstrated a concomitant ethanol-induced increase in the activity of ventral tegmental dopaminergic neurons (Brodie 1990, 1999). On the other hand, systemic administration of NMDAR antagonists increases burst firing of dopaminergic neurons (Zhang 1992) and can stimulate release of DA from dopaminergic terminal areas suggesting that glutamate – acting through NMDARs – exerts a tonic inhibitory action on DA release in the nucleus accumbens (Hoffman 1996; Kretschmer 1999). According to the model of Fadda and Rossetti (1998), blockade of the NMDARs by acute ethanol treatment disinhibits dopaminergic neurons via GABAergic interneurons possessing NMDARs. Withdrawal of alcohol, similarly to withdrawal of opiates or cocaine, has been found associated with decreased DA release in the limbic forebrain areas (Rossetti 1992) due to decreased firing rate of dopaminergic neurons (Diana 1993). Thus, reduced dopaminergic functions seen after ethanol withdrawal may arise as a consequence of enhanced NMDA responses induced by chronic ethanol exposure.

All the above-discussed findings suggest the possibility that increased NMDA mediated neurotransmission may constitute the basis of both the motor signs (e.g. tremor, seizures etc.) and the affective or emotional disturbances (e.g. craving, dysphoria) associated with alcohol withdrawal. Bearing in mind that besides the glutamatergic system other transmitter mechanisms are also involved in the adaptive changes induced by chronic ethanol treatment and that *in vitro* experiments allow only limited conclusions to deduce for a whole organism, it is clear that alterations in NMDAR function play a critical role in these processes leading to the development of alcohol dependence and withdrawal symptoms. According to this view of the pathomechanism of alcohol withdrawal syndrome (AWS), the NMDAR may be a possible target and NMDAR antagonists may be useful agents for the treatment of both the physical and the psychical signs of alcohol withdrawal.

NMDAR Antagonists in Pharmacotherapy for Alcohol Withdrawal

In the early 90s, it was already hypothesized that NMDAR antagonists can block alcohol withdrawal induced seizures in ethanol dependent animals. Since then, extensive literature on animal experimental and preliminary clinical data suggest that NMDAR antagonists are promising candidates for the treatment of AWS, inasmuch as these compounds may attenuate not only the physical but also the affective and motivational components of AWS.

Competitive NMDA Receptor Antagonists

So far, mostly the classic competitive and channel blocking NMDAR antagonists were tested and proved to be useful in *in vitro* or animal models of alcoholism. Early experiments showed that competitive NMDA receptor antagonists acting at the glutamate site (e.g. CGP 39551, D-CPP-ene) decreased handling-induced hyperactivity after withdrawal from chronic ethanol treatment in mice (Liljequist 1991; Lamblin 1993; Ripley 1995). These compounds increased the threshold for population spikes in hippocampal slices from the same animals. NMDA antagonists acting within the ion channel (e.g. ketamine, MK-801 and ADCI) were also shown to suppress withdrawal-induced seizures effectively in both rats and mice (Fidecka 1989; Morrisett 1990; Grant 1996). Unfortunately, preclinical studies indicated that most of these compounds produce psychotomimetic or sedative effects, ataxia, muscle relaxation, neuronal damages (e.g. in the cingulate cortex) as well as motor and learning impairment.

These serious side effects impeded their introduction to the human therapy (Koek 1990; Murata 1993; Carter 1994; Parsons 1998; Yenari 1998; Breese 2002). However, due to immense therapeutic promise of NMDA antagonists in acute and/or chronic neurodegenerative and psychiatric disorders efforts have been made to develop compounds lacking these side effects. More encouraging approaches were performed with low affinity channel blockers like memantine or with NMDA antagonists acting on the glycine binding site (L-701,324) having more tolerable side effect profiles.

Channel Blockers

Novel channel blockers like memantine (1-amino-3,5-dimethyl-adamantane) and its analogue neramexane (MRZ 2/579, 1-amino-1,3,3,5,5-pentamethyl-cyclohexane hydrochloride) have improved side effect profile probably due to their moderate potency and rapid, strongly voltage-dependent blocking kinetics (Parsons 1999). These compounds greatly inhibited alcohol consumption without affecting water or food intake during relapse in a rat craving model (Hölter 1996, 2000; Shelton 1997; Piasecki 1998). In addition, neramexane as well as memantine effectively suppressed ethanol withdrawal induced seizures in alcohol dependent rats (Bienkowski 2001).

Glycine Site Antagonists

There are several observations that glycine and D-cycloserine can reverse the inhibitory effects of and rapid tolerance to ethanol (Rabe 1990; Woodward 1990; Buller 1995; Khanna 1995; Moraes-Ferreira 1997). Glycine-B site antagonists also attenuated the expression of withdrawal symptoms. A member of this type of NMDA receptor antagonists, L-701,324 (7-chloro-4-hydroxy-3-3-phenoxyphenyl-2-1H–quinolone) produced a dose-dependent inhibition of audiogenic seizures associated with alcohol withdrawal (Kotlinska and Liljequist 1996) and potently blocked the acquisition of ethanol-induced conditioned place preference (Biala and Kotlinska 1999).

Preliminary problems with glycine site antagonists included poor systemic availability has now been overcome with agents like GV196771A ((E)-4,6-Dichloro-3-(2-oxo-1-phenylpyrrolidin-3-ylidenemethyl)-1H-indole-2-carboxylic acid sodium salt) or SM-31900 (3(S)-(2- (4- (Amino methyl) -2- (1(R)-carboxyethoxy)phenylamino) -2- oxoethyl) -7- chloro -1,3,4,5- tetrahydrobenzo(c,d)indole -2- carboxylic acid hydrochloride) which are not only of very high affinity, but also have improved pharmacokinetic and physicochemical properties, such as good brain permeation and solubility (Jansen and Dannhardt 2003). Although these compounds were not tested in animal models related to alcoholism, the facts that SM-31900 has a potent anticonvulsant activity (Katayama 2001), GV196771A can inhibit the development of morphine tolerance (Quartaroli 2001) and both compounds are devoid of behavioral side effects (hyperactivity, motor dysfunction) make these compounds promising candidates also in the treatment of alcoholism.

Acamprosate

The American Food and Drug Administration (FDA) recently granted the approval of a novel anti-alcohol medication Campral (acamprosate calcium) to maintain abstinence in patients with alcohol dependence. In Europe, more than 4 million people have been treated with this agent since 1989 when it became commercially available in France. According to several human studies, acamprosate has a consistent effect on prolonging abstinence and reducing the rate of relapse, in conjunction with an equally consistent absence of effect on self-reported craving, suggesting that it can be used as a relapse-prevention medication.

In previous animal studies, it was observed that acamprosate dose-dependently reduced alcohol consumption with no effects on food and water intake. It could also lessen the increase in extracellular glutamate level and hypermotility during ethanol withdrawal (Dahchour 1998; Dahchour and De Witte 1999; Rammes 2001). The ability of acamprosate to reduce the glutamate level in microdialysates from nucleus accumbens during ethanol withdrawal was an important finding for the therapeutic use of this agent.

The exact mechanism of action of acamprosate, originally developed as a GABA analogue, was intensively investigated in the past years. According to several studies, acamprosate had no direct effect on NMDA responses in primary neuronal cultures (Popp and Lovinger 2000; Mayer 2002a; Nagy 2004). In spite of these findings, acamprosate effectively reduced the ethanol-withdrawal evoked neuronal cell death in ethanol pre-treated organotypic hippocampal (Mayer 2002b) or primary cortical cultures from the rat (Figure 8) (al-Qatari 2001; Nagy 2004). On the other hand, acamprosate was reported by several other authors to have an inhibitory effect on the native or recombinant NMDA-receptors (Spanagel 1997b; Allgaier 2000; Rammes 2001). alQatari et al. (1998) argued that acamprosate may have an excitatory or inhibitory effect on NMDA receptors depending upon the experimental conditions indicating that acamprosate, at least partly, acts as a `partial agonist' at the NMDA receptor (alQatari 1998). In addition, acamprosate was shown to bind to a spermidine-sensitive site on the NMDAR and to modulate the function of NMDARs (Naassila 1998). Based on these observations a glutamatergic hypothesis for the action of acamprosate and a causal relationship between NMDA antagonism and the anticraving properties of acamprosate has been suggested (Lhuintre 1985, 1990; Spanagel and Zieglgansberger 1997a; Holter 1997; Heyser 1998; Dahchour 1998; Bolo 1998).

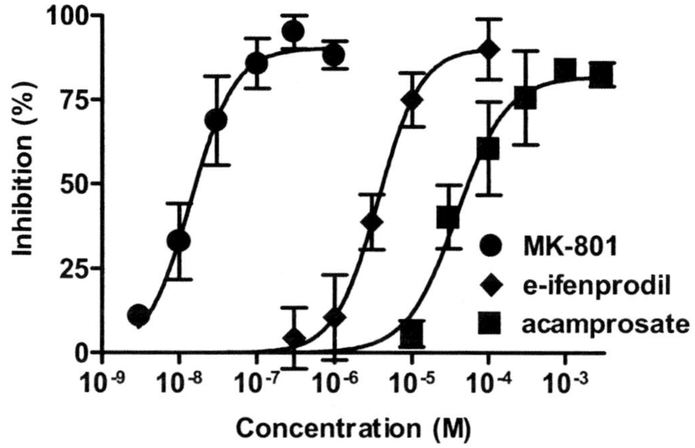

Figure 8. Inhibitory effect of different compounds on ethanol-withdrawal-induced neurotoxicity Neuronal cell death caused by 24-hour ethanol-withdrawal in primary cultures of rat cortical neurones pre-treated with 100 mM ethanol for 3 consecutive days was quantified by measuring LDH-release. Different concentrations of MK-801, erythro-ifenprodil and acamprosate were present during the withdrawal period. Each point represents the percentage of inhibition (mean ± S.E. (error bars)). *From:* Nagy, J., Horváth, C., Farkas, S., Kolok, S., Szombathelyi, Z. (2004a) NR2B subunit selective NMDA antagonists inhibit neurotoxic effect of alcohol-withdrawal in primary cultures of rat cortical neurons. *Neurochem. Int., 44(1),* 17-23.

NR2B Subunit Selective Antagonists

In recent years, a novel type of NMDAR antagonists, the NR2B subunit selective ones have emerged and received considerable attention. This type of compounds showed potency in models of neurodegeneration (Kew 1998), Parkinson disease (Nash 1999, 2000; Wright 2000; Steece-Collier 2000), and hyperalgesia (Boyce 1999; Carter 2000; Fillhard 2000; Chizh 2001). It was also realized that this type of compounds lacks the serious side effects of the classic NMDAR antagonists' (Nikam and Meltzer 2002). Although, like other un-competitive NMDAR antagonists they may have some adverse effect on learning and memory, it was proved that they have a wider separation between doses that are effective in seizure or stroke models and those that disrupt learning and memory. The limited information on the novel NR2B SSNAs such as CP-101,606 (traxoprodil) (Chenard 1995; Kemp and McKernan 2002), Ro25-6981 (Fischer 1997), Co-101244 (Zhou 1999), CI-1041 (Chizh 2002) and RG-1103 (Borza 2003) (Figure 9, Table 3) also suggests that these drugs are better tolerated and are largely devoid of adverse CNS effects at antinociceptive doses, at least with respect to psychotomimetic, ataxic and sedative effects (Taniguchi 1997; Boyce 1999; Carter 2000; Fillhard 2000; Murray 2000; Chizh 2001).

Table 3. Selectivity of Novel NR2B Subunit Selective NMDA Antagonists

Compound	Inhibition of NMDA evoked Ca^{2+} responses (IC_{50} values)		Selectivity (NR2B vs. 2A)
	NR1/NR2A	NR1/NR2B	
CP-101,606	> 30 μM	30.0 ± 3.6 nM	> 1000
Ro-25-6981	> 10 μM	57 ± 10 nM	> 175
Co-101244	> 30 μM	4.8 ± 0.7 nM	> 6250
CI-1041	> 50 μM	8.4 ± 1.4 nM	> 5950
RG-13579	> 30 μM	19.9 ± 4.5 nM	> 1600
RG-1103	> 30 μM	2.1 ± 0.4 nM	> 13600

Compounds were tested for their inhibitory activity on 100 μM NMDA induced calcium elevation in HEK 293 cells stably expressing recombinant NMDA receptor (Nagy 2003b). IC_{50} values were determined and degrees of selectivity were evaluated as the ratio of IC_{50} values obtained in cells expressing NR1/NR2A and NR1/NR2B receptors.

From: József Nagy, Csilla Horváth, Sándor Farkas, Sándor Kolok, Zsolt Szombathelyi (2004) NR2B subunit selective NMDA antagonists inhibit neurotoxic effect of alcohol-withdrawal in primary cultures of rat cortical neurons. *Neurochemistry International 44(1),* 17-23.

The initial chemical leads of NR2B SSNAs – ifenprodil and eliprodil – were found effective in animal models of alcohol dependence. According to Malinowska et al. (1999), ifenprodil potently reduced the severity of withdrawal-induced seizures. Oral administration of increasing doses of eliprodil produced a dose-dependent and almost complete inhibition of ethanol withdrawal produced audiogenic seizures in alcohol dependent Sprague-Dawley rats (Kotlinska and Liljequist 1996). This effect of eliprodil was achieved at doses that by themselves did not alter the basal locomotor activity of untreated control animals. Similar results were observed in alcohol dependent male C57BL/6J mice (Narita 2000). The expression of spontaneous ethanol withdrawal signs (piloerection, jerk, tremor) occurring 6 – 12 hours after the discontinuation of ethanol treatment was suppressed by ifenprodil and was accompanied by a marked up-regulation of the NR2B subunit in the limbic forebrain of ethanol treated mice. According to Kotlinska and Liljequist (1997), eliprodil effectively reversed the reduction in extracellular DA level during ethanol withdrawal but only partially and dose-independently substituted ethanol indicating that it has no discriminative stimulus properties similar to those produced by ethanol.

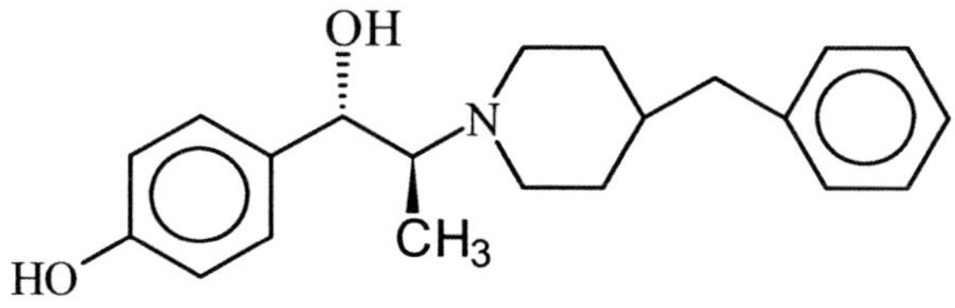

ifenprodil

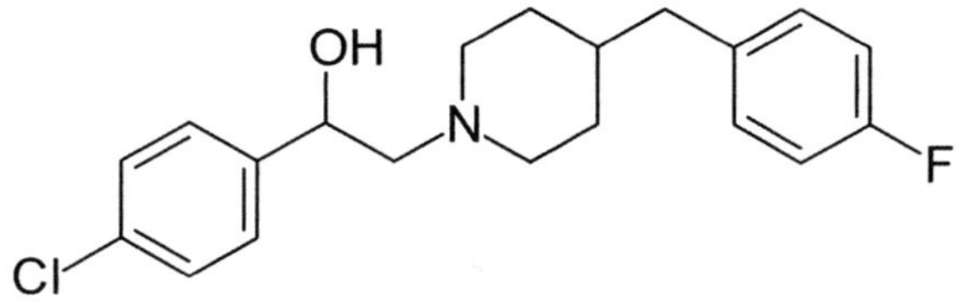

eliprodil

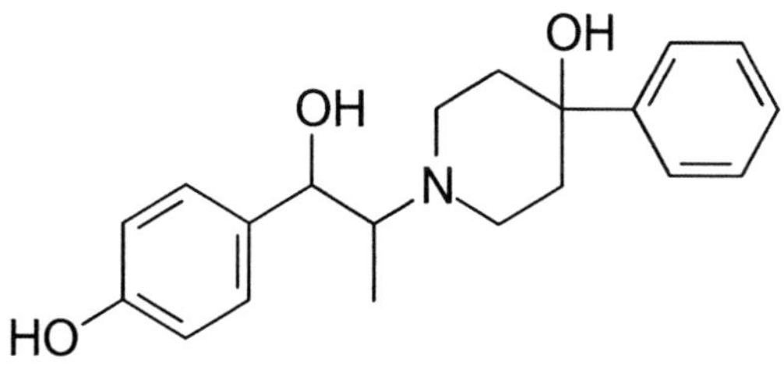

CP-101,606

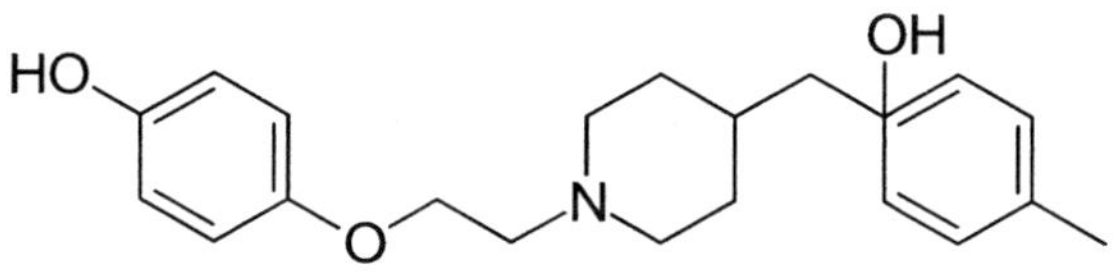

Co-101244

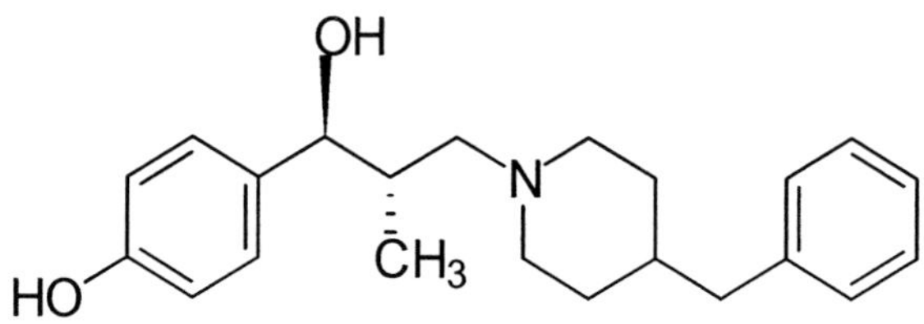

Ro-25-6981

Figure 9. NR2B subunit selective NMDAR antagonists

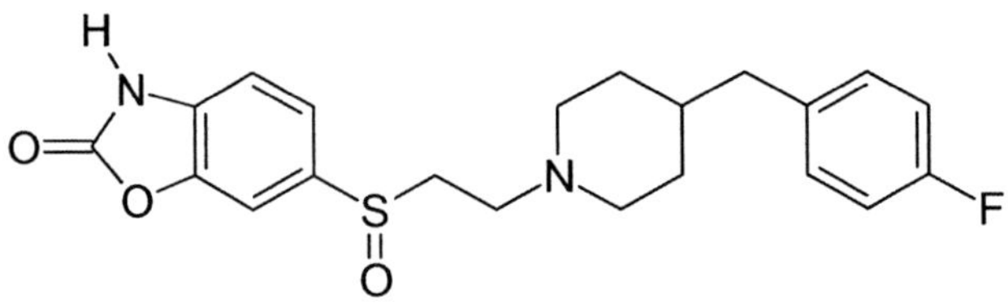

CI-1041

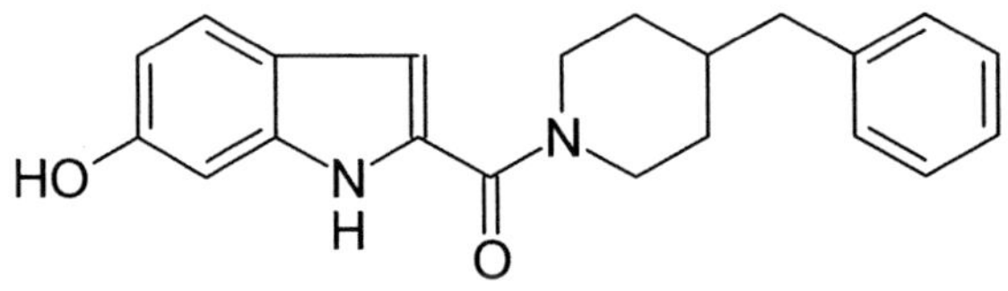

RGH-13579

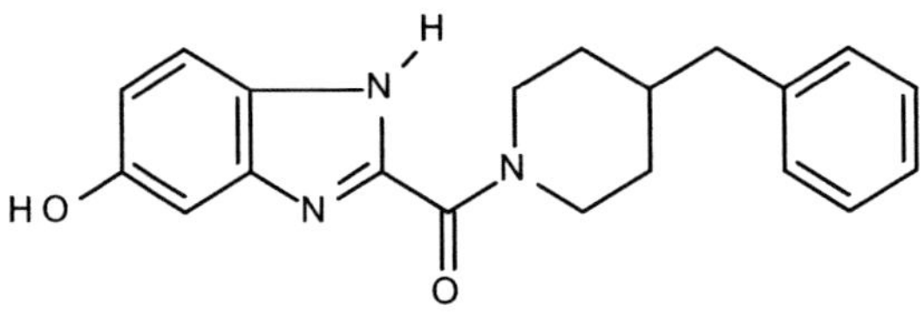

RGH-1103

Figure 9. NR2B subunit selective NMDAR antagonists (Continued)

In primary cultures of cortical neurons from rats pretreated with ethanol intermittently for 3 days, some of the novel indole-2-carboxamide derivative NR2B SSNAs (RGH-13579 and RGH-1103 (Borza 2003)) as well as CP-101,606, Co-101244 and CI-1041 potently and dose-dependently reduced the withdrawal-evoked LDH release (Figure 10). One of the novel compounds (RGH-1103) was as effective as MK-801, the most potent but not subunit selective, channel blocker NMDAR antagonist. The inhibitory potencies of the NR2B SSNAs for withdrawal-induced toxicity was in good linear relation with their effectiveness for inhibition of NMDA induced cytosolic calcium elevation (Table 4) (Nagy 2004a).

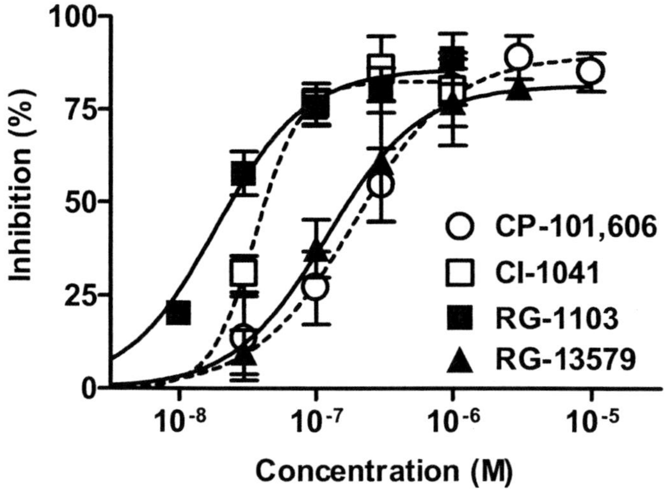

Figure 10. Inhibitory effect of NR2B SSNAs on ethanol-withdrawal-induced neurotoxicity Neuronal cell death caused by 24-hour ethanol-withdrawal in primary cultures of rat cortical neurones pre-treated with 100 mM ethanol for 3 consecutive days was quantified by measuring LDH-release. Different concentrations of some known (open symbols, dashed lines) and novel (filled symbols, straight lines) NR2B SSNAs were present during the withdrawal period. Each point represents the percentage of inhibition (mean ± S.E. (error bars)).

From: Nagy, J., Horváth, C., Farkas, S., Kolok, S., Szombathelyi, Z. (2004) NR2B subunit selective NMDA antagonists inhibit neurotoxic effect of alcohol-withdrawal in primary cultures of rat cortical neurons. *Neurochem. Int., 44(1),* 17-23.

Table 4. Inhibitory Effect of NMDA Antagonists on Cytosolic Calcium Rises and Ethanol Withdrawal Induced Neurotoxicity

Compound	NMDA evoked Ca^{2+} elevation		withdrawal induced toxicity	
	I_{max} (%)	IC_{50} (μM)	I_{max} (%)	IC_{50} (μM)
MK-801	100 ± 5	0.037 ± 0.002	90 ± 5	0.020 ± 0.004
CP-101,606	58 ± 2	0.041 ± 0.005	89 ± 9	0.201 ± 0.061
Co-101244	67 ± 2	0.024 ± 0.003	101 ± 5	0.206 ± 0.045
CI-1041	70 ± 3	0.007 ± 0.001	82 ± 6	0.037 ± 0.008
RGH-13579	77 ± 3	0.018 ± 0.004	81 ± 18	0.137 ± 0.067
RGH-1103	74 ± 2	0.002 ± 0.0004	87 ± 6	0.019 ± 0.005
Acamprosate	−1.2 ± 2.9	> 300	82 ± 8	40 ± 13

Compounds were tested for their inhibitory effect on cytosolic calcium elevation evoked by 40 μM NMDA and on ethanol-withdrawal produced neurotoxicity in primary cultures of rat cortical neurons.

From: József Nagy, Csilla Horváth, Sándor Farkas, Sándor Kolok, Zsolt Szombathelyi (2004) NR2B subunit selective NMDA antagonists inhibit neurotoxic effect of alcohol-withdrawal in primary cultures of rat cortical neurons. *Neurochemistry International 44(1)*, 17-23.

Summary

According to the recently emerged glutamatergic theory of the pathomechanism of alcohol withdrawal syndrome, increased NMDA receptor function plays a central role in the development of alcohol dependence and manifestation of the withdrawal symptoms. Furthermore, at least partly, an increase in the expression of the NR2B subunit underlies the alcohol-induced alterations (Nagy 2004b, 2004c). Since NMDA receptor antagonists can reduce hyperactivity, seizures, and neuronal cell loss associated with alcohol withdrawal, novel NMDA receptor antagonists, especially NR2B subunit selective ones, may be effective in the treatment of the physical symptoms of withdrawal. Since NMDAR antagonists can also restore the reduced DA brain levels associated with ethanol withdrawal and thereby can reduce psychical symptoms of alcohol dependence, novel type of NMDAR antagonists could be promising agents to prevent relapse and maintain abstinence as well.

In the last decade, great achievements have been made in research for novel, subunit selective NMDA antagonists. Several such agents potent in animal models of a number of CNS diseases and showing improved side-effect profile have been identified and characterized. Although, only few of them have been examined and found effective in animal models of alcoholism, their *in vitro* and – according to the available preliminary data – *in vivo* efficiency is encouraging. To answer the question whether these agents will be useful in the pharmacotherapy for alcohol dependence is a major challenge for the forthcoming years.

References

Akinshola, B.E. (2001) Straight-chain alcohols exhibit a cutoff in potency for the inhibition of recombinant glutamate receptor subunits. *Br. J. Pharmacol., 133(5),* 651-658.

Allgaier, C., Franke, H., Sobottka, H., Scheibler, P. (2000) Acamprosate inhibits Ca2+ influx mediated by NMDA receptors and voltage-sensitive Ca2+ channels in cultured rat mesencephalic neurones. *Naunyn Schmiedebergs Arch. Pharmacol., 362(4-5),* 440-443.

Allgaier, C. (2002) Ethanol sensitivity of NMDA receptors. *Neurochem. Int.,41(6),* 377-382.

Alvestad, R.M., Grosshans, D.R., Coultrap, S.J., Nakazawa, T., Yamamoto, T., Browning, M.D. (2003) Tyrosine dephosphorylation and ethanol inhibition of N-Methyl-D-aspartate receptor function. *J. Biol. Chem., 278(13),* 11020-11025.

Anders, D.L., Blevins, T., Sutton, G., Chandler, L.J., Woodward, J.J. (1999) Effects of c-Src tyrosine kinase on ethanol sensitivity of recombinant NMDA receptors expressed in HEK 293 cells. *Alcohol Clin. Exp. Res., 23(2),* 357-362.

Anders, D.L., Blevins, T., Smothers, C.T., Woodward, J.J. (2000) Reduced ethanol inhibition of N-methyl-D-aspartate receptors by deletion of the NR1 C0 domain or overexpression of alpha-actinin-2 proteins. *J. Biol. Chem., 275(20),* 5019-5024.

Avenet, P., Leonardon, J., Besnard, F., Graham, D., Frost, J., Depoortere, H., Langer, S.Z., Scatton, B. (1996) Antagonist properties of the stereoisomers of ifenprodil at NR1A/NR2A and NR1A/NR2B subtypes of the NMDA receptor expressed in Xenopus oocytes. *Eur. J. Pharmacol., 296(2),* 209-213.

Avenet, P., Leonardon, J., Besnard, F., Graham, D., Depoortere, H., Scatton, B. (1997) Antagonist properties of eliprodil and other NMDA receptor antagonists at rat NR1A/NR2A and NR1A/NR2B receptors expressed in Xenopus oocytes. *Neurosci. Lett., 223(2),* 133-136.

Ballenger, J.C. & Post, R.M. (1978) Kindling as a model for alcohol withdrawal syndromes. *Br. J. Psychiatry., 133,* 1-14.

Becker, H.C. & Hale, R.L. (1993) Repeated episodes of ethanol withdrawal potentiate the severity of subsequent withdrawal seizures: an animal model of alcohol withdrawal "kindling". *Alcohol Clin. Exp. Res., 17(1),* 94-98.

Biala, G. & Kotlinska, J. (1999) Blockade of the acquisition of ethanol-induced conditioned place preference by N-methyl-D-aspartate receptor antagonists. *Alcohol Alcohol., 34(2),* 175-182.

Bienkowski, P., Krzascik, P., Koros, E., Kostowski, W., Scinska, A., Danysz, W. (2001) Effects of a novel uncompetitive NMDA receptor antagonist, MRZ 2/579 on ethanol self-administration and ethanol withdrawal seizures in the rat. *Eur. J. Pharmacol., 413(1),* 81-89.

Blevins, T., Mirshahi, T., Woodward, J., (1995) Increased agonist and antagonist sensitivity of N-methyl-D-aspartate stimulated calcium flux in cultured neurons following chronic ethanol exposure. *Neurosci. Lett., 200,* 214-218.

Blevins, T., Mirshahi, T., Chandler, L.J., Woodward, J.J. (1997) Effects of acute and chronic ethanol exposure on heteromeric N-methyl-D-aspartate receptors expressed in HEK 293 cells. *J. Neurochem., 69(6),* 2345-2354.

Borza, I., Kolok, S., Gere, A., Agai-Csongor, E., Agai, B., Tarkanyi, G., Horvath, C., Barta-Szalai, G., Bozo, E., Kiss, C., Bielik, A., Nagy, J., Farkas, S., Domany, G. (2003) Indole-2-carboxamides as novel NR2B selective NMDA receptor antagonists. *Bioorganic & Medicinal Chemistry Letters, 13,* 3859–3861.

Boyce, S., Wyatt, A., Webb, J.K., O'Donnell, R., Mason, G., Rigby, M., Sirinathsinghji, D., Hill, R.G., Rupniak, N.M. (1999) Selective NMDA NR2B antagonists induce antinociception without motor dysfunction: correlation with restricted localisation of NR2B subunit in dorsal horn. *Neuropharmacology, 38(5),* 611-623.

Breese, G.R., Knapp, D.J., Moy, S.S. (2002) Integrative role for serotonergic and glutamatergic receptor mechanisms in the action of NMDA antagonists: potential relationships to antipsychotic drug actions on NMDA antagonist responsiveness. *Neurosci. Biobehav. Rev., 26(4),* 441-455.

Brimecombe, J.C., Boeckman, F.A. and Aizenman E. (1997) Functional consequences of NR2 subunit composition in single recombinant N-methyl-D-aspartate receptors. *Proc. Natl. Acad. Sci. USA, 94,* 11019-11024.

Brodie, M.S., Shefner, S.A., Dunwiddie, T.V. (1990) Ethanol increases the firing rate of dopamine neurons of the rat ventral tegmental area in vitro. *Brain Res., 508(1),* 65-69.

Brodie, M.S., Pesold, C., Appel, S.B. (1999) Ethanol directly excites dopaminergic ventral tegmental area reward neurons. *Alcohol Clin. Exp. Res., 23(11),* 1848-1852.

Buller, A.L., Larson, H.C., Morrisett, R.A., Monaghan, D.T., (1995) Glycine modulates ethanol inhibition of heteromeric N-methyl-D-aspartate receptors expressed in Xenopus oocytes. *Mol. Pharmacol., 48(4),* 717-723.

Carpenter-Hyland E.P., Woodward J.J., Chandler L.J. (2004) Chronic Ethanol Induces Synaptic But Not Extrasynaptic Targeting of NMDA Receptors. *J. Neurosci., 24(36),* 7859 –7868.

Carter, A.J., (1994) Many agents that antagonize the NMDA receptor-channel complex in vivo also cause disturbances of motor coordination. *J. Pharmacol. Exp. Ther., 269(2),* 573-580.

Carter, R.B., Wilent, W., Huber, M., Xu, Z., Vanover, K.E., Woodward R.M. (2000) Anti-allodynic and anti-hyperalgesic effects of the nr2b subunit-selective nmda antagonist ci-1041 in rat models of inflammation and neuropathy *Soc. Neurosci. Abstr., 26,* 617.3.

Cebere, A., Cebers, G., Liljequist, S. (1999) Enhancement of NMDA-induced functional responses without concomitant NMDA receptor changes following chronic ethanol exposure in cerebellar granule cells. *Naunyn Schmiedebergs Arch. Pharmacol., 360(6),* 623-632.

Chandler, L.J., Norwood, D., Sutton, G. (1999) Chronic ethanol upregulates NMDA and AMPA, but not kainate receptor subunit proteins in rat primary cortical cultures. *Alcohol Clin. Exp. Res., 23(2),* 363-370.

Chen, C, Leonard, J.P. (1996) Protein tyrosine kinase-mediated potentiation of currents from cloned NMDA receptors. *J. Neurochem., 67(1),* 194-200.

Chenard, B.L., Bordner, J., Butler, T.W., Chambers, L.K., Collins, M.A., De Costa, D.L., Ducat, M.F., Dumont, M.L., Fox, C.B., Mena, E.E., et al. (1995) (1S,2S)-1-(4-hydroxyphenyl)-2-(4-hydroxy-4-phenylpiperidino)-1-propanol: a potent new neuroprotectant which blocks N-methyl-D-aspartate responses. *J. Med. Chem., 38(16),* 3138-3145.

Chizh, B.A., Headley, P.M., Tzschentke, T.M. (2001) NMDA receptor antagonists as analgesics: focus on the NR2B subtype. *Trends Pharmacol. Sci., 22(12),* 636-642.

Chizh, B.A., (2002) Novel approaches to targeting glutamate receptors for the treatment of chronic pain: Review article *Amino Acids, 23,* 169-176.

Christine, C.W., Choi, D.W. (1990) Effect of zinc on NMDA receptor-mediated channel currents in cortical neurons. *J. Neurosci., 10(1),* 108-116.

Chu, B., Anantharam, V., Treistman, S.N. (1995) Ethanol inhibition of recombinant heteromeric NMDA channels in the presence and absence of modulators. *J. Neurochem., 65(1),* 140-148.

Costa, E.T., Savage, D.D., Valenzuela, C.F. (2000) Acute effects of ethanol on kainate receptors in cultured hippocampal neurons. *Alcohol Clin. Exp. Res., 24(5),* 706-715.

Cull-Candy, S., Brickley, S., Farrant, M. (2001) NMDA receptor subunits: diversity, development and disease. *Current Opinion in Neurobiology, 11,* 327–335.

Dahchour, A., De Witte, P., Bolo, N., Nedelec, J.F., Muzet, M., Durbin, P., Macher, J.P. (1998) Central effects of acamprosate: part 1. Acamprosate blocks the glutamate increase in the nucleus accumbens microdialysate in ethanol withdrawn rats. *Psychiatry Res., 82(2),* 107-114.

Dahchour, A. & De Witte, P. (1999) Effect of repeated ethanol withdrawal on glutamate microdialysate in the hippocampus. *Alcohol Clin. Exp. Res., 23(10),* 1698-1703.

Dahchour, A. & De Witte, P. (2000) Ethanol and amino acids in the central nervous system: assessment of the pharmacological actions of acamprosate. *Pro. Neurobiol., 60(4),* 343-362.

Danysz, W. & Parsons, C.G. (1998) Glycine and N-methyl-D-aspartate receptors: physiological significance and possible therapeutic applications. *Pharmacological Reviews, 50(4),* 597-664.

Darstein, M.B., Landwehrmeyer, G.B., Feuerstein, T.J. (2000) Changes in NMDA receptor subunit gene expression in the rat brain following withdrawal from forced long-term ethanol intake. *Naunyn Schmiedebergs Arch. Pharmacol., 361(2),* 206-213.

Das, S., Sasaki, Y F., Rothe, T., Premkumar, L S., Takasu, M., Crandall, J E., Dikkes, P., Conner, D A., Rayudu, P V., Cheung, W., Chen, H S., Lipton, S A., Nakanishi, N. (1998) Increased NMDA current and spine density in mice lacking the NMDA receptor subunit NR3A. *Nature, 393(6683),* 377-381.

Davies, M. (2003) The role of GABAA receptors in mediating the effects of alcohol in the central nervous system. *J. Psychiatry Neurosci., 28(4),* 263-274.

Di Chiara, G., Tanda G., Cadoni, C., Aquas E., Bassareo, V., Carboni, E. (1998) Homologies and differences in the action of drugs of abuse and a conventional reinforcer (food) on dopamine transmission: an interpretative framework of the mechanism of drug dependenc *Adv. Pharmacol., 42,* 983-987.

Diana, M., Pistis, M., Carboni, S., Gessa, G L., Rossetti, Z L. (1993) Profound decrement of mesolimbic dopaminergic neuronal activity during ethanol withdrawal syndrome in rats: electrophysiological and biochemical evidence. *Proc. Natl. Acad. Sci. USA, 90(17),* 7966-7969.

Dildy, J.E. & Leslie, S.W. (1989) Ethanol inhibits NMDA-induced increases in free intracellular Ca2+ in dissociated brain cells. *Brain Res., 499,* 383-387.

Editorial (2004) Time for coordinated action on alcohol. *The Lancet, 363,* 9414.

Engberg, G. & Hajos, M. (1992) Ethanol attenuates the response of locus coeruleus neurons to excitatory amino acid agonists in vivo. *Naunyn Schmiedebergs Arch. Pharmacol., 345(2),* 222-226

Engberg, G. & Hajos, M. (1992b) Alcohol withdrawal reaction as a result of adaptive changes of excitatory amino acid receptors. *Naunyn Schmiedebergs Arch. Pharmacol., 346(4),* 437-441.

Fadda, F. & Rossetti, Z.L. (1998) Chronic ethanol consumption: from neuroadaptation to neurodegeneration. *Prog. Neurobiol., 56(4),* 385-431.

Ferrani-Kile, K., Randall, P.K., Leslie, S.W. (2003) Acute ethanol affects phosphorylation state of the NMDA receptor complex: implication of tyrosine phosphatases and protein kinase A. *Brain Res. Mol. Brain Res., 115(1),* 78-86.

Ferreira, V.M., Frausto, S., Browning, M.D., Savage, D.D., Morato, G.S., Valenzuela, C.F. (2001) Ionotropic glutamate receptor subunit expression in the rat hippocampus: lack of an effect of a long-term ethanol exposure paradigm. *Alcohol Clin. Exp. Res., 25(10),* 1536-1541.

Fidecka, S., Langwinski, R. (1989) Interaction between ketamine and ethanol in rats and mice. *Pol. J. Pharmacol. Pharm., 41(1),* 23-32.

Fillhard, J.A., Kinsora J.J., Meltzer L.T. (2000) The effects of ci-1041 in two tests of analgesia: acetic acid-induced writing test and formalin foot pad test. Soc. *Neurosci. Abstr., 26,* 617.4.

Fink, K. & Göthert, M. (1996) Both ethanol and ifenprodil inhibit NMDA-evoked release of various neurotransmitters at different, yet proportional potency: potential relation to NMDA receptor subunit composition. *Naunyn Schmiedebergs Arch. Pharmacol., 354(3),* 312-319.

Fischer, G., Mutel, V., Trube, G., Malherbe, P., Kew, J.N., Mohacsi, E., Heitz, M.P., Kemp, J.A. (1997) Ro 25-6981, a highly potent and selective blocker of N-methyl-D-aspartate receptors containing the NR2B subunit. Characterization in vitro. *J. Pharmacol. Exp. Ther., 283(3),* 1285-1292.

Floyd, D.W., Jung, K.Y., McCool, B.A. (2003) Chronic ethanol ingestion facilitates N-methyl-D-aspartate receptor function and expression in rat lateral/basolateral amygdala neurons. *J. Pharmacol. Exp. Ther., 307(3),* 1020-1029.

Follesa, P. & Ticku, M.K. (1996a) NMDA receptor upregulation: molecular studies in cultured mouse cortical neurons after chronic antagonist exposure. *J. Neurosci., 16,* 2172-2178.

Follesa, P. & Ticku, M.K. (1996b) Chronic ethanol-mediated up-regulation of the N-methyl-D-aspartate receptor polypeptide subunits in mouse cortical neurons in culture. *J. Biol. Chem., 271(23),* 13297-13299.

Franks, N.P. & Lieb, W.R. (1994) Molecular and cellular mechanisms of general anaesthesia. *Nature, 367(6464),* 607-614.

Gallagher, M.J., Huang, H., Pritchett, D.B., Lynch, D.R. (1996) Interactions between ifenprodil and the NR2B subunit of the N-methyl-D-aspartate receptor. *J. Biol. Chem., 271(16),* 9603- 9611

Gallagher, M J, Huang, H, Grant, E R, Lynch, D R. (1997) The NR2B-specific interactions of polyamines and protons with the N-methyl-D-aspartate receptor. *J. Biol. Chem., 272(40),* 24971-24979.

Godden, E.L., Harris, R.A., Dunwiddie, T.V. (2001). Correlation between molecular volume and effects of n-alcohols on human neuronal nicotinic acetylcholine receptors expressed in Xenopus oocytes. *J. Pharmacol. Exp. Ther., 296(3),* 716-722.

Gonzalez, L.P., Veatch, L.M., Ticku, M.K., Becker, H.C. (2001) Alcohol withdrawal kindling: mechanisms and implications for treatment. *Alcohol Clin. Exp. Res., 25,* 197S-201S.

Grant, K.A., Valverius, P., Hudspith, M., Tabakoff, B. (1990) Ethanol withdrawal seizures and the NMDA receptor complex. *Eur. J. Pharmacol., 176,* 289-296.

Grant, K.A., Lovinger, D.M. (1995) Cellular and behavioral neurobiology of alcohol: receptor-mediated neuronal processes. *Clin. Neurosci., 3(3),* 155-164.

Grant, K.A., Colombo, G., Grant, J., Rogawski, M.A. (1996) Dizocilpine-like discriminative stimulus effects of low-affinity uncompetitive NMDA antagonists. *Neuropharmacology, 35(12),* 1709-1719.

Grant, B.F., Dawson, D.A., Stinson, F.S., Chou, S.P., Dufour, M.C., Pickering, R.P. (2004). The 12-month prevalence and trends in DSM-IV alcohol abuse and dependence: United States, 1991–1992 and 2001–2002. *Drug and Alcohol Dependence, 74,* 223–234.

Hardy, P.A., Chen, W., Wilce, P.A., (1999) Chronic ethanol exposure and withdrawal influence NMDA receptor subunit and splice variant mRNA expression in the rat cerebral cortex. *Brain Res., 819(1-2),* 33-39.

Harris, B.R., Gibson, D.A., Prendergast, M.A., Blanchard, J.A., Holley, R.C., Hart, S.R., Scotland, R.L., Foster, T.C., Pedigo, N.W., Littleton, J.M. (2003) The neurotoxicity induced by ethanol withdrawal in mature organotypic hippocampal slices might involve cross-talk between metabotropic glutamate type 5 receptors and N-methyl-D-aspartate receptors. *Alcohol Clin. Exp. Res., 27(11),* 1724-1735.

Henniger, M.S., Wotjak, C.T., Holter, S.M. (2003) Long-term voluntary ethanol drinking increases expression of NMDA receptor 2B subunits in rat frontal cortex. *Eur. J. Pharmacol., 470(1-2),* 33-36.

Heyser, C.J., Schulteis, G., Durbin, P., Koob, G.F. (1998) Chronic acamprosate eliminates the alcohol deprivation effect while having limited effects on baseline responding for ethanol in rats. *Neuropsychopharmacology*, *18(2),* 125-133.

Hoffman, P.L., Rabe, C.S., Moses, F., Tabakoff, B. (1989) N-methyl-D-aspartate receptors and ethanol: inhibition of calcium flux and cyclic GMP production. *J. Neurochem., 52,* 61937-61940.

Hoffman, P.L., Rabe, C.S., Grant, K.A., Valverius, P., Hudspith, M., Tabakoff, B. (1990) Ethanol and the NMDA receptor *Alcohol., 7(3),* 229-231.

Hoffman, P.L. & Tabakoff, B. (1996) Alcohol dependence: a commentary on mechanisms. *Alcohol Alcohol., 31(4),* 333-340.

Hollmann, M. Structure of ionotropic glutamate receptors. In Ionotropic Glutamate Receptors in the CNS. Edited by Jonas, P., Monyer, H., Berlin: Springer, 1999, 1-98.

Hölter, S.M., Danysz, W., Spanagel, R. (1996) Evidence for alcohol anti-craving properties of memantine. *Eur. J. Pharmacol., 314,* R1–R2.

Hölter, SM., Landgraf, R., Zieglgansberger, W., Spanagel, R. (1997) Time course of acamprosate action on operant ethanol self-administration after ethanol deprivation. *Alcohol Clin. Exp. Res., 21(5),* 862-868.

Hölter, S.M., Danysz, W., Spanagel, R. (2000) Novel Uncompetitive N-Methyl-D-Aspartate (NMDA)-Receptor Antagonist MRZ 2/579 Suppresses Ethanol Intake in Long-Term Ethanol-Experienced Rats and Generalizes to Ethanol Cue in Drug Discrimination Procedure. *J. Pharmacol. Exp. Ther., 292,* 545–552.

Honer, M., Benke, D., Laube, B., Kuhse, J., Heckendorn, R., Allgeier, H., Angst, C., Monyer, H., Seeburg, P H., Betz, H., Mohler, H. (1998) Differentiation of glycine antagonist sites of N-methyl-D-aspartate receptor subtypes. Preferential interaction of CGP 61594 with NR1/2B receptors. *J. Biol. Chem., 273(18),* 11158-11163.

Hu, X.J., Follesa, P., Ticku, M.K. (1996) Chronic ethanol treatment produces a selective upregulation of the NMDA receptor subunit gene expression in mammalian cultured cortical neurons. *Brain Res. Mol. Brain Res., 36,* 211-218.

Hu, X.J. & Ticku, M.K. (1997) Functional characterization of a kindling-like model of ethanol withdrawal in cortical cultured neurons after chronic intermittent ethanol exposure. *Brain Res., 767(2),* 228-234.

Husi, H. & Grant, S.G. (2001) Isolation of 2000-kDa complexes of N-methyl-D-aspartate receptor and postsynaptic density 95 from mouse brain. *J. Neurochem., 77(1),* 281-291.

Ibbotson, T., Field, M.J., Boden, P.R. (1997) Effect of chronic ethanol treatment in vivo on excitability in mouse cortical neurones in vitro. *Br. J. Pharmacol., 122(5),* 956-962.

Imperato, A. & Di Chiara, G. (1986) Preferential stimulation of dopamine release in the nucleus accumbens of freely moving rats by ethanol. *J. Pharmacol. Exp. Ther., 239(1),* 219-228.

Jansen, M., Dannhardt, G. (2003) Antagonists and agonists at the glycine site of the NMDA receptor for therapeutic interventions. *European Journal of Medicinal Chemistry, 38,* 661-670.

Kalluri, H.S., Mehta, A.K., Ticku, M.K. (1998) Up-regulation of NMDA receptor subunits in rat brain following chronic ethanol treatment. *Brain Res. Mol. Brain Res., 58(1-2),* 221-224.

Kashiwagi, K., Fukuchi, J., Chao, J., Igarashi, K., Williams, K. (1996) An aspartate residue in the extracellular loop of the N-methyl-D-aspartate receptor controls sensitivity to spermine and protons. *Mol. Pharmacol., 49(6),* 1131-1141.

Katayama, S., Ae, N., Nagata, R. (2001) Synthesis of tricyclic indole-2-carboxylic [correction of caboxylic] acids as potent NMDA-glycine antagonists. *J. Org. Chem., 66,* 3474-3483.

Kemp, J.A. & McKernan, R.M. (2002) NMDA receptor pathways as drug targets. *Nature Neurosci. Supp., 5,* 1039-1042.

Kew, J.N., Trube, G., Kemp, J.A. (1998) State-dependent NMDA receptor antagonism by Ro 8-4304, a novel NR2B selective, non-competitive, voltage-independent antagonist. *Br. J. Pharmacol., 123(3),* 463-472.

Khanna, J.M., Morato, G.S., Chau, A., Shah, G. (1995) D-cycloserine enhances rapid tolerance to ethanol motor incoordination. *Pharmacol Biochem Behav., 52(3),* 609-614.

Koek, W., Colpaert, F.C. (1990) Selective blockade of N-methyl-D-aspartate (NMDA)-induced convulsions by NMDA antagonists and putative glycine antagonists: relationship with phencyclidine-like behavioral effects. *J. Pharmacol. Exp. Ther., 252(1),* 349-357.

Kohr, G., Seeburg, P.H. (1996) Subtype-specific regulation of recombinant NMDA receptor-channels by protein tyrosine kinases of the src family. *J. Physiol., 492 (Pt 2),* 445-452.

Koob, G.F. (1992) Drugs of abuse: anatomy, pharmacology and function of reward pathways. *Trends Pharmacol. Sci., 13(5),* 177-184.

Kotlinska, J. & Liljequist, S. (1996) Oral administration of glycine and polyamine receptor antagonists blocks ethanol withdrawal seizures. *Psychopharmacology (Berl), 127(3),* 238-244.

Kotlinska, J. & Liljequist, S. (1997) The NMDA/glycine receptor antagonist, L-701,324, produces discriminative stimuli similar to those of ethanol. *Eur. J. Pharmacol., 332(1),* 1-8.

Kretschmer, B.D. (1999) Modulation of the mesolimbic dopamine system by glutamate: role of NMDA receptors. *J. Neurochem., 73(2),* 839-848.

Kumari, M., Ticku, M.K. (2000) Regulation of NMDA receptors by ethanol. *Prog. Drug Res., 54,* 152-189.

Kuner, T., Schoepfer, R., Korpi, E.R. (1993) Ethanol inhibits glutamate-induced currents in heteromeric NMDA receptor subtypes. *Neuroreport, 5(3),* 297-300.

Lamblin, F., Deuceuninck, D., DeWitte, P. (1993) Modulation of alcohol preference by NMDA antagonists in male rats. *Alcohol Alcohol., 28(6),* 639-647.

Laube, B., Kuhse, J., Betz, H. (1998) Evidence for a tetrameric structure of recombinant NMDA receptors. *J. Neurosci., 18(8),* 2954-2961.

Lerma, J. (1992) Spermine regulates N-methyl-D-aspartate receptor desensitization. *Neuron, 8(2),* 343-352.

Lhuintre, J.P., Daoust, M., Moore, N.D., Chretien, P., Saligaut, C., Tran, G., Boismare, F., Hillemand, B. (1985) Ability of calcium bis acetyl homotaurine, a GABA agonist, to prevent relapse in weaned alcoholics. *Lancet, 1,* 1014-1016.

Lhuintre, J.P., Moore, N., Tran, G., Steru, L., Langrenon, S., Daoust, M., Parot, P., Ladure, P., Libert, C., Boismare, F., et al. (1990) Acamprosate appears to decrease alcohol intake in weaned alcoholics. *Alcohol Alcohol., 25,* 613-622.

Li, C., Peoples, R.W., Weight, F.F. (1994) Alcohol action on a neuronal membrane receptor: evidence for a direct interaction with the receptor protein. *Proc. Natl. Acad. Sci. USA, 91(17),* 8200-8204.

Liljequist, S. (1991) The competitive NMDA receptor antagonist, CGP 39551, inhibits ethanol withdrawal seizures. *Eur. J. Pharmacol., 192(1),* 197-198.

Lovinger, D.M., White G. and Weight, F.F. (1989) Ethanol inhibits NMDA-activated ion current in hippocampal neurons. *Science, 243,* 1721-1724.

Lovinger, D.M., White, G., Weight, F.F. (1990) NMDA receptor-mediated synaptic excitation selectively inhibited by ethanol in hippocampal slice from adult rat. *J. Neurosci., 10,* 41372-41379.

Lovinger, D.M. (1993) Excitotoxicity and alcohol-related brain damage. *Alcohol Clin. Exp. Res., 17(1),* 19-27.

Lovinger, D.M. (1995) Developmental decrease in ethanol inhibition of N-methyl-D-aspartate receptors in rat neocortical neurons: relation to the actions of ifenprodil. *J. Pharmacol. Exp. Ther., 274(1),* 164-172.

Maldve, R.E., Zhang, T.A., Ferrani-Kile, K., Schreiber, S.S., Lippmann, M.J., Snyder, G.L., Fienberg, A.A., Leslie, S.W., Gonzales, R.A., Morrisett, R.A. (2002) DARPP-32 and regulation of the ethanol sensitivity of NMDA receptors in the nucleus accumbens. *Nat. Neurosci., 5(7),* 641-648.

Malinowska, B., Napiorkowska-Pawlak, D., Pawlak, R., Buczko, W., Gothert, M. (1999) Ifenprodil influences changes in mouse behaviour related to acute and chronic ethanol administration. *Eur. J. Pharmacol., 377(1),* 13-19.

Masood, K., Wu, C., Brauneis, U., Weight, F.F. (1994) Differential ethanol sensitivity of recombinant N-methyl-D-aspartate receptor subunits. *Mol. Pharmacol., 45,* 324–329.

Mayer, S., Harris, B., Gibson, D.A., Blanchard, J., Prendergast, M.A., Holley, R.C., Littleton, J. (2002a) Acamprosate has no effect on NMDA-induced toxicity but reduces toxicity induced by spermidine or by changing the medium in organotypic hippocampal slice cultures from rat. *Alcohol. Clin. Exp. Res., 26(5),* 655-662.

Mayer, S., Harris, B.R., Gibson, D.A., Blanchard, J.A., Prendergast, M.A., Holley, R.C., Littleton, J. (2002b) Acamprosate, MK-801, and ifenprodil inhibit neurotoxicity and calcium entry induced by ethanol withdrawal in organotypic slice cultures from neonatal rat hippocampus. *Alcohol. Clin. Exp. Res., 26(10),* 1468-1478.

Meguro, H., Mori, H., Araki, K., Kushiya, E., Kutsuwada, T., Yamazaki, M., Kumanishi, T., Arakawa, M., Sakimura, K., Mishina, M. (1992) Functional characterization of a heteromeric NMDA receptor channel expressed from cloned cDNAs. *Nature, 357(6373),* 70-74.

Meyer, H. (1901) Zur theorie der alkolnarkose der einfluss wechselnder temperatur auf wirkungsstarke und theilungscoefficient der narcotica. *Naunyn-Schmiedeberg's Arch. Exp. Pathol. Pharmacol., 46,* 338–346.

Mihic, S.J., Ye, Q., Wick, M.J., Koltchine, V.V., Krasowski, M.D., Finn, S.E., Mascia, M.P., Valenzuela, C.F., Hanson, K.K., Greenblatt, E.P., Harris, R.A., Harrison, N.L. (1997) Sites of alcohol and volatile anaesthetic action on GABA(A) and glycine receptors. *Nature, 389(6649),* 385-389.

Mirshahi, T., & Woodward, J. (1995) Ethanol sensitivity of heteromeric NMDA receptors: effects of subunit assembly, glycine and NMDAR1 Mg(2+)-insensitive mutants. *J. Neuropharmacology, 34,* 347–355.

Miyakawa, T., Yagi, T., Kitazawa, H., Yasuda, M., Kawai, N., Tsuboi, K., Niki, H. (1997) Fyn-kinase as a determinant of ethanol sensitivity: relation to NMDA-receptor function. *Science, 278(5338),* 698-701.

Monyer, H., Sprengler, R., Schoepfer, R., Herb, A., Higuchi, M., Lomeli, H., Burnashev, N., Sakmann, B., Seeburg. P.H. (1992) Heteromeric NMDA receptors: molecular and functional distinction of subtypes. *Science, 256,* 1217-1221.

Monyer, H., Burnashev, N., Laurie, D.J., Sakmann, B., Seeburg, P.H. (1994) Developmental and regional expression in the rat brain and functional properties of four NMDA receptors. *Neuron, 12(3),* 529-40.

Moraes-Ferreira, V.M., Morato, G.S. (1997) D-cycloserine blocks the effects of ethanol and HA-966 in rats tested in the elevated plus-maze. *Alcohol. Clin. Exp. Res., 21(9),* 1638-1642.

Moriyoshi, K., Masu, M., Ishii, T., Shigemoto, R., Mizuno, N., Nakanishi, S. (1991) Molecular cloning and characterization of the rat NMDA receptor. *Nature, 354(6348),* 31-37.

Morrisett, R.A., Rezvani, A.H., Overstreet, D., Janowsky, D.S., Wilson, W.A., Swartzwelder, H.S. (1990) MK-801 potently inhibits alcohol withdrawal seizures in rats. *Eur. J. Pharmacol., 176(1),* 103-105.

Morrow, A.L., Devaud L.L., Bucci, D., Smith F.D., (1994) GABAA and NMDA receptor subunit mRNA expression in ethanol dependent rats. *Alcohol Alcohol., 2,* 89-95.

Mothet, J.P., Parent, A.T., Wolosker, H., Brady, R.O. Jr,, Linden, D.J., Ferris, C.D., Rogawski, M.A., Snyder, S.H. (2000) D-serine is an endogenous ligand for the glycine site of the N-methyl-D-aspartate receptor. *Proc. Natl. Acad. Sci. USA., 97(9),* 4926-4931.

Murata, S., Kawasaki, K. (1993) Common and uncommon behavioural effects of antagonists for different modulatory sites in the NMDA receptor/channel complex. *Eur. J. Pharmacol., 239(1-3),* 9-15.

Murray, F. Kennedy, J., Hutson, P.H., Elliot, J., Huscroft, I., Mohnen, K., Russell, M.G., Grimwood, S. (2000) Modulation of [3H]MK-801 binding to NMDA receptors in vivo and in vitro. *Eur. J. Pharmacol., 397,* 263–270.

Nagy, J., Müller, F., László L. (2001) Cytotoxic effect of alcohol-withdrawal on primary cultures of cortical neurones. *Drug Alcohol Depend., 61,* 155–162.

Nagy, J. & László L. (2002) Increased sensitivity to NMDA is involved in alcohol-withdrawal induced cytotoxicity observed in primary cultures of cortical neurones chronically pre-treated with ethanol. *Neurochem. Int., 40,* 585–591.

Nagy, J., Kolok, S., Dezső, P., Boros, A. (2003) Szombathelyi, Z. Differential alterations in the expression of NMDA receptor subunits following chronic ethanol treatment in primary cultures of rat cortical and hippocampal neurones. *Neurochem. Int., 42(1),* 35-43.

Nagy J., Boros, A., Dezső, P., Kolok, S., Fodor L. (2003b) Inducible expression and pharmacology of recombinant NMDA receptors, composed of rat NR1a/NR2B subunits. *Neurochemistry International 43,* 19–29.

Nagy, J., Horváth, C., Farkas, S., Kolok, S., Szombathelyi, Z. (2004a) NR2B subunit selective NMDA antagonists inhibit neurotoxic effect of alcohol-withdrawal in primary cultures of rat cortical neurones. *Neurochem. Int., 44(1),* 17-23.

Nagy, J. (2004b) Renaissance of NMDA receptor antagonists: Do they have a role in the pharmacotherapy for alcoholism? *IDrugs: The Investigational Drugs Journal, 7(4),* 339-350.

Nagy, J. (2004c) NR2B type of NMDA receptor subunits: a possible target in treatment for alcohol dependence. *Current Drug Targets CNS Neurol Disord. 3(3),* 169-179.

Narita, M., Soma, M., Narita, M., Mizoguchi, H., Tseng, L F., Suzuki, T. (2000) Implications of the NR2B subunit-containing NMDA receptor localized in mouse limbic forebrain in ethanol dependence. *Eur. J. Pharmacol., 401(2),* 191-195.

Nash, J.E., Hill, M.P., Brotchie, J.M. (1999) Antiparkinsonian actions of blockade of NR2B-containing NMDA receptors in the reserpine-treated rat. *Exp. Neurol., 155(1),* 42-48.

Nash, J.E., Fox, S.H., Henry, B., Hill, M.P., Peggs, D., McGuire, S., Maneuf, Y., Hille, C., Brotchie, J.M., Crossman, A.R. (2000) Antiparkinsonian actions of ifenprodil in the MPTP-lesioned marmoset model of Parkinson's disease. *Exp. Neurol., 165(1),* 136-142.

Nestler, E.J. (1993) Cellular responses to chronic treatment with drugs of abuse. *Crit. Rev. Neurobiol., 7(1),* 23-39.

Nevo, I., Hamon, M. (1995) Neurotransmitter and neuromodulatory mechanisms involved in alcohol abuse and alcoholism. *Neurochem. Int., 26(4),* 305-336.

Nikam, S.S., Meltzer, L.T., (2002) NR2B selective NMDA receptor antagonists. *Current Pharmaceutical Design, 8,* 845-855.

Overton, E. (1896) Uber die osmotischem eigenschaften der zelle in ihrer betdeurtung dur die toxikologie und pharmakolgie. *Z. Physikal Chem., 22,* 189–209.

Parsons, C.G., Danysz, W., Hesselink, M., Hartmann, S., Lorenz, B., Wollenburg, C., Quack, G. (1998) Modulation of NMDA receptors by glycine--introduction to some basic aspects and recent developments. *Amino-Acids., 14(1-3),* 207-216.

Parsons, C.G., Danysz, W., Bartmann, A., Spielmanns, P., Frankiewicz, T., Hesselink, M., Eilbacher, B., Quack, G. (1999) Amino-alkyl-cyclohexanes are novel uncompetitive NMDA receptor antagonists with strong voltage-dependency and fast blocking kinetics: in vitro and in vivo characterization. *Neuropharmacology, 38(1),* 85-108.

Peoples, R.W., Weight, F.F. (1995) Cutoff in potency implicates alcohol inhibition of N-methyl-D-aspartate receptors in alcohol intoxication. *Proc. Natl. Acad. Sci. USA., 92(7),* 2825-2829.

Peoples, R.W., White, G., Lovinger, D.M., Weight, F.F. (1997) Ethanol inhibition of N-methyl-D-aspartate-activated current in mouse hippocampal neurones: whole-cell patch-clamp analysis. *Br. J. Pharmacol., 122(6),* 1035-1042.

Perez-Otano, I., Schulteis, C.T., Contractor, A., Lipton, S.A.., Trimmer, J S., Sucher, N.J., Heinemann, S.F. (2001) Assembly with the NR1 subunit is required for surface expression of NR3A-containing NMDA receptors. *J. Neurosci., 21(4),* 1228-1237.

Piasecki, J., Koros, E., Dyr, W., Kostowski, W., Danysz, W., Bienkowski, P. (1998) Ethanol-reinforced behaviour in the rat: effects of uncompetitive NMDA receptor antagonist, memantine. *Eur. J. Pharmacol., 354(2-3),* 135-143.

Popp, R.L Lickteig, R., Browning, M.D., Lovinger, D.M. (1998) Ethanol sensitivity and subunit composition of NMDA receptors in cultured striatal neurons. *Neuropharmacology, 37(1),* 45-56.

Popp, R.L & Lovinger, D.M. (2000) Interaction of acamprosate with ethanol and spermine on NMDA receptors in primary cultured neurons. *Eur. J. Pharmacol., 394(2-3),* 221-231.

Premkumar, L.S. & Auerbach, A.J. (1997) Subconductance states of a mutant NMDA receptor channel kinetics, calcium, and voltage dependence. *Gen. Physiol., 110(5),* 485-502.

Quartaroli, M., Fasdelli, N., Bettelini, L., Maraia, G., Corsi, M. (2001) GV196771A, an NMDA receptor/glycine site antagonist, attenuates mechanical allodynia in neuropathic rats and reduces tolerance induced by morphine in mice. *Eur. J. Pharmacol., 430(2-3),* 219-227.

alQatari, M.; Bouchenafa, O., Littleton, J. (1998) Mechanism of action of acamprosate. Part II. Ethanol dependence modifies effects of acamprosate on NMDA receptor binding in membranes from rat cerebral cortex. *Alcohol. Clin. Exp. Res. 22(4),* 810-814.

al-Qatari, M., Khan, S., Harris, B., Littleton, J. (2001) Acamprosate is neuroprotective against glutamate-induced excitotoxicity when enhanced by ethanol withdrawal in neocortical cultures of fetal rat brain. *Alcohol. Clin. Exp. Res. 25(9),* 1276-1283.

Rabe, C.S. & Tabakoff, B. (1990) Glycine site-directed agonists reverse the actions of ethanol at the N-methyl-D-aspartate receptor. *Mol. Pharmacol., 38(6),* 753-757.

Rameau, A.G., Akaneya, Y., Chiu, L.Y., Ziff, E.B. (2000) Role of NMDA receptor functional domains in excitatory cell death. *Neuropharmacology, 39,* 2255-2266.

Rammes, G., Mahal, B., Putzke, J., Parsons, C., Spielmanns, P., Pestel, E., Spanagel, R., Zieglgansberger, W., Schadrack, J., (2001) The anti-craving compound acamprosate acts as a weak NMDA-receptor antagonist, but modulates NMDA-receptor subunit expression similar to memantine and MK-801. *Neuropharmacology, 40(6),* 749-760.

Ripley, T.L. & Little, H.J. (1995) Effects on ethanol withdrawal hyperexcitability of chronic treatment with a competitive N-methyl-D-aspartate receptor antagonist. *J. Pharmacol. Exp. Ther., 272(1),* 112-118.

Rock, D.M. & Macdonald, R.L. (1992) The polyamine spermine has multiple actions on N-methyl-D-aspartate receptor single-channel currents in cultured cortical neurons. *Mol. Pharmacol., 41(1),* 83-88.

Ronald, K.M., Mirshahi, T., Woodward J.J. (2001) Ethanol inhibition of N-methyl-D-aspartate receptors is reduced by site-directed mutagenesis of a transmembrane domain phenylalanine residue. *J. Biol. Chem., 276(48),* 44729–44735.

Rossetti, Z.L., Hmaidan, Y., Gessa, G.L. (1992) Marked inhibition of mesolimbic dopamine release: a common feature of ethanol, morphine, cocaine and amphetamine abstinence in rats. *Eur. J. Pharmacol., 221(2-3),* 227-234.

Rossetti, Z.L. & Carboni, S. (1995) Ethanol withdrawal is associated with increased extracellular glutamate in the rat striatum. *Eur. J. Pharmacol., 283(1-3),* 177-183.

Rudolph, J.G., Walker, D.W., Iimuro, Y., Thurman, R.G., Crews, F.T. (1997) NMDA receptor binding in adult rat brain after several chronic ethanol treatment protocols. *Alcohol Clin Exp Res., 21(8),* 1508-1519.

Rumbaugh, G., Prybylowski, K., Wang, J.F., Vicini, S.J. (2000) Exon 5 and spermine regulate deactivation of NMDA receptor subtypes. *Neurophysiol., 83(3),* 1300-1306.

Salter, M.W. & Kalia, L.V. (2004) Src kinases: a hub for NMDA receptor regulation. *Nat. Rev. Neurosci., 5,* 317-328.

Schorge, S. & Colquhoun, D. (2003) Studies of NMDA receptor function and stoichiometry with truncated and tandem subunits. *J. Neurosci., 23(4),* 1151–1158.

Shelton, K.L. & Balster, R.L. (1997) Effects of gamma-aminobutyric acid agonists and N-methyl-D-aspartate antagonists on a multiple schedule of ethanol and saccharin self-administration in rats. *J. Pharmacol. Exp. Ther., 280(3),* 1250-1260.

Smothers, C.T., Mrotek, J.J., Lovinger, D.M. (1997) Chronic ethanol exposure leads to a selective enhancement of N-methyl-D-aspartate receptor function in cultured hippocampal neurons. *J. Pharmacol. Exp. Ther., 283,* 1214-1222.

Smothers, C.T., Clayton, R., Blevins, T., Woodward, J.J. (2001) Ethanol sensitivity of recombinant human N-methyl-D-aspartate receptors. *Neurochem. Int., 38(4),* 333-340.

Snell, L.D., Bhave, S.V., Tabakoff, B., Hoffman, P.L. (2001) Chronic ethanol exposure delays the 'developmental switch' of the NMDA receptor 2A and 2B subunits in cultured cerebellar granule neurons.. *J. Neurochem., 78(2),* 396-405.

Spanagel, R. & Zieglgansberger, W. (1997a) Anti-craving compounds for ethanol: new pharmacological tools to study addictive processes. *Trends Pharmacol. Sci., 18(2),* 54-59.

Spanagel, R., Holter, S., Putzke, J., Pestel, E., Wetzel, C., Zieglgansberger, W., (1997b). Behavioral, molecular and electrophysiological characterization of the action of acamprosate. *Alcohol. Clin. Exp. Res. Suppl. 21,* 637.

Steece-Collier, K., Chambers, L.K., Jaw-Tsai, S.S., Menniti, F.S., Greenamyre, J.T. (2000) Antiparkinsonian actions of CP-101,606, an antagonist of NR2B subunit-containing N-methyl-d-aspartate receptors. *Exp. Neurol., 163(1),* 239-243.

Stork, O., Kojima, N., Stork, S., Kume, N., Obata, K. (2002) Resistance to alcohol withdrawal-induced behaviour in Fyn transgenic mice and its reversal by ifenprodil. *Brain Res. Mol. Brain Res., 105(1-2),* 126-135.

Tang, C.M., Dichter, M., Morad, M. (1990) Modulation of the N-methyl-D-aspartate channel by extracellular H+. *Proc. Natl. Acad. Sci. USA., 87(16),* 6445-6449.

Taniguchi, K., Shinjo, K., Mizutani, M., Shimada, K., Ishikawa, T., Menniti, F.S., Nagahisa, A. (1997) Antinociceptive activity of CP-101,606, an NMDA receptor NR2B subunit antagonist. *Br. J. Pharmacol., 122(5),* 809–812.

The World Health Report (2003) World Health Organization, Geneva, Switzerland.

Thomas, M.P., Davis, M.I., Monaghan, D.T., Morrisett, R.A. (1998) Organotypic brain slice cultures for functional analysis of alcohol-related disorders: novel versus conventional preparations. *Alcohol Clin. Exp. Res., 22(1),* 51-59.

Thomas, M.P. & Morrisett, R.A. (2000) Dynamics of NMDAR-mediated neurotoxicity during chronic ethanol exposure and withdrawal. *Neuropharmacology, 39(2),* 218-226.

Traynelis, S.F. & Cull-Candy, S.G. (1990) Proton inhibition of N-methyl-D-aspartate receptors in cerebellar neurons. *Nature, 345(6273),* 347-350.

Trevisan, L., Fitzgerald, L.W., Brose, N., Gasic, G.P., Heinemann, S.F., Duman, R.S., Nestler, E.J. (1994) Chronic ingestion of ethanol up-regulates NMDAR1 receptor subunit immunoreactivity in rat hippocampus. *J. Neurochem., 62(4),* 1635-1638.

Tsai, G., Gastfriend, D.R., Coyle, J.T. (1995) The glutamatergic basis of human alcoholism. *Am. J. Psychiatry, 152(3),* 332-340.

Tsai, G. & Coyle, J.T. (1998) The role of glutamatergic neurotransmission in the pathophysiology of alcoholism. *Annu. Rev. Med., 49,* 173-184.

Ueno, S., Lin, A., Nikolaeva, N., Trudell, J.R., Mihic, S.J., Harris, R.A., Harrison, N.L. (2000) Tryptophan scanning mutagenesis in TM2 of the GABA(A) receptor alpha subunit: effects on channel gating and regulation by ethanol. *Br. J. Pharmacol., 131(2),* 296-302.

Wang, Y.T., Salter, M.W. (1994) Regulation of NMDA receptors by tyrosine kinases and phosphatases. *Nature, 369(6477),* 233-235.

Whittington, M.A., Lambert, J.D., Little, H.J. (1995) Increased NMDA receptor and calcium channel activity underlying ethanol withdrawal hyperexcitability. *Alcohol Alcohol., 30(1),* 105-114.

Williams, K. (1993) Ifenprodil discriminates subtypes of the N-methyl-D-aspartate receptor: selectivity and mechanisms at recombinant heteromeric receptors. *Mol. Pharmacol., 44(4),* 851-859.

Williams, K., Kashiwagi, K., Fukuchi, J., Igarashi, K. (1995) An acidic amino acid in the N-methyl-D-aspartate receptor that is important for spermine stimulation. *Mol. Pharmacol., 48(6),* 1087-1098.

Williams, K, Pahk, A J, Kashiwagi, K, Masuko, T, Nguyen, N D, Igarashi, K. (1998) The selectivity filter of the N-methyl-D-aspartate receptor: a tryptophan residue controls block and permeation of Mg2+. *Mol. Pharmacol., 53(5),* 933-941.

Wirkner, K., Poelchen, W., Koles, L., Muhlberg, K., Scheibler, P., Allgaier, C., Illes, P. (1999) Ethanol-induced inhibition of NMDA receptor channels. *Neurochem. Int., 35,* 153-162.

Woodward, J.J. & Gonzales, R.A. (1990) Ethanol inhibition of N-methyl-D-aspartate-stimulated endogenous dopamine release from rat striatal slices: reversal by glycine. *J. Neurochem., 54(2),* 712-715.

Wright, J.L., Gregory, T.F., Kesten, S.R., Boxer, P.A., Serpa, K.A., Meltzer, L.T., Wise, L.D., Espitia, S.A., Konkoy, C.S., Whittemore, E.R., Woodward, R.M. (2000) Parallel synthesis of a series of subtype-selective NMDA receptor antagonists. *J. Med. Chem., 43,* 3408-3419.

Yaka, R., Phamluong, K., Ron, D. (2003a) Scaffolding of Fyn kinase to the NMDA receptor determines brain region sensitivity to ethanol. *J. Neurosci. 23(9),* 3623-3632.

Yaka, R., Tang, K.C., Camarini, R., Janak, P.H., Ron, D. (2003b) Fyn kinase and NR2B-containing NMDA receptors regulate acute ethanol sensitivity but not ethanol intake or conditioned reward. *Alcohol Clin Exp Res., 27(11),* 1736-1742.

Yenari, M.A., Bell, T.E., Kotake, A.N., Powell, M., Steinberg, G.K. (1998) Dose escalation safety and tolerance study of the competitive NMDA antagonist selfotel (CGS 19755) in neurosurgery patients. *Clin Neuropharmacol., 21(1),* 28-34.

Zhang, J., Chiodo, L A., Freeman, A S. (1992) Electrophysiological effects of MK-801 on rat nigrostriatal and mesoaccumbal dopaminergic neurons. *Brain Res., 590(1-2),* 153-163.

Zhou, Z.L., Cai, S.X., Whittemore, E.R. (1999) 4-Hydroxy-1-[2-(4-hydroxyphenoxy)ethyl]-4-(4-methylbenzyl)piperidine: a novel, potent, and selective NR1/2B NMDA receptor antagonist. *J. Med. Chem., 42,* 2993-3000.

In: Focus on Neurochemistry Research
Editor: Robert M. Coleman, pp.101-115

ISBN 1-59454-417-4

Chapter III

Benzodiazepines as Modulators of Neurotransmission

M. Palmada
Department of Physiology I, University of Tuebingen, Tuebingen, Germany
R.K.H. Kinne
Department of Epithelial Cell Physiology, Max Planck Institute of Molecular Physiology, Dortmund, Germany

Abstract

Benzodiazepines are widely prescribed anxiolytic, anticonvulsant and sedative drugs. Their pharmacological effects are mediated by a variety of mechanisms that converge in the alteration of synaptic transmission in the central nervous system. Initially, benzodiazepines were thought to modulate synaptic transmission by interacting with postsynaptic neurotransmitter receptors. High-affinity benzodiazepine binding sites were described on neuronal $GABA_A$ receptors. The current contribution focuses on the newly identified benzodiazepine targets in the modulation of neurotransmission.

Introduction

Benzodiazepines are widely prescribed anxiolytic, anticonvulsant and sedative drugs. Their pharmacological effects are mediated by a variety of mechanisms that converge in the alteration of synaptic transmission in the central nervous system. Initially, benzodiazepines were thought to modulate synaptic transmission by interacting with postsynaptic neurotransmitter receptors. High-affinity benzodiazepine binding sites were described on neuronal $GABA_A$ receptors (Costa and Guidotti 1979) as well as in glutamate (AMPA) receptors (May et al. 1999). Benzodiazepines enhance the effect of the inhibitory neurotransmitter GABA and attenuate the excitatory amino acid-mediated excitotoxicity. Recently, benzodiazepines were also reported to alter the activity of neurotransmitter transporters responsible for terminating the action of neurotransmitters in the synaptic cleft. Diazepam inhibits norepinephrine transporters and thus enhances the action of endogenous catecholamines (Hara et al. 2000). The glial and neuronal high affinity excitatory amino acid transporters are differently modulated by benzodiazepines (Palmada et al. 2002; Palmada et al. 1999). Depending on which transporter is most abundant in a specific brain area, an inhibition or a potentiation of excitatory neurotransmission may occur. The latter might be responsible of the undesired side-effects of these drugs. Thus, their action not only on postsynaptic neurotransmitter receptors but also on pre/postsynaptic neurotransmitter transporters account for the depressant effects of benzodiazepines.

The molecular mechanisms in the receptor action of benzodiazepines, the allosteric modulation of $GABA_A$ receptors by these drugs and the role of $GABA_A$ receptors and benzodiazepine structure diversities in this modulation have been reviewed (Costa and Guidotti 1979; Costa et al. 2002). The synthesis and analysis of the neuroprotective efficacy of competitive and non-competitive glutamate receptors antagonists have also been the subject of several studies (May et al. 1999). The current contribution will therefore focus on the newly identified benzodiazepine targets in the modulation of neurotransmission.

New Molecular Targets of Benzodiazepines

The reason benzodiazepines are effective in the treatment of such different symptoms might be that they exert their effects through several mechanisms.

GABAergic and glutamatergic neurotransmission might be regulated by modulating neurotransmitter release, neuronal/glial reuptake, binding at the postsynaptic receptor site(s) and/or metabolism. Several reports have indicated that benzodiazepines do not act presynaptically by modifying the release, reuptake or synthesis of GABA (Choi et al. 1981; Olsen et al. 1978; Pericic et al. 1977; Iversen and Johnston 1971). The drugs also do not change GABA metabolism (Sawaya et al. 1975). Thus, they must act postsynaptically by facilitating the action of GABA on its receptors ($GABA_A$ receptors) (Costa and Guidotti 1979). The effect of benzodiazepines on glutamatergic neurotransmission was initially investigated only at the receptor level (May et al. 1999). Benzodiazepines antagonize glutamate receptors thereby attenuating excitatory amino acid-mediated neurotoxicity. Recently, several studies have explored the role of pre/postsynaptic glutamate transporters as benzodiazepine targets.

Excitatory Amino Acid Transporters (EAATs)

High-affinity glutamate transporters play an essential role in regulating glutamate/aspartate concentrations in the synaptic cleft to avoid excessive neurotoxic concentrations of the neurotransmitter. The transport is mediated predominantly by the neuronal EAAT3 (Kanai and Hediger 1992) and the astroglial EAAT1 (Storck et al. 1992) and EAAT2 (Pines et al. 1992; Arriza et al. 1994) transporters. Although EAAT3 and EAAT2 have also been shown to be present in astrocytes and neurons (Conti et al. 1998; Mennerick et al. 1998), they are denoted as neuronal and glial respectively, since they are primarily expressed in these cell types. All three transporters use the electrical potential difference and the Na^+ and K^+ gradient across the plasma membrane as the driving force to translocate the neurotransmitter against a steep concentration gradient. Due to the increasing evidence that effective removal of excessive glutamate/aspartate from the synaptic cleft is crucial for neuroprotection (Rothstein et al. 1992; Masliah et al. 1996) these transporters have rapidly developed into pharmacological targets of great interest. As mentioned above, these transporters are modulated by benzodiazepines.

The Neuronal Glutamate Transporter EAAT3

The activity of the neuronal excitatory amino acid transporter EAAT3 is enhanced at low benzodiazepine concentrations (10 μM) and inhibited at higher concentrations (1 mM). The stimulatory and inhibitory effects are in turn modulated by the transporter substrate concentration (Figure 1) (Palmada et al. 1999). A reciprocal modification between benzodiazepines and target substrate has also been observed with $GABA_A$ receptors. Benzodiazepines potentiate GABA effects and GABA increases the affinity of the benzodiazepine binding sites (Tallman et al. 1978; Gallager et al. 1978).

The neuronal transporter has a variety of regulatory sites (see Figure 2A) including a PKC phosphorylation site (illustrated by P) that up-regulates transport activity by an increase in the catalytic rate as well as in the number of transporters available on the cell surface (Davis et al. 1998), a N-glycosylation site (indicated by a star) that regulates targeting of the transporter (Ferrer-Martinez et al. 1995), sites for EAAs and analogues which can act as substrates or as non-transportable inhibitors and a binding site for Zn^{2+} that inhibits transport in a non-competitive manner (Vandenberg et al. 1998). Two novel sites for benzodiazepines have also been found in the neuronal transporter, one stimulatory and the other inhibitory which are in turn subject of modulatory control by several compounds like the transporter substrates, the transmitters norepinephrine and L-DOPA and the unrelated amino acid, L-lysine (Palmada 2001). The difference in relative potency observed, i.e., oxazepam>diazepam≈medazepam≈lorazepam for stimulation versus diazepam≈lorazepam >oxazepam>medazepam for inhibition, supports the assumption of the presence of two benzodiazepine binding sites. The stimulatory site is desentisized by all mentioned compounds whereas the inhibitory site is sentisized specifically by the transporter substrates and the transmitters norepinephrine and L-DOPA.

$GABA_A$ receptors are pentameric and consist of three of at least sixteen different subunits, grouped in seven classes: alpha, beta, gamma, delta, sigma, epsilon and pi. The subunit composition determines receptor kinetics and the specificity of modulation by protein kinases, zinc, barbiturates, steroids, and benzodiazepines. Benzodiazepines modulate $GABA_A$ receptors containing an α1, α2, α3 or α5 subunit together with a gamma subunit (Mehta and Ticku 1999). These receptors contain two active sites for GABA and one site for benzodiazepine binding (see Figure 2B). The latter may affect the dissociation and/or isomerization constants of the channel (MacDonald et al. 1989) thereby increasing the

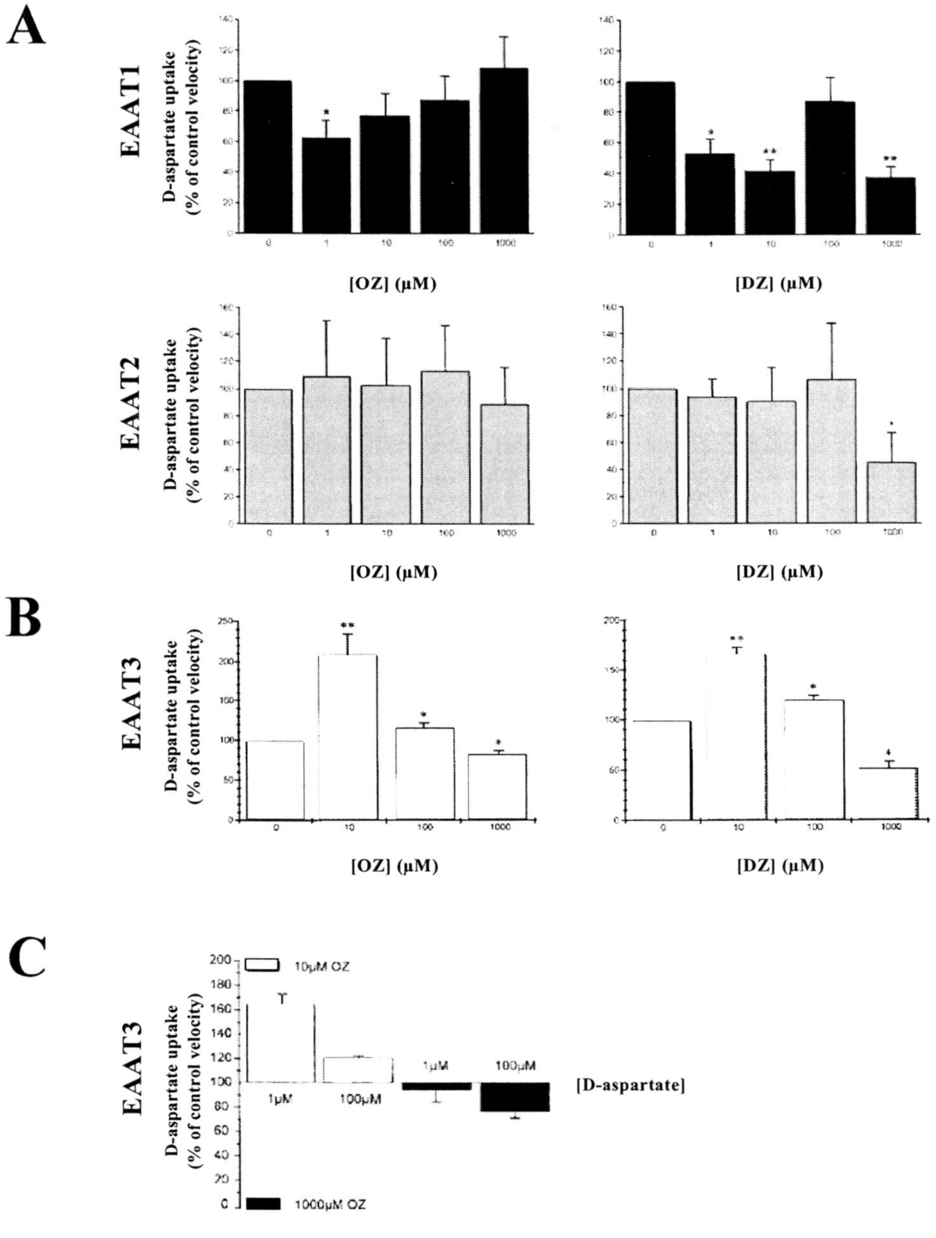

Figure 1. Modulation of excitatory amino acid transporters by benzodiazepines.
Effect of oxazepam (OZ) and diazepam (DZ) on D-[^{3}H]aspartate uptake in CHO cells transfected with glial EAAT1 or EAAT2 (**A**) or neuronal EAAT3 (**B**) excitatory amino acid transporters. Labelled aspartate uptake was measured for 2 min at 37 °C in control buffer or buffer containing OZ or DZ at the indicated concentrations. Asterisks denote that the uptake velocity was significantly different from the control value. (**C**) Effects of D-aspartate concentration on the EAAT3 stimulation and inhibition mediated by oxazepam. Uptake velocity was measured at low (1 μM) and high (100 μM) D-aspartate concentrations at a fixed concentration of OZ. [Data from 4 and 5].

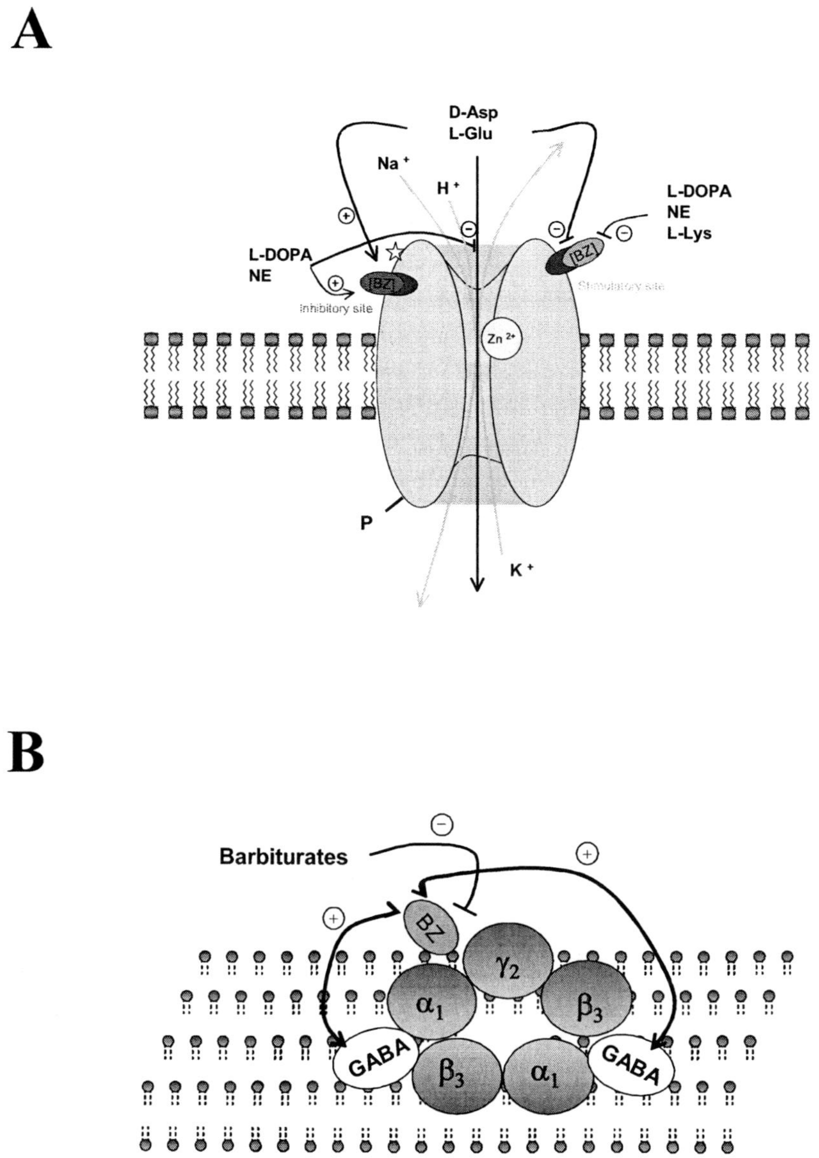

Figure 2. Schematic diagram of the EAAT3/EAAC1 transporter (**A**) and $GABA_A$ receptor (**B**) regulatory sites. Abbreviations: BZ (benzodiazepine), D-Asp (D-Aspartate), DOPA (L-3,4-dihydroxyphenylalanine), GABA (γ-aminobutyric acid), L-Glu (L-Glutamate), L-Lys (L-Lysine), NE (Norepinephrine), and Zn^{2+} (Zinc). The phosphorylation site is illustrated by P and the N-glycosylation site is indicated by a star.

affinity of $GABA_A$ receptors for GABA and enhancing the receptor open probability. GABA, in turn, increases the affinity of the benzodiazepine for its binding site. In contrast to what has been observed for the neuronal glutamate transporter, the modulation of the benzodiazepine effect by GABA cannot be mimicked by any other amino acid transmitter tested (Tallman et al. 1978). Barbiturates have been shown to decrease the affinity of beta-carboline-3-carboxylic acid methyl ester to benzodiazepine binding sites (Wong et al. 1984).

The Glial Glutamate Transporters EAAT1 and EAAT2

Further studies have also investigated the effect of benzodiazepines (oxazepam and diazepam) on the glial excitatory amino acid transporters. Regarding the distinct physiological role of glutamate transporter subtypes, glial neuroprotection versus neuronal neurotransmission and GABA metabolism, benzodiazepines might exert different effects on the transporters. Similar to the neuronal transporter, the glial ones are inhibited by high diazepam concentrations. However, at micromolar concentrations the benzodiazepines exclusively affect EAAT1 (Figure 1). These effects are modified by the substrate concentration (Palmada et al. 2002).

Although EAAT2 is not regulated by oxazepam and diazepam at micromolar concentrations, it is modulated by midazolam. The anesthetic midazolam reduces glutamate release mediated by the reversed operation of EAAT2 during cerebral ischemia/anoxia (Sakai and Amaha 2000).

Monoamine Transporters and Receptors

The biogenic amine neurotransmitters noradrenaline, 5-hydroxytryptamine (5-HT, serotonin) and dopamine control a variety of functions including locomotion, hormone secretion and behaviours associated with affect, emotion and reward. Transporters for noradrenaline, serotonin and dopamine (named NET, SERT and DAT respectively) control the intensity of monoamine-mediated signalling by the reuptake of neurotransmitters released by presynaptic neurons. These proteins mediate monoamine transport by sequential binding and cotransport of Na^+ and Cl^- ions. NET and SERT cotransport their substrates with one Na^+ and one Cl^- ion whereas DAT transports dopamine with two Na^+ ions and one Cl^- ion (Gu et al. 1994). These transporters are well-known targets for psychotropic drugs and neurotoxins (Amara and Sonders 1998; Miller et al. 1999). The availability of genetic animal models with targeted disruption of these transporters has provided an opportunity to investigate the physiological role of these proteins *in vivo* and to resolve important aspects on specificity and mechanism of certain pharmacological agents (for review see (Torres et al. 2003)). Once released into the synaptic cleft, monoamines act on monoamine receptors that are coupled to G-proteins. Adrenergic receptors are classified as α_1, α_2, β_1 or β_2. Serotoninergic (5-HT) receptors are classified as 5-HT_1 (with four subtypes), 5-HT_2 and 5-HT_3. 5-HT_{1A} receptors are expressed presynaptically in the raphe nucleus and inhibit 5-HT uptake, and postsynaptically in the hippocampus. 5-HT_2 receptors are found in the fourth layer of the cortex and 5-HT_3 receptors presynaptically in the nucleus tractus solitarius. Dopaminergic receptors are classified as D_1, D_2, D_3, D_4, and D_5. D_3 and D_4 limit negative symptoms of schizophrenic processes whereas D_2 receptor controls the extrapyramidal system.

Some of these receptors have been reported to be sensitive to psychotropic drugs (Bourin et al. 2002). Benzodiazepines have recently been shown to regulate monoamine transporters and receptors.

Norepinephrine Transporters (NETs) and Adrenergic Receptors

An increasing number of recent studies demonstrate that the general anesthetics are effective not only by affecting postsynaptic neurotransmitter receptors or presynaptic neurotransmitter release but also by modulating presynaptic transporters (Mantz et al. 1995; el Maghrabi and Eckenhoff 1993; Martin et al. 1990). In an attempt to explore the interaction between NET and anesthetics including diazepam, the authors found that short-term treatment with intravenous anesthetics inhibit norepinephrine uptake noncompetitively in a concentration-dependent manner (Hara et al. 2000). The role of other neurotransmitters in modulating benzodiazepine effects on NET activity was not investigated.

Competition binding studies have demonstrated that benzodiazepines can also modulate the function of α_1-adrenergic receptors. Diazepam inhibits phenylephrine binding at the α_{1A} subtype. The inhibitory profile of diazepam reveals a noncompetitive interaction between phenylephrine and diazepam at this receptor (Waugh et al. 1999). Lorazepam and midazolam potentiate the responses of full and partial agonists at the α_{1B} and α_{1D} subtypes. To potentiate the agonists response, the drugs must bind simultaneously with the agonist to the receptor at different binding sites. Thus, similar to what has been described for $GABA_A$ receptors, benzodiazepines act as allosteric modulators of α_1-adrenergic receptors. Aromatic and hydrophobic residues on the $\gamma1$ and $\gamma2$ subunits of $GABA_A$ receptors mediate benzodiazepine binding (Wieland et al. 1992; Sigel and Buhr 1997; Sigel et al. 1998). α_1-adrenergic receptors contain various aromatic and hydrophobic residues in the extracellular loops and in the juxtamembrane regions of the transmembrane helices and thus these residues might also mediate benzodiazepine binding (Waugh et al. 1999). Benzodiazepines are unable to activate $GABA_A$ receptors by themselves, however, at α_1-adrenergic receptors the drugs alone display partial agonism.

Further experiments supported the role of α_1-adrenergic receptors as benzodiazepine targets. Experiments performed to study the tocolytic potencies of benzodiazepines (diazepam, midazolam and nitrazepam) in a postpartum rat model and their interactions with alpha-adrenoreceptors indicated that there is a correlation between the tocolytic potencies and inhibition constants of the tested drugs, indicating that the smooth muscle relaxing effect of the studied benzodiazepines is due to a modulation of the α_1-adrenergic receptors (Zupko et al. 2003).

Serotonin Transporters (SERTs) and Receptors

Long-term administration of antidepressant drugs increase serotonergic activity by desensitization of presynaptic $5\text{-}HT_{1A}$ receptors. The atypical neuroleptic clozapine (dibenzodiazepine) acts as an agonist for G protein activation at postsynaptic $5\text{-}HT_{1A}$ receptors (Newman-Tancredi et al. 2003). The use of selective serotonin reuptake inhibitors has been reported to lead to desensitization of $5\text{-}HT_{1A}$ autoreceptors (Blier et al. 1984; Blier and de Montigny 1994). In addition, substance P receptor (NK1R) antagonism results in decreased anxiety, enhanced firing of 5-HT neurons and reduced presynaptic 5-HT_{1A} receptor function (Santarelli et al. 2001). The selective NK1R antagonist, RP67580, acts as a potent anxiolytic compound in wild type mice and is ineffective in animals lacking the NK1R receptor whereas diazepam is effective in wild type and knockout mice. However, the diazepam anxiolytic effect in knockout mice is not as pronounced as in wild type animals, suggesting that NK1R is not required but participates in the benzodiazepine-mediated reduced anxiety.

Voltage-gated Sodium Channels

An additional site of benzodiazepines action are the voltage-dependent sodium channels. As mentioned before, benzodiazepines enhance the inhibitory effect of GABA. Upon binding to $GABA_A$ receptors, Cl^- channels open and hyperpolarize neurons and therefore prevent the generation of action potentials. Thus benzodiazepines could exert their effects by inhibiting voltage-gated sodium channels that are responsible for the generation and propagation of action potentials. The voltage-gated sodium channels consists of an alpha subunit which constitutes the sodium selective pore. In the central nervous system this subunit is accompanied by two beta subunits ($\beta 1$ and $\beta 2$) that control gating kinetics and voltage dependency. The peak sodium current and inactivation kinetics can be modulated by phosphorylation. Drugs including several anaesthetics and antiepileptics also modulate channel properties.

As a matter of fact, several studies have shown that the benzodiazepines carbamazepine and diazepam inhibit voltage-dependent sodium currents. This reduction is voltage dependent and is due to a shift of the inactivation curve to hyperpolarizing potentials. Benzodiazepines bind to the inactivated state of the sodium channel, in the inner pore of the channel. Carbamazepine has been reported to reduce sodium inward currents in isolated hippocampal neurons from epileptic patients or rats (Vreugdenhil and Wadman 1999; Reckziegel et al. 1999; Kuo et al. 1997), in mammalian cell line heterologously expressing $Na_v1.2$ (Ragsdale et al. 1991), mammalian and amphibian nerve fibers (Courtney and Etter 1983; Schwarz and Grigat 1989) and snail neurons (Brown et al. 1992). In isolated rat hippocampal neurons and in cultured mouse neuroblastoma cells (Wakamori et al. 1989), diazepam reduces I_{Na}. In nanomolar concentrations diazepam, nitrazepam, lorazepam and clonazepam decrease repetitive firing in spinal cord neurons (McLean and MacDonald 1988). In micromolar concentrations diazepam reduces action potentials in rat optic nerves.

Mechanisms Involved in the Development of Tolerance and Physical Dependence

Receptors hypo- and hypersensitization influence the development of tolerance and physical dependence. Protracted daily doses of benzodiazepines lead to tolerance to the acute effects of these drugs (Greenblatt and Shader 1978; Ryan and Boisse 1983; Ryan and Boisse

1984), physical dependence (Ryan and Boisse 1983; McNicholas et al. 1983), and induce withdrawal reactions when the benzodiazepine treatment is stopped abruptly (Ryan and Boisse 1984; Pevnick et al. 1978). Tolerance and dependence to benzodiazepines is due to changes in GABAergic and glutamatergic systems.

The tolerance to benzodiazepines is associated with an increase in the turnover rate of DBI (Diazepam-Binding Inhibitor), which is a negative allosteric regulator of $GABA_A$ receptors, and might be responsible for the GABA receptor desensitization that takes place after protracted benzodiazepine administration (Miyata et al. 1987). Tolerance is also associated with changes in the expression of $GABA_A$ receptor subunits (decrease in γ_2 and α_1 and increase in α_5), with an increase of glutamic acid decarboxylase (GAD_{67}) transcripts (Izzo et al. 2001) and with an uncoupling of the benzodiazepine and GABA binding sites. This uncoupling (decline in the GABA enhancement of benzodiazepine binding) is related to phosphorylation and internalization processes. Chronic benzodiazepine exposure leads to enhanced $GABA_A$ receptor internalization, an effect requiring phosphorylation of a hitherto unknown protein, into intracellular acidic vesicles where benzodiazepines still bind to the receptor but potentiation by GABA is impaired (Ali and Olsen 2001).

Excitatory amino acid-related processes might be involved in the physical dependence after benzodiazepine withdrawal. AMPA receptor GluR1 subunit mRNA and protein expression levels are increased in cortex and hippocampus 96 h after benzodiazepine withdrawal (Izzo et al. 2001). Thus, GABAergic and glutamatergic systems are involved in the development of tolerance and dependence to benzodiazepines, respectively.

Conclusion

The reason benzodiazepines exert a broad spectrum of effects such as sedation, inducing sleep, preventing seizures, relieving anxiety and muscle spasms is that they exert their effects through several mechanisms (Figure 3). Originally, benzodiazepines were thought to enhance inhibitory GABAergic neurotransmission and inhibit excitatory glutamatergic neurotransmission by modulating only postsynaptic receptors ($GABA_A$ and AMPA receptors, respectively). Recently, several pre/postsynaptic neurotransmitter transporters and receptors have been identified as targets for benzodiazepines. Neuronal and glial excitatory amino acid transporters (EAATs) are differently modulated by benzodiazepines. Depending on which transporter is most abundant in a specific brain area, an inhibition or a potentation of excitatory neurotransmission may occur. Neuronal and glial EAA transporters exist on the contraluminal membrane of the blood-brain barrier where they restrict aspartate/glutamate entry to the brain and remove them from the extracellular fluid in conjunction with facilitative transporters on the luminal membrane (O'Kane et al. 1999). In this region, the relative activity of the three transporters is approximately 1:3:6 for EAAT1, EAAT2, and EAAT3, respectively. Hence, low benzodiazepine concentrations would have a neuroprotective effect. Monoamine transporters and receptors are also regulated by benzodiazepines. The drugs inhibit the reuptake of monoamines (i.e. NET), act as agonists of monoamine receptors or desensitize inhibitory $5\text{-}HT_{1A}$ receptors. Benzodiazepines further

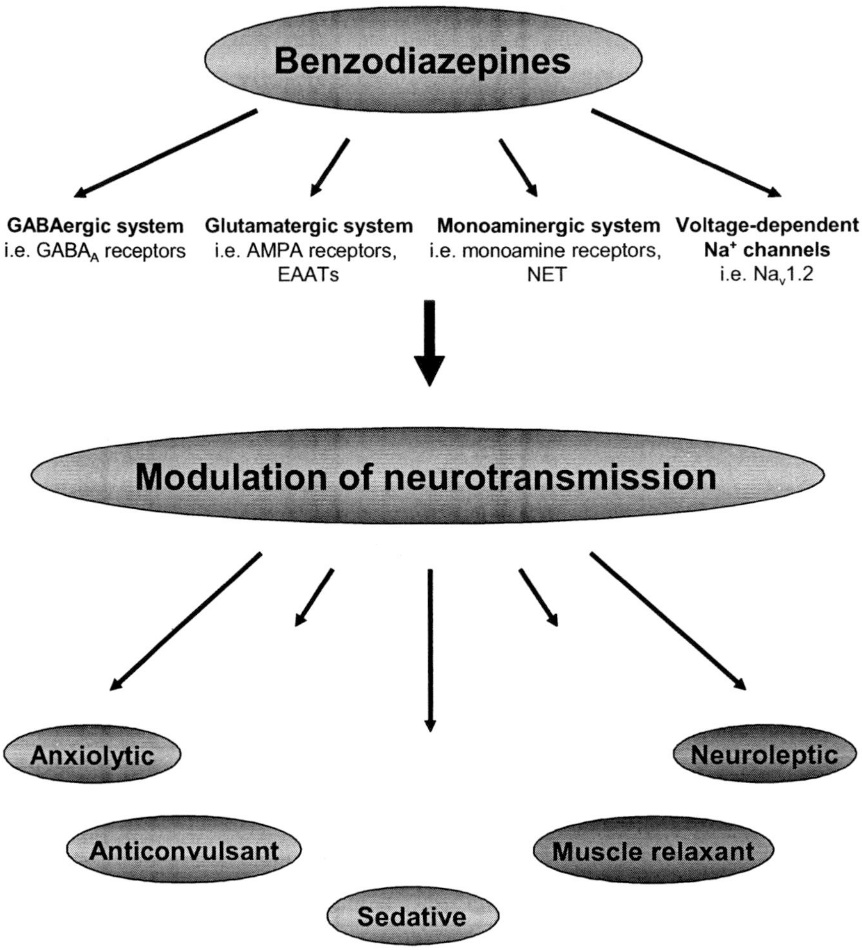

Figure 3. Benzodiazepines exert their effects through several mechanisms including GABAergic, glutamatergic, monoaminergic systems and voltage-dependent sodium channels.

inhibit voltage-dependent Na^+ channels responsible for the generation of action potentials. Thus, benzodiazepines are effective through different mechanisms that converge in the alteration of synaptic transmission in the central nervous system. Advances in neuroscience and in understanding the origin of neuropsychiatric disorders will lead to discern even more targets of benzodiazepines action.

Acknowledgements

This chapter is dedicated to the late Prof. Dr. E. Kinne-Saffran and Dr. J. J. Centelles for their valuable advice and experimental support in the elucidation of the excitatory amino acid transporters as benzodiazepine targets.

References

[1] Costa, E. and Guidotti, A. 1979. Molecular mechanisms in the receptor action of benzodiazepines. *Annu. Rev. Pharmacol. Toxicol. 19:* 531-545.

[2] May, P.C., Robison, P.M., and Fuson, K.S. 1999. Stereoselective neuroprotection by novel 2, 3-benzodiazepine non-competitive AMPA antagonist against non-NMDA receptor-mediated excitotoxicity in primary rat hippocampal cultures. *Neurosci. Lett. 262:* 219-221.

[3] Hara, K., Yanagihara, N., Minami, K., Hirano, H., Sata, T., Shigematsu, A., and Izumi, F. 2000. Dual effects of intravenous anesthetics on the function of norepinephrine transporters. *Anesthesiology 93:* 1329-1335.

[4] Palmada, M., Kinne-Saffran, E., Centelles, J.J., and Kinne, R.K. 2002. Benzodiazepines differently modulate EAAT1/GLAST and EAAT2/GLT1 glutamate transporters expressed in CHO cells. *Neurochem. Int. 40:* 321-326.

[5] Palmada, M., Bohmer, C., Centelles, J.J., and Kinne, R.K. 1999. Effect of benzodiazepines on the epithelial and neuronal high-affinity glutamate transporter EAAC1. *J. Neurochem. 73:* 2389-2396.

[6] Costa, E., Auta, J., Grayson, D.R., Matsumoto, K., Pappas, G.D., Zhang, X., and Guidotti, A. 2002. GABAA receptors and benzodiazepines: a role for dendritic resident subunit mRNAs. *Neuropharmacology 43:* 925-937.

[7] Choi, D.W., Farb, D.H., and Fischbach, G.D. 1981. Chlordiazepoxide selectively potentiates GABA conductance of spinal cord and sensory neurons in cell culture. *J. Neurophysiol. 45:* 621-631.

[8] Olsen, R.W., Ticku, M.K., Van Ness, P.C., and Greenlee, D. 1978. Effects of drugs on gamma-aminobutyric acid receptors, uptake, release and synthesis in vitro. *Brain Res. 139:* 277-294.

[9] Pericic, D., Walters, J.R., and Chase, T.N. 1977. Effect of diazepam and pentobarbital on aminooxyacetic acid-induced accumulation on GABA. *J. Neurochem. 29:* 839-846.

[10] Iversen, L.L. and Johnston, G.A. 1971. GABA uptake in rat central nervous system: comparison of uptake in slices and homogenates and the effects of some inhibitors. *J. Neurochem. 18:* 1939-1950.

[11] Sawaya, M.C., Horton, R.W., and Meldrum, B.S. 1975. Effects of anticonvulsant drugs on the cerebral enzymes metabolizing GABA. *Epilepsia 16:* 649-655.

[12] Kanai, Y. and Hediger, M.A. 1992. Primary structure and functional characterization of a high-affinity glutamate transporter. *Nature 360:* 467-471.

[13] Storck, T., Schulte, S., Hofmann, K., and Stoffel, W. 1992. Structure, expression, and functional analysis of a Na(+)-dependent glutamate/aspartate transporter from rat brain. *Proc. Natl. Acad. Sci. U. S. A 89:* 10955-10959.

[14] Pines, G., Danbolt, N.C., Bjoras, M., Zhang, Y., Bendahan, A., Eide, L., Koepsell, H., Storm-Mathisen, J., Seeberg, E., and Kanner, B.I. 1992. Cloning and expression of a rat brain L-glutamate transporter. *Nature 360:* 464-467.

[15] Arriza, J.L., Fairman, W.A., Wadiche, J.I., Murdoch, G.H., Kavanaugh, M.P., and Amara, S.G. 1994. Functional comparisons of three glutamate transporter subtypes cloned from human motor cortex. *J. Neurosci. 14:* 5559-5569.

[16] Conti, F., DeBiasi, S., Minelli, A., Rothstein, J.D., and Melone, M. 1998. EAAC1, a high-affinity glutamate tranporter, is localized to astrocytes and gabaergic neurons besides pyramidal cells in the rat cerebral cortex. *Cereb. Cortex 8:* 108-116.

[17] Mennerick, S., Dhond, R.P., Benz, A., Xu, W., Rothstein, J.D., Danbolt, N.C., Isenberg, K.E., and Zorumski, C.F. 1998. Neuronal expression of the glutamate transporter GLT-1 in hippocampal microcultures. *J. Neurosci. 18:* 4490-4499.

[18] Rothstein, J.D., Martin, L.J., and Kuncl, R.W. 1992. Decreased glutamate transport by the brain and spinal cord in amyotrophic lateral sclerosis. *N. Engl. J. Med. 326:* 1464-1468.

[19] Masliah, E., Alford, M., DeTeresa, R., Mallory, M., and Hansen, L. 1996. Deficient glutamate transport is associated with neurodegeneration in Alzheimer's disease. *Ann. Neurol. 40:* 759-766.

[20] Tallman, J.F., Thomas, J.W., and Gallager, D.W. 1978. GABAergic modulation of benzodiazepine binding site sensitivity. *Nature 274:* 383-385.

[21] Gallager, D.W., Thomas, J.W., and Tallman, J.F. 1978. Effect of GABAergic drugs on benzodiazepine binding site sensitivity in rat cerebral cortex. *Biochem Pharmacol. 27:* 2745-2749.

[22] Davis, K.E., Straff, D.J., Weinstein, E.A., Bannerman, P.G., Correale, D.M., Rothstein, J.D., and Robinson, M.B. 1998. Multiple signaling pathways regulate cell surface expression and activity of the excitatory amino acid carrier 1 subtype of Glu transporter in C6 glioma. *J. Neurosci. 18:* 2475-2485.

[23] Ferrer-Martinez, A., Felipe, A., Nicholson, B., Casado, J., Pastor-Anglada, M., and McGivan, J. 1995. Induction of the high-affinity Na(+)-dependent glutamate transport system XAG- by hypertonic stress in the renal epithelial cell line NBL-1. *Biochem J. 310 (Pt 2):* 689-692.

[24] Vandenberg, R.J., Mitrovic, A.D., and Johnston, G.A. 1998. Molecular basis for differential inhibition of glutamate transporter subtypes by zinc ions. *Mol. Pharmacol. 54:* 189-196.

[25] Palmada, M. 2001. *Modulation of brain glutamate transporter activities by neurotransmitters and benzodiazepines in heterologous expression systems.* Max-Planck-Institute of molecular physiology, PhD thesis, Dortmund.

[26] Mehta, A.K. and Ticku, M.K. 1999. An update on GABAA receptors. *Brain Res. Brain Res. Rev. 29:* 196-217.

[27] MacDonald, R.L., Rogers, C.J., and Twyman, R.E. 1989. Barbiturate regulation of kinetic properties of the GABAA receptor channel of mouse spinal neurones in culture. *J. Physiol 417:* 483-500.

[28] Wong, E.H., Snowman, A.M., Leeb-Lundberg, L.M., and Olsen, R.W. 1984. Barbiturates allosterically inhibit GABA antagonist and benzodiazepine inverse agonist binding. *Eur. J. Pharmacol. 102:* 205-212.

[29] Sakai, F. and Amaha, K. 2000. Midazolam and ketamine inhibit glutamate release via a cloned human brain glutamate transporter. *Can. J. Anaesth. 47:* 800-806.

[30] Gu, H., Wall, S.C., and Rudnick, G. 1994. Stable expression of biogenic amine transporters reveals differences in inhibitor sensitivity, kinetics, and ion dependence. *J. Biol. Chem. 269:* 7124-7130.

[31] Amara, S.G. and Sonders, M.S. 1998. Neurotransmitter transporters as molecular targets for addictive drugs. *Drug Alcohol Depend. 51:* 87-96.

[32] Miller, G.W., Gainetdinov, R.R., Levey, A.I., and Caron, M.G. 1999. Dopamine transporters and neuronal injury. *Trends Pharmacol. Sci. 20:* 424-429.

[33] Torres, G.E., Gainetdinov, R.R., and Caron, M.G. 2003. Plasma membrane monoamine transporters: structure, regulation and function. *Nat. Rev. Neurosci. 4:* 13-25.

[34] Bourin, M., David, D.J., Jolliet, P., and Gardier, A. 2002. [Mechanism of action of antidepressants and therapeutic perspectives]. *Therapie 57:* 385-396.

[35] Mantz, J., Lecharny, J.B., Laudenbach, V., Henzel, D., Peytavin, G., and Desmonts, J.M. 1995. Anesthetics affect the uptake but not the depolarization-evoked release of GABA in rat striatal synaptosomes. *Anesthesiology 82:* 502-511.

[36] el Maghrabi, E.A. and Eckenhoff, R.G. 1993. Inhibition of dopamine transport in rat brain synaptosomes by volatile anesthetics. *Anesthesiology 78:* 750-756.

[37] Martin, D.C., Introna, R.P., and Aronstam, R.S. 1990. Inhibition of neuronal 5-HT uptake by ketamine, but not halothane, involves disruption of substrate recognition by the transporter. *Neurosci. Lett. 112:* 99-103.

[38] Waugh, D.J., Gaivin, R.J., Damron, D.S., Murray, P.A., and Perez, D.M. 1999. Binding, partial agonism, and potentiation of alpha(1)-adrenergic receptor function by benzodiazepines: A potential site of allosteric modulation. *J. Pharmacol. Exp. Ther. 291:* 1164-1171.

[39] Wieland, H.A., Luddens, H., and Seeburg, P.H. 1992. A single histidine in GABAA receptors is essential for benzodiazepine agonist binding. *J. Biol. Chem. 267:* 1426-1429.

[40] Sigel, E. and Buhr, A. 1997. The benzodiazepine binding site of GABAA receptors. *Trends Pharmacol. Sci. 18:* 425-429.

[41] Sigel, E., Schaerer, M.T., Buhr, A., and Baur, R. 1998. The benzodiazepine binding pocket of recombinant alpha1beta2gamma2 gamma-aminobutyric acidA receptors: relative orientation of ligands and amino acid side chains. *Mol. Pharmacol. 54:* 1097-1105.

[42] Zupko, I., Janossy, K., Maul, K., Marki, A., and Falkay, G. 2003. Alpha-adrenergic blockade: a possible mechanism of tocolytic action of certain benzodiazepines in a postpartum rat model in vivo. *Life Sci. 72:* 1093-1102.

[43] Newman-Tancredi, A., Rivet, J.M., Cussac, D., Touzard, M., Chaput, C., Marini, L., and Millan, M.J. 2003. Comparison of hippocampal G protein activation by 5-HT(1A) receptor agonists and the atypical antipsychotics clozapine and S16924. Naunyn Schmiedebergs Arch. *Pharmacol. 368:* 188-199.

[44] Blier, P., de Montigny, C., and Tardif, D. 1984. Effects of the two antidepressant drugs mianserin and indalpine on the serotonergic system: single-cell studies in the rat. *Psychopharmacology (Berl) 84:* 242-249.

[45] Blier, P. and de Montigny, C. 1994. Current advances and trends in the treatment of depression. *Trends Pharmacol. Sci. 15:* 220-226.

[46] Santarelli, L., Gobbi, G., Debs, P.C., Sibille, E.T., Blier, P., Hen, R., and Heath, M.J. 2001. Genetic and pharmacological disruption of neurokinin 1 receptor function decreases anxiety-related behaviors and increases serotonergic function. *Proc. Natl. Acad. Sci. U. S. A 98:* 1912-1917.

[47] Vreugdenhil, M. and Wadman, W.J. 1999. Modulation of sodium currents in rat CA1 neurons by carbamazepine and valproate after kindling epileptogenesis. *Epilepsia 40:* 1512-1522.

[48] Reckziegel, G., Beck, H., Schramm, J., Urban, B.W., and Elger, C.E. 1999. Carbamazepine effects on Na+ currents in human dentate granule cells from epileptogenic tissue. *Epilepsia 40:* 401-407.

[49] Kuo, C.C., Chen, R.S., Lu, L., and Chen, R.C. 1997. Carbamazepine inhibition of neuronal Na+ currents: quantitative distinction from phenytoin and possible therapeutic implications. *Mol. Pharmacol. 51:* 1077-1083.

[50] Ragsdale, D.S., Scheuer, T., and Catterall, W.A. 1991. Frequency and voltage-dependent inhibition of type IIA Na+ channels, expressed in a mammalian cell line, by local anesthetic, antiarrhythmic, and anticonvulsant drugs. *Mol. Pharmacol. 40:* 756-765.

[51] Courtney, K.R. and Etter, E.F. 1983. Modulated anticonvulsant block of sodium channels in nerve and muscle. *Eur. J. Pharmacol. 88:* 1-9.

[52] Schwarz, J.R. and Grigat, G. 1989. Phenytoin and carbamazepine: potential- and frequency-dependent block of Na currents in mammalian myelinated nerve fibers. *Epilepsia 30:* 286-294.

[53] Brown, A.M., McCrohan, C.R., and Pamplin, P. 1992. Inhibition of slow TTX-insensitive inward current by the anticonvulsant carbamazepine in an identified neuron of Lymnaea stagnalis. *Comp Biochem Physiol C 103:* 549-551.

[54] Wakamori, M., Kaneda, M., Oyama, Y., and Akaike, N. 1989. Effects of chlordiazepoxide, chlorpromazine, diazepam, diphenylhydantoin, flunitrazepam and haloperidol on the voltage-dependent sodium current of isolated mammalian brain neurons. *Brain Res. 494:* 374-378.

[55] McLean, M.J. and MacDonald, R.L. 1988. Benzodiazepines, but not beta carbolines, limit high frequency repetitive firing of action potentials of spinal cord neurons in cell culture. *J. Pharmacol. Exp. Ther. 244:* 789-795.

[56] Greenblatt, D.J. and Shader, R.I. 1978. Dependence, tolerance, and addiction to benzodiazepines: clinical and pharmacokinetic considerations. *Drug Metab Rev. 8:* 13-28.

[57] Ryan, G.P. and Boisse, N.R. 1983. Experimental induction of benzodiazepine tolerance and physical dependence. *J. Pharmacol. Exp. Ther. 226:* 100-107.

[58] Ryan, G.P. and Boisse, N.R. 1984. Benzodiazepine tolerance, physical dependence and withdrawal: electrophysiological study of spinal reflex function. *J. Pharmacol. Exp. Ther. 231:* 464-471.

[59] McNicholas, L.F., Martin, W.R., and Cherian, S. 1983. Physical dependence on diazepam and lorazepam in the dog. *J. Pharmacol. Exp. Ther. 226:* 783-789.

[60] Pevnick, J.S., Jasinski, D.R., and Haertzen, C.A. 1978. Abrupt withdrawal from therapeutically administered diazepam. Report of a case. *Arch. Gen. Psychiatry 35:* 995-998.

[61] Miyata, M., Mocchetti, I., Ferrarese, C., Guidotti, A., and Costa, E. 1987. Protracted treatment with diazepam increases the turnover of putative endogenous ligands for the benzodiazepine/beta-carboline recognition site. *Proc. Natl. Acad. Sci. U. S. A 84:* 1444-1448.

[62] Izzo, E., Auta, J., Impagnatiello, F., Pesold, C., Guidotti, A., and Costa, E. 2001. Glutamic acid decarboxylase and glutamate receptor changes during tolerance and dependence to benzodiazepines. *Proc. Natl. Acad. Sci. U. S. A 98:* 3483-3488.

[63] Ali, N.J. and Olsen, R.W. 2001. Chronic benzodiazepine treatment of cells expressing recombinant GABA(A) receptors uncouples allosteric binding: studies on possible mechanisms. *J. Neurochem. 79:* 1100-1108.

[64] O'Kane, R.L., Martinez-Lopez, I., DeJoseph, M.R., Vina, J.R., and Hawkins, R.A. 1999. Na(+)-dependent glutamate transporters (EAAT1, EAAT2, and EAAT3) of the blood-brain barrier. A mechanism for glutamate removal. *J. Biol. Chem. 274:* 31891-31895.

In: Focus on Neurochemistry Research
Editor: Robert M. Coleman, pp.117-140

ISBN 1-59454-417-4

Chapter IV

Expression of Cytokines and Co-stimulatory Molecules by Activated Human Microglia

Camilla Reali, Franca Scintu, Rita Pillai, Monica Curto, Giuseppina Palmieri[1], Maria Antonietta Marcialis[1], Herbert Schwarz[2] and Valeria Sogos***

Department of Cytomorphology and [1]Department of Sciences and Biomedical Technologies, University of Cagliari, Italy

[2]Department of Pathology, University of Regensburg, Germany

Abstract

Microglia are the principal immune cells in the central nervous system (CNS) and have a critical role in host defense against invading microorganisms and neoplastic cells.

These cells are known to interact with peripheral immunocompetent cells via production of cytokines and co-stimulatory molecules. In fact, like macrophage, microglia have an array of cellular defense mechanisms, including secretion of inflammatory cytokine, production of complement proteins and receptors. Thus, we activated in vitro human fetal microglia either by infection with *Mycobacterium avium* (SmD, SmT) and *Mycobacterium tuberculosis* (Myc) or by stimulation with myelin basic protein (MBP) and estrogens. At different times, cells were fixed or collected for RNA extraction. The expression of cytokines (TNF-α, IL-1, IL-10), chemokines (MIP, MCP-1, RANTES, IL-8) and other co-stimulatory molecules, such as CD137, a member of TNF-α receptor family, was investigated. Infected cells showed a correlation between the bacterial pathogenicity and increased inflammatory factors expression or decreased anti-

* Address correspondence to: Valeria Sogos, Dept. of Cytomorphology, Cittadella Universitaria, 09042 Monserrato (CA). Tel: (39) 070-6754063; Fax: (39) 070-6754003; e-mail: sogos@unica.it

inflammatory molecules production. The co-stimulatory molecule CD137 production was also affected by infection and MBP. On the other hand, pretreatment of microglial cells with estrogens may prevents the effect of mycobacteria infection on cytokine expression. These results suggest that microglia inflammatory response as well as their interaction with immunocompetent cells can be modulated by pathogenic factors such as bacteria or MBP (multiple sclerosis). They also help in understanding the mechanism(s) of brain damage in inflammatory disease.

Introduction

Microglia represent the resident immune cells of the brain and have phenotypic and functional features similar to the monocytes or tissue macrophages of peripheral organs. These cells were first described by Del Rio Ortega (1932) as a distinct population within the Central Nervous System (CNS), with distinctive characteristics differentiating them from other glial cells and neurons. Microglia may adopt different morphological states throughout their life cycle, two principal forms of which have been described. In the normal adult brain, they display ramified morphology, with a small cell body and long branched processes (Ling, 1981). This ramified form is called resting microglia, since it represents functional quiescent cells devoid of monocytic properties. During the embryonic period and after brain lesions, they are quickly transformed into reactive or activated microglia, with a large cell body and short processes, closely resembling macrophages (Giulian and Baker, 1986; Giulian,1987). This form of microglia is believed to be involved in the phagocytic removal of debris from naturally-occurring cell death during brain remodelling and maturation (Ferrer et al., 1990; Killackey, 1984). Like other tissue macrophages, activated microglia can synthesize and secrete a great variety of soluble factors, including growth factors, cytokines, chemokines, complement factors, fatty acid metabolites, proteolitic enzymes and free radicals (Minghetti and Levi, 1998; Aloisi, 2001; Hanish, 2002). A number of these factors exercise trophic functions, promoting neuronal survival and regulating neuronal growth, function and regeneration. However, the majority of substances produced by activated microglia are proinflammatory factors. In fact, when microglia are activated by brain lesions or the presence of pathogens, they migrate to the site of injury, where they proliferate and release inflammatory factors or neurotoxic substances such as tumour necrosis factor alpha (TNF-α), which may potentially damage the CNS. For these reasons, it has been suggested that activated microglia may play a pivotal role in the pathogenesis of several neurodegenerative diseases (for a review see Gonzalez-Scarano and Baltuch, 1999). The constitutive expression of cytokines is generally low in the CNS. High levels of cytokines have been reported in patients suffering from various infections including HIV (Stanley et al., 1994) and in several neurological disorders such as Alzheimer's disease, Down's syndrome (Griffin et al., 1989), multiple sclerosis (Hofman, 1989), cerebral ischemia or brain injury (Feuerstein et al., 1998). The early response of glial cells to nervous system injury and their ability to secrete pro-inflammatory cytokines have been proposed as an early signalling mechanism responsible for subsequent neuronal degeneration (Benveniste, 1992).

Chemokines are a family of structurally-related cytokines with potent leukocyte activation and/or chemotactic activity, involved in cell-cell interaction (Rollins,1997; Glabinsky, 1995). Chemokines and their receptors are present in the CNS in the form of neurons and glial cells (Bajetto et al., 2001; Ragozzino, 2002) and their levels increase in certain neurodegenerative disorders, such as Alzheimer's disease (Streit et al., 2001; Halks-Miller et al., 2003), multiple sclerosis (Ransohoff, 2002; Hulshof et al., 2003), amyotrophic lateral sclerosis (Wilms, 2003) or inflammatory processes (De Groot and Woodroofe, 2001). It has been hypothesized that the temporal expression and combination of chemokines may contribute to a cascade of events resulting in specific patterns of neurodegeneration. These molecules represent important mediators in cell interaction among neurons, glia and leucocytes (Bajetto et al., 2002). Chemokines have been studied principally for their ability to draw immunocompetent cells, macrophages, etc. from the bloodstream into the damaged brain.

The major objective of this study was to investigate how microglia would respond to pathogenic factors. For this reason, human fetal microglia cultures were either infected with Mycobacterium avium (SmD, SmT) and Mycobacterium tuberculosis (Myc) or treated with myelin basic protein (MBP) or estrogens, and the expression of cytokines, chemokines and the co-stimulatory molecule CD 137 was investigated.

Materials and Methods

Microglia Cultures

Microglia cell cultures were obtained from long-term cultures of human fetal brain, as previously described (Lauro et al., 1995). These cultures were characterized and their purity and absence of previous contamination by other cells assessed, along with their functions (Presta et al., 1995; Jaranowska et al., 1995; Gremo et al., 1997)

Infection of Microglia Cells

Microglia cultures were infected with MAC 905 type 1 M. avium domed-opaque variant (SmD) and transparent variant (SmT), kindly provided by Dr. Orefici, Istituto Superiore di Sanità, Rome, Italy, and with N.C. 0741708 M.tuberculosis (from Central Public Health Laboratory Service, National Collection of Type Cultures, London)

Bacteria were grown and colonial variants were isolated as previously described (Curto et al., 2004). Microglial cells (10^6 cells per Petri dish, 56.7 cm^2) were incubated at 37°C in 5% CO_2 with the inoculum prepared at a bacterial concentration of 10^6-10^7 CFU/ml, as determined by OD and as previously described (Curto et al., 2004). At different times (30 min to 8 days after infection), cells were washed and fixed for immunocytochemistry or RNA extracted for RT-PCR, as described below.

Since a considerable number of bacteria remained stuck to the dish or cells and counts might not have reflected internalized bacteria, mild trypsinization was carried out (0.2% of trypsin for 5 min) in order to detach excess bacteria after the infection peak was reached (2 hours and 16 hours respectively for M. avium and M. tuberculosis).

Culture Treatment

Myelin Basic Protein was added to microglia cell cultures (final concentration 1, 5 or 10 μg/ml) in DMEM plus 10% fetal calf serum; after 48 h, cells were collected with Trizol (Gibco) for total RNA extraction, while parallel cells on coverslips were fixed at –20°C with cold methanol for 4 min for immunocytochemistry.

For estrogen treatments, cells were grown for 24 hr before replacing the medium with serum-free, phenol red-free DMEM (Life Technologies). Then, cells were incubated in the presence of 10^{-8}M 17β-estradiol (Sigma) in freshly added serum-free, phenol red-free DMEM.

Reverse Transcriptase- Polymerase Chain Reaction

Total RNA was isolated from control and treated microglial cell cultures using Trizol reagent (Gibco), according to Chomczynski, 1993. RNA concentration was measured with a spectrophotometer at 260 nm. Identical amounts of RNA were reverse transcribed into cDNA, which was subsequently amplified by polymerase chain reaction (PCR). PCR primers for TNF-α were 5'GAG CTG AGA GAT AAC CAG CTG GTG3' and 5'CAG ATA GAT GGG CTC ATA CCA GGG3' ; IL-1 were 5'TTC GAG CCA ATG ATC AGT ACC TCA3' and 5'TTT GGG TAT CTC AGG CAT CTC3' ; IL-10 were 5'GAT CTC CGA GAT GCC TTC AGC3' and 5'GGA AGA AAT CGA TGA CAG CGC3'; CD 137 were ATC ATG GGA AAC AGC TGT TAC AAC and TGG TCC ACA GAC CAC GTC CCT CTC; glyceraldehyde-3-phosphate-dehydrogenase (GAPDH) were 5'GCC AAA AGG GTC ATC ATC TCT G3' and 5'CAT GCC AGT GAG CTT CCC GT3' (Gibco BRL). Polymerase chain reaction was carried out using thermal cycle 2400 (Applied Biosystems).

According to Powell and Kroon (1992) for quantitative PCR with a non-radioactive label, digoxigenin-11-dUTP (DIG) (Roche Mannheim) was incorporated during the PCR reaction. An aliquot of the PCR reaction mixture was electrophoresed in 2% agarose gel Sigma in TAE (Tris-acetate/EDTA) buffer, to allow adequate separation of DNA. DNA was transferred to a nylon membrane by blotting for 18h in 10X SSC.

The blot was washed briefly in buffer A (maleic acid, 0.1 M; sodium chloride 0.15 M, pH 7.5) plus 0.3% Tween 20 and incubated for 30 min at room temperature in blocking buffer (1% blocking reagent in buffer A, Roche Mannheim, Germany). This was followed by a 30 min incubation with an antidigoxigenin-IgG conjugated to alkaline phosphatase (Roche Mannheim, Germany) diluted in blocking buffer. The blot was washed and chemiluminescent detection performed by incubating the membrane for 5 min in a CSPD solution (Roche Mannheim, Germany). The membrane was sealed in a clear plastic bag and exposed to X-ray film (KODAK X-OMAT) for 10-20 min in an X-ray cassette at room temperature. The signals corresponding to each cytokine and GAPDH mRNA were scanned by a densitometer (Hoefer Scientific Instruments, San Francisco, CA, U.S.A.). mRNA quantities were normalised by comparison with GAPDH levels.

Immunocytochemistry

Cells on coverslips were removed from dishes and fixed at -20° C with cold methanol for 4 min or with paraformadehyde (4% in phosphate buffer) for 15 min. For staining, samples were washed three times (20 min each wash) with Tris-HCl (50 mM, pH 7.4)-Triton X-100 (0.2%) and incubated with normal goat serum (1:10, Vector, Burlingame, CA, U.S.A.) for 30 min at room temperature. Subsequently, they were overlaid with monoclonal antibody against human CD137 (3 μg/ml, Biosource, Camarillo, CA, U.S.A.) and incubated for 1h. Slides were then rinsed three times in Tris-HCl-Triton and antibody visualized using the APAAP method (kit from DAKO, Carpinteria, CA, U.S.A.) according to the manufacturer's instructions and colour-developed with BCIP/NBT or Fast Red. Negative controls were incubated with non-immune serum or with murine IgG (MOPC21).

To study chemokine immunolocalization, sheep polyclonal antibodies against IL-8, MCP-1, MIP, RANTES (1:100; a gift from Dr. A. Mire-Sluis) were used. After incubation for an hour with the primary antibody, slides were washed three times and overlaid with a biotinylated anti-sheep (1:80; Chemicon) IgG (Vector, 1/200) and subsequently incubated with Texas Red conjugated avidin (1:80; Chemicon) for 30 min. Cells were viewed using an Olympus BX41 fluorescence microscope.

Statistical Analysis

Data were analyzed using the paired *t* test. Values were considered significant at $p < 0.05$. Data shown represent the mean ± SE of individual experiments.

Results

MBP Effect on CD137 Expression in Microglia Cells

Microglia cells were treated with MBP at concentrations ranging from 1 μg/ml to 10 μg/ml for 48 hours. CD137, a member of the TNF receptor family, is expressed by activated T-lymphocytes and monocytes. This protein was not detected in microglia cells before treatment or those treated with 1 μg/ml MBP, as shown in Figure 1 a,b. When cells were exposed to higher concentrations of MBP (5-10 μg/ml), immunostaining for CD 137 was evident (Figure 1 c,d). On the other hand, RT-PCR showed the presence of CD137 mRNA in untreated microglia, as already observed (Reali et al., 2003; Curto et al., 2004). Low concentrations of MBP (1 μg/ml) had no effect, while 5-10 μg/ml induced a significant increase in CD137 mRNA expression (Figure 2).

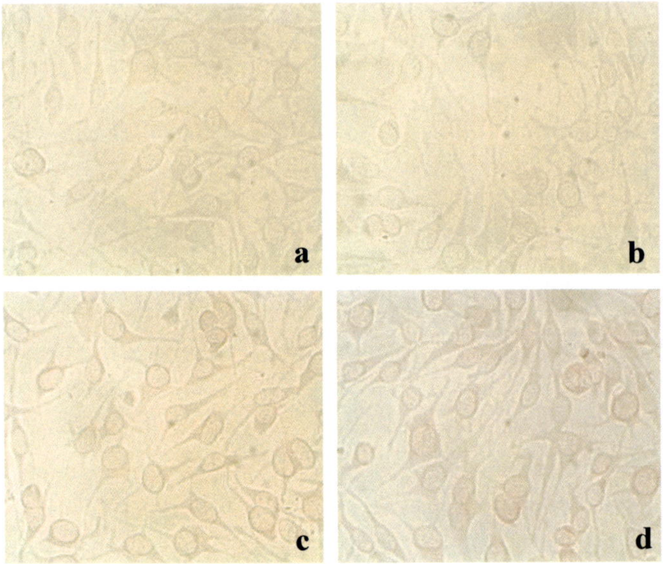

Figure 1: Immunolocalization of CD137 in microglia treated with MBP. (a) Control; (b) 1μg/ml MBP; (c) 5μg/ml MBP; (d) 10μg/ml MBP.

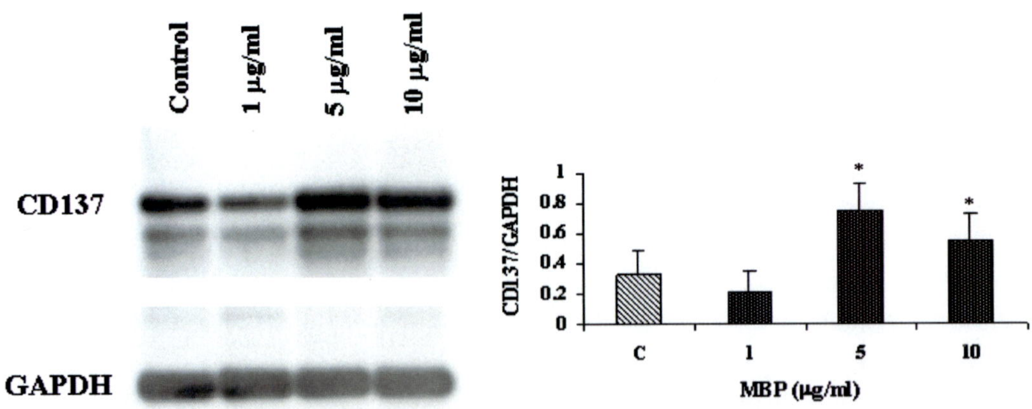

Figure 2: RT-PCR for CD137 in microglia. MBP (5-10 μg/ml) treatment significantly increased CD137 expression. Results are the mean of four experiments. (*P< 0.05)

Cytokine Expression in Microglia Infected by Mycobacteria

Microglia were incubated with the two variants of Mycobacterium avium (domed-opaque SmD, and transparent SmT) and with M. tuberculosis. Microglia were in effect infected by mycobacteria growing inside the cells, as already shown in a previous work (Curto et al., 2004). Using RT-PCR, we examined how mycobacteria infection modified the expression of three different cytokines mRNAs in microglia: tumour necrosis factor-alpha (TNF-α), interleukin-1 (IL-1) and interleukin-10 (IL-10). A basal expression of all three of these cytokines was observed in uninfected cells.

TNF-α mRNA expression was not significantly modified by infection with the less pathogenic variant of M. avium (SmD), whereas SmT induced a peak at 30 minutes, followed by a decrease (Figure 3 a). M. tuberculosis infection determined a significant increase in TNF-α expression between 2 and 8 hours, after which it decreased and returned to basal conditions (Figure 3 b).

A significant increase in IL-1 mRNA expression was observed only at 16 hours in SmD-infected cells. On the contrary, a transient decrease was observed between 4 and 16 hours in SmT-infected microglia (Figure 4 a). Moreover, M. tuberculosis induced a lasting, significant inhibition of this cytokine expression 16 hours after infection (Figure 4 b).

In both SmD and SmT infected cells, IL-10 mRNA expression was transiently inhibited: SmD induced a significant decrease only 30 minutes after infection, while SmT inhibited IL-10 expression between 2 and 16 hours (Fig 5 a). Moreover, M. tuberculosis induced a strong lasting decrease in IL-10 expression 24 hours after infection (Figure 5 b).

CD137 Expression in Infected Microglia

CD137 expression was assayed by immunocytochemistry and RT-PCR in M. avium and tuberculosis- infected microglia cells. As shown in Figure 6, uninfected cells were not immunostained by an anti-CD137 antibody. SmD infection induced very low labeling after 4 hours, which rapidly decreased, whereas after only 2 hours from infection with SmT, stronger staining was observed and lasted a little longer. M. tuberculosis infection induced early significant immunolabeling (after 30 minutes) that was maintained for a longer period.

RT-PCR showed no significant difference between infected and control microglia (Fig 7).

Chemokine Expression in Infected Microglia

Immunocytochemistry for four different chemokines (RANTES, IL-8, MIP, MCP-1) was performed in controls and in mycobacteria-infected microglia. Result are summarized in Table 1. Uninfected cells showed no immunoreactivity for any of the assayed chemokines. Faint immunostaining for RANTES was observed after SmD infection, whereas a progressive increase in labeling was induced between 4 and 16 hours from infection with SmT, followed by a return to basal levels. M. tuberculosis determined an early strong increase in RANTES expression between 30 minutes and 16 hours, as shown in Figure 8.

a)

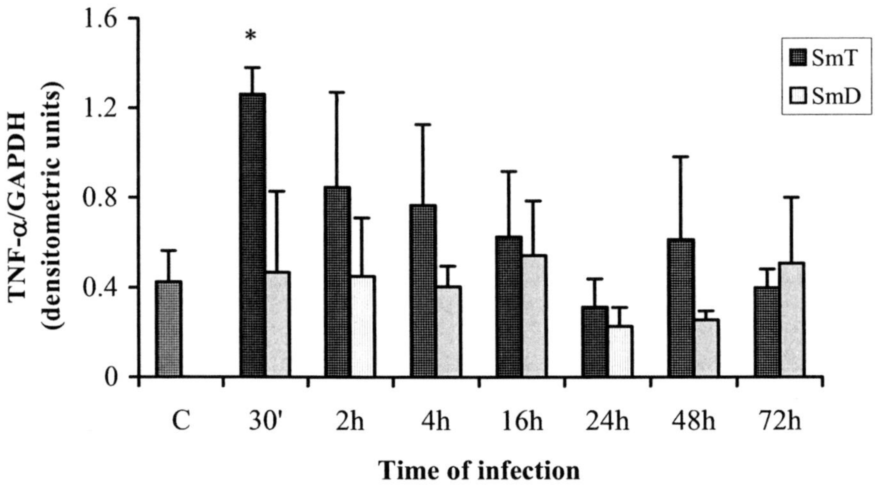

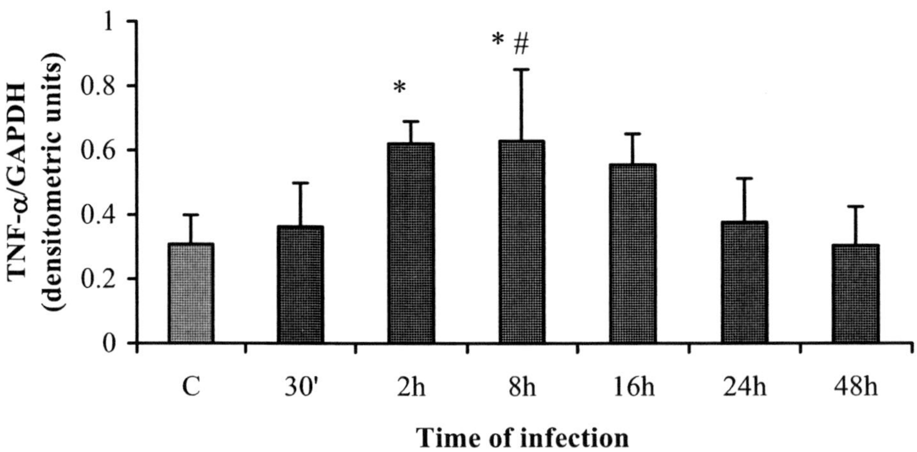

b)

Figure 3: GAPDH normalized results of RT-PCR for TNF-α at different times from infection with SmD, SmT (a) and M. tuberculosis (b). Results are the mean of four experiments. *P<0.05 vs. control; #P <0.05 vs. 30 min. (with permission from Neurochem. Int., Curto et al., 2004).

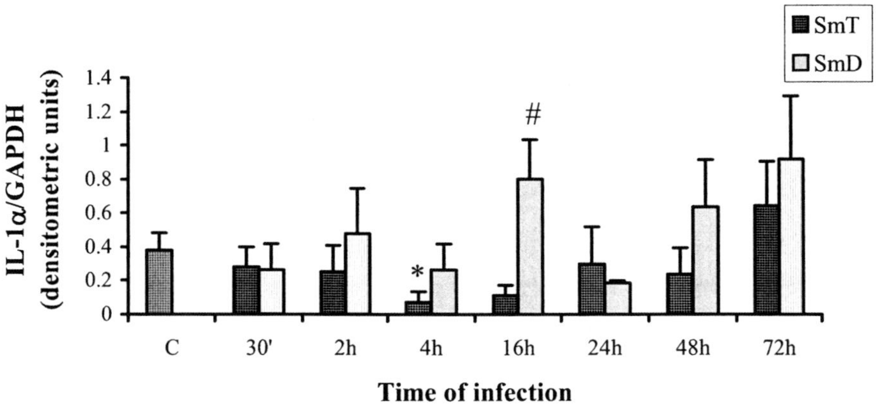

a)

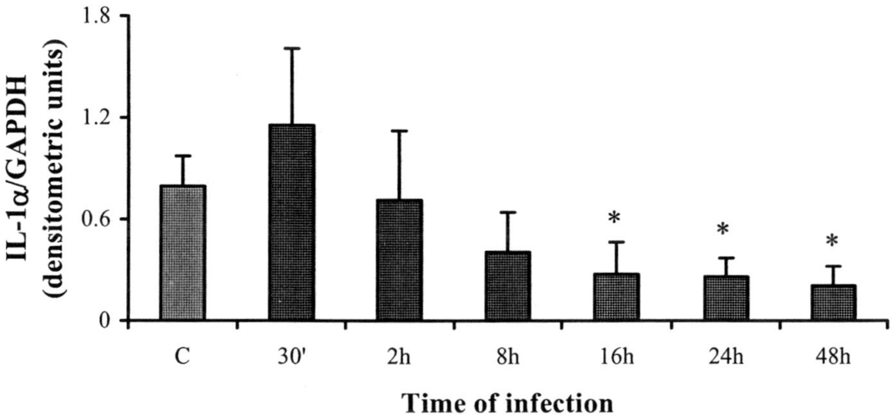

b)

Figure 4: GAPDH normalized results of RT-PCR for IL-1 at different times from infection with SmD, SmT (a) and M. tuberculosis (b). Results are the mean of four experiments. *P<0.05 vs. control; [#]P <0.01 vs. 30 min. (with permission from Neurochem. Int., Curto et al., 2004).

In SmD-infected microglia, weak IL-8 immunostaining was observed after 30 minutes; it increased between 2 and 4 hours, but progressively decreased after 24 hours. SmT infection determined a rise in this chemokine immunoreactivity between 2 and 24 hours, followed by a decrease and a return to basal levels after four days. On the other hand, we observed strong immunolabeling for IL-8 in *M. tuberculosis*-infected cells after 30 minutes, lasting for the whole period of infection (Figure 9).

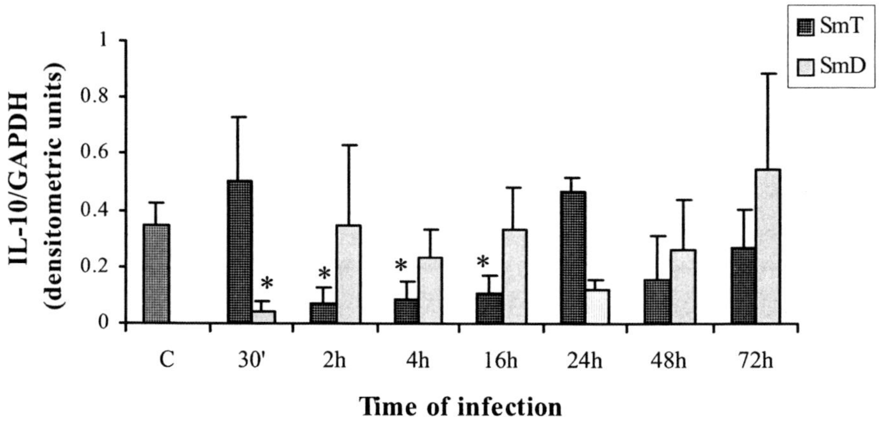

a)

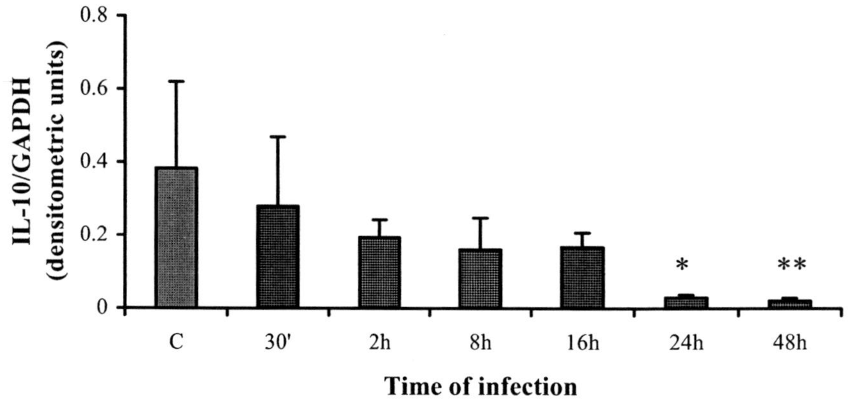

b)

Figure 5: GAPDH normalized results of RT-PCR for IL-10 at different times from infection with SmD, SmT (a) and M. tuberculosis (b). Results are the mean of four experiments. *P<0.05; **P <0.01 vs. control. (with permission from Neurochem. Int., Curto et al., 2004).

MIP immunoreactivity was not induced by SmD, whereas SmT infection determined transient, faint labeling for this chemokine. On the contrary, *M. tuberculosis* induced strong immunostaining, at least until 48 hours after infection (Figure 10).

MCP-1 was present at low levels in both SmD- and SmT-infected cells, whereas *M. tuberculosis* infection determined early strong immunoreactivity for this chemokine (Figure 11).

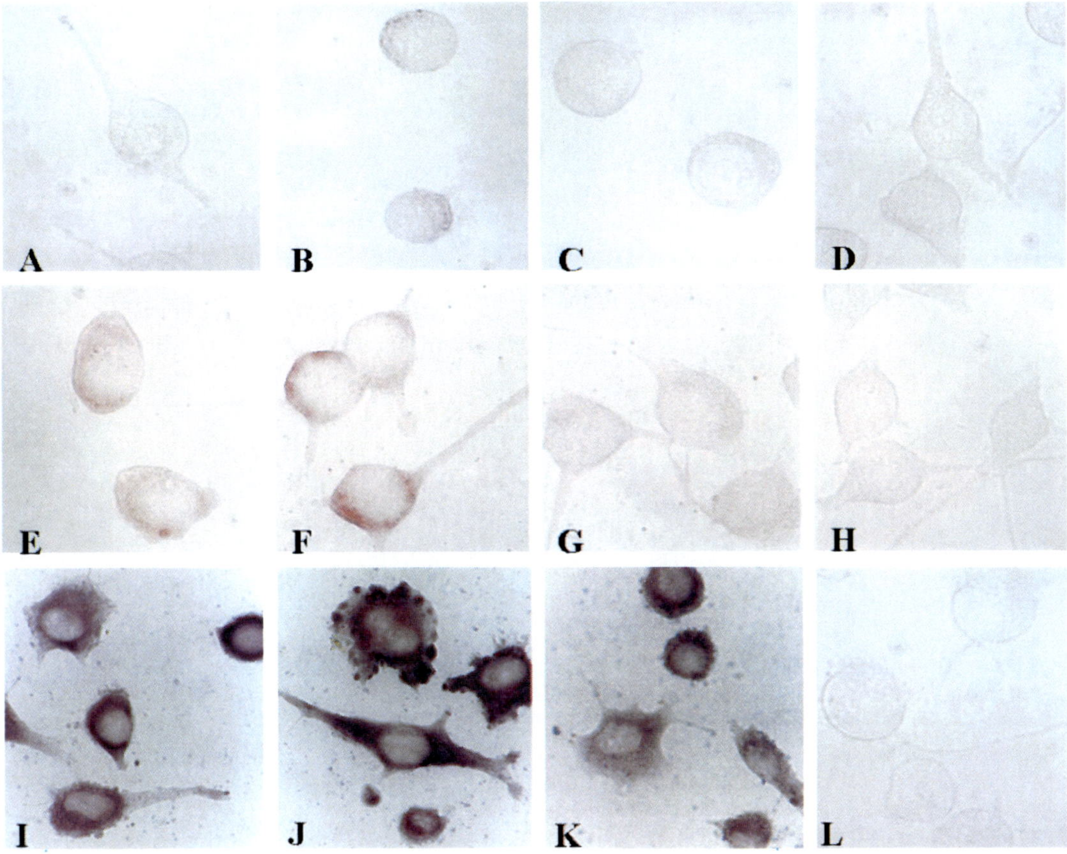

Figure 6: CD137 immunostaining in microglia infected with mycobacteria. (a) Non-infected cells; (b) at 4h, SmD; (c) at16h, SmD; (d) at 48h, SmD; (e) at 2h, SmT; (f) at 24h, SmT; (g) at 48h, SmT; (h) at 72h, SmT; (i) at 30 min, M. tuberculosis; (j) at 2 h, M. tuberculosis; (k) at 48h, M. tuberculosis; (l) negative control. Bar = 10μ (with permission from Neurochem. Int., Curto et al., 2004).

Effect of Estrogens on Cytokine Expression by Microglia

Microglia cells were treated with 10^{-8} M β-estradiol for 30 minutes to 2 hours and mRNA extracted and assayed by RT-PCR for IL-1 and IL-10 expression. Figure 12 a shows that a slight increase in IL-1 expression occurred 30 minutes after treatment, but a significant increase was observed only after 2 hours. On the other hand, β-estradiol induced a significant increase in IL-10 as early as 30 minutes after treatment, which was still present after 2 hours (Figure 12 b).

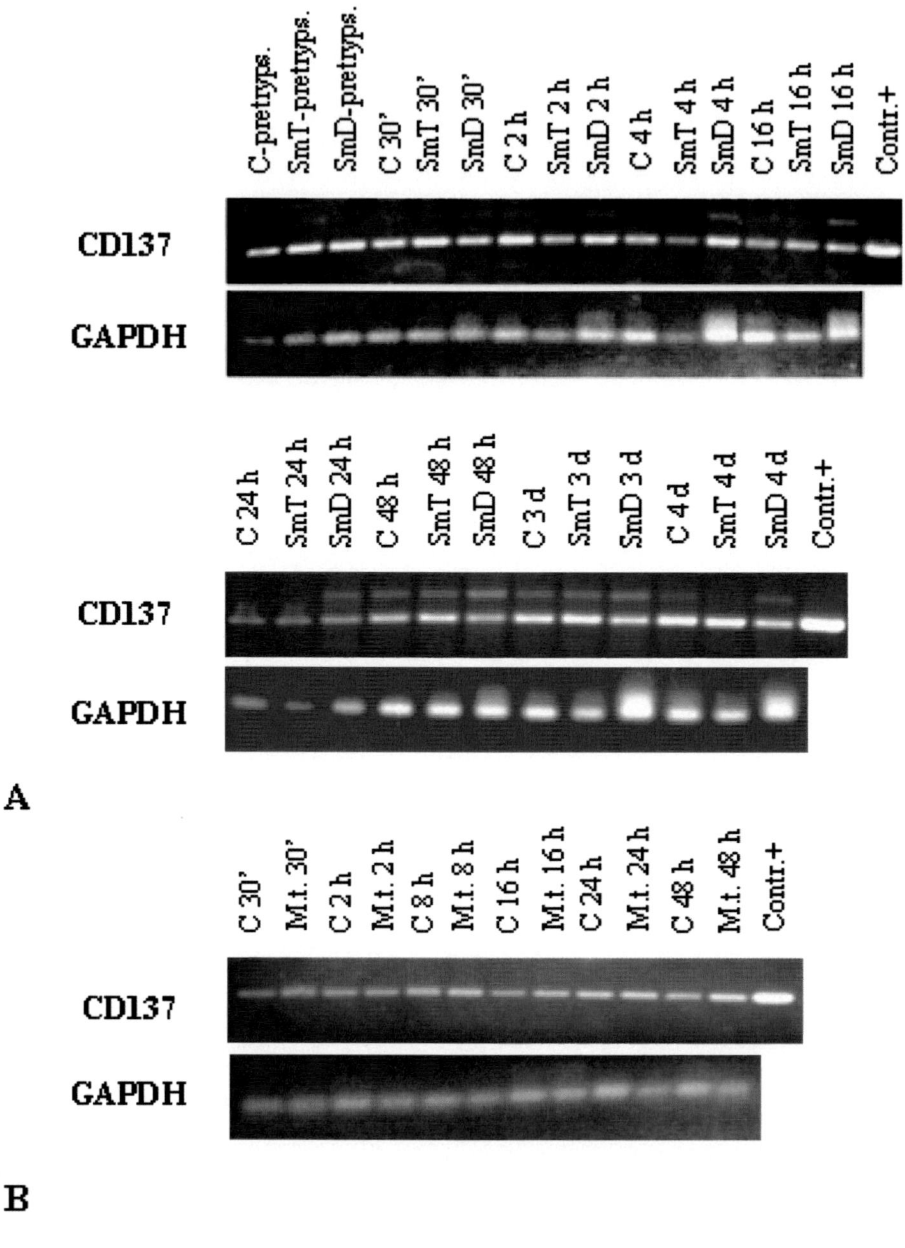

Figure 7: RT-PCR for CD137 in microglia at different times from infection with (a) M. avium and (b) M. tuberculosis. Contr.+: positive control

Table 1. Comparison of Chemokine Expression in Microglia Infected with Mycobacteria

	RANTES			IL-8			MIP			MCP-1		
	SmD	**SmT**	**M.t.**	**SmD**	**SmT**	**M.t.**	**SmD**	**SmT**	**M.t.**	**SmD**	**SmT**	**M.t.**
C	–	–	–	–	–	–	–	–	–	–	–	–
30'	±	±	++	+	±	+++	–	+	+++	+	+	+++
2h	+	±	+++	+±	+	+++	–	±	++++	±	±	++++
4h	±	+		++	+		±	±	.	±	±	.
8h			+++			+++			+++			+++
16h	±	++	+++	±	+	+++	–	±	++	±	±	++
24h	+	+	++	+	+	++±	–	±	++	±	±	++
48h	±	±	+	+	±	++±	–	±	++	–	±	++
3d	±	±		+	±		–	–		–	–	
4d	±	±		±	±		–	–		–	–	

SmD: *Mycobacterium avium*, Domed-opaque variant; SmT: *Mycobacterium avium*, Transparent variant; M.t.: *Mycobacterium tuberculosis*

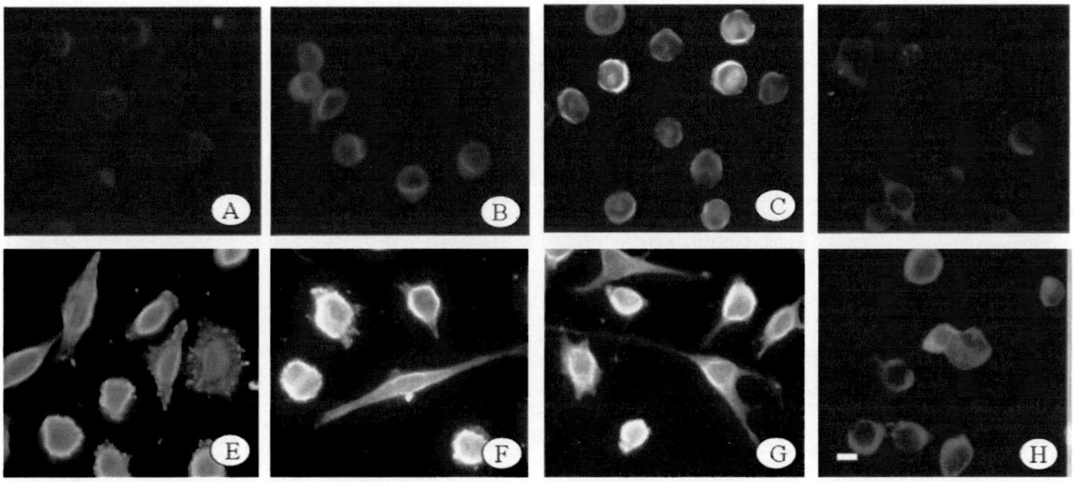

Figure 8:RANTES immunolocalization in microglia infected with mycobacteria: (a) Non-infected cells; (b) at 2h, SmD; (c) at 16h, SmT; (d) at 48h, SmT; (e) at 30 min, M. tuberculosis; (f) at 2h, M. tuberculosis; (g) at 24h, M. tuberculosis; (h) at 48h, M. tuberculosis. Bar = 10μ.

Conclusions

In this study, we investigated the effects of pathogenic factors such as mycobacteria or MBP on cultured human microglia. Our results demonstrate that the expression of inflammatory and anti-inflammatory cytokines and chemokines, as well as the co-stimulatory molecule CD137, can be modulated in microglia by different stimuli. These data confirm the key role played by microglia in many pathological brain dysfunctions.

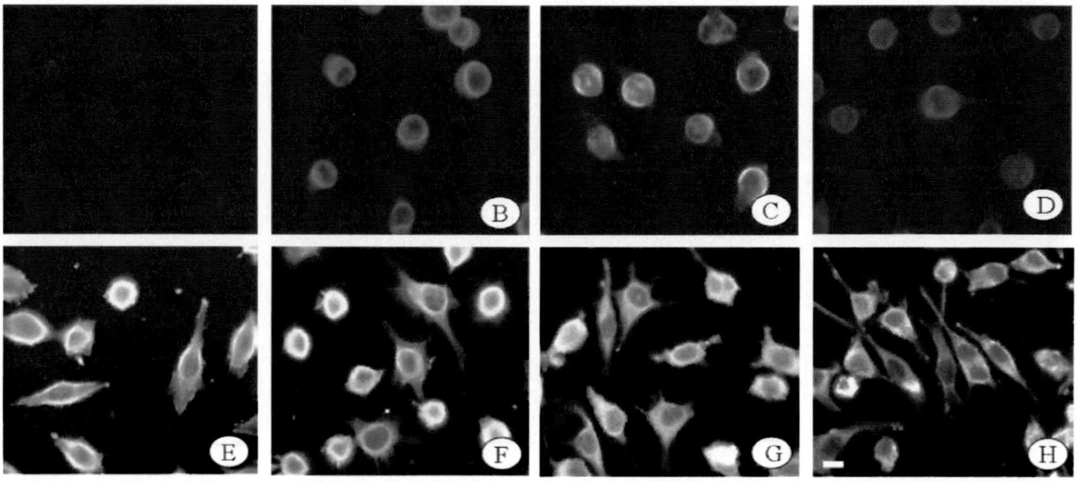

Figure 9: IL-8 immunolocalization in microglia infected with mycobacteria: (a) Non-infected cells; (b) at 2h, SmD; (c) at 4h, SmD; (d) at 16h, SmT; (e) at 30 min, M. tuberculosis; (f) at 2h, M. tuberculosis; (g) at 16h, M. tuberculosis; (h) at 48h, M. tuberculosis. Bar = 10μ.

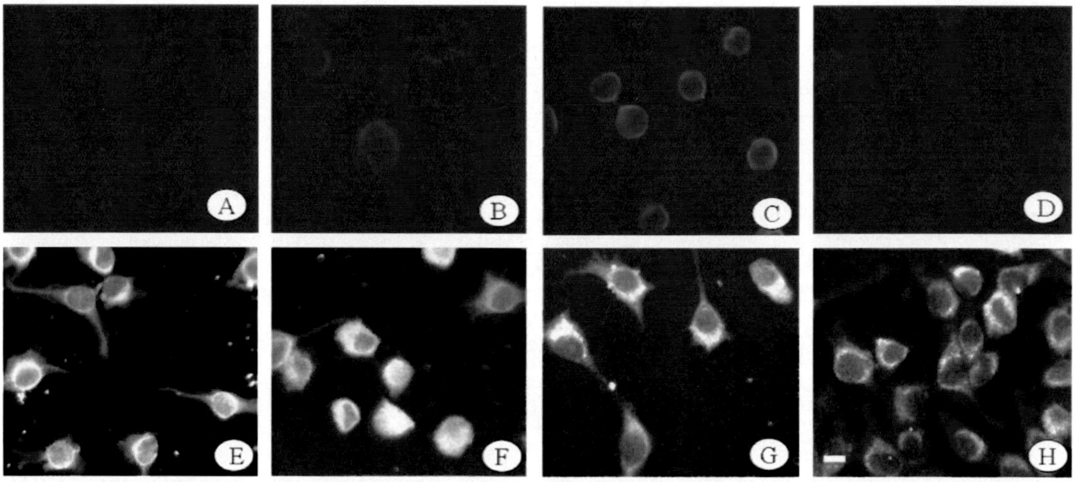

Figure 10: MIP immunolocalization in microglia infected with mycobacteria: (a) Non-infected cells; (b) at 4h, SmD; (c) at 30 min, SmT; (d) at 48h, SmT; (e) at 30 min, M. tuberculosis; (f) at 2h, M. tuberculosis; (g) at 8h, M. tuberculosis; (h) at 48h, M. tuberculosis. Bar = 10μ.

MBP is a major component of the myelin sheath in the central nervous system, and its immunological destruction is the major pathology of multiple sclerosis (MS). In this disease, MBP is found in cerebral spinal fluid (CSF). CD137 is a member of the TNF receptor superfamily and is expressed on activated T lymphocytes (Kwon and Weissman,1989; Pollock et al., 1993) as well as myeloid cells, including monocytes, neutrophils and dendritic cells (Kwon et al., 1997; Heinisch et al., 2000; Futagawa et al., 2002; Wilcox et al., 2002). Its expression in primary cells is strictly activation-dependent (Schwarz et al., 1995; Kienzle and von Kempis, 2000) and regulates the activation and proliferation of immune cells, possibly

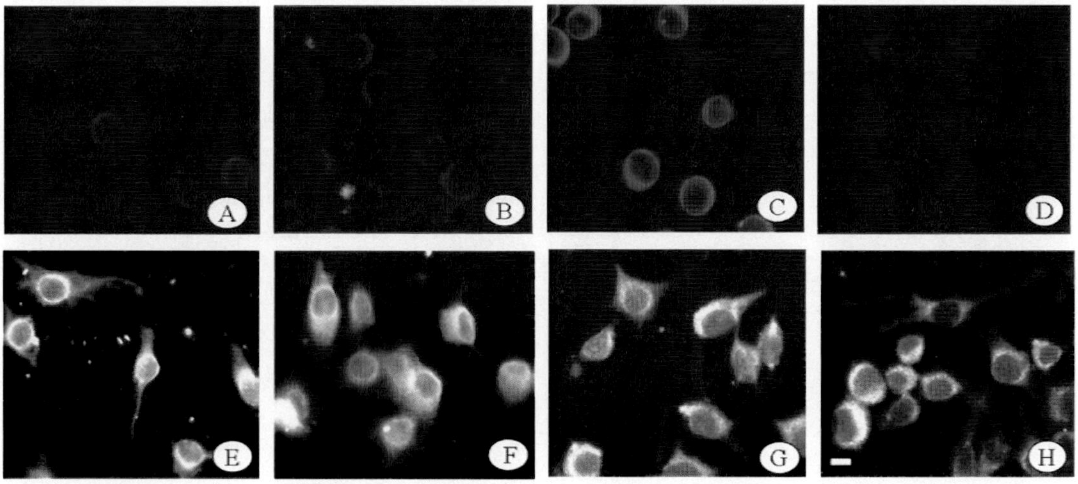

Figure 11:MCP-1 immunolocalization in microglia infected with mycobacteria: (a) Non-infected cells; (b) at 30 min, SmD; (c) at 30 min, SmT; (d) at 48h, SmT; (e) at 30 min, M. tuberculosis; (f) at 2h, M. tuberculosis; (g) at 16h, M. tuberculosis; (h) at 48h, M. tuberculosis. Bar = 10μ.

inducing the apoptosis of activated lymphocytes. Several reports have indicated evidence of CD137 interaction with a specific ligand (CD137-L), which may have biological effects like stimulating cell proliferation. In a recent work, we showed that human microglia, neurons and astrocytes express and synthesize CD137, whereas ligand expression is restricted to neurons and astrocytes, suggesting that these molecules may play a role in interaction among human brain cells (Reali et al., 2003). In this chapter, we have demonstrated that treatment of microglia cultures with MBP induces an increase in CD137 expression. These results are consistent with previous work showing increased intrathecal release of CD137 in patients with multiple sclerosis (Sharief, 2002). The involvement of microglia cells in multiple sclerosis has already been demonstrated. An increase in the number of microglial cells occurs in and around the demyelinating plaques causing inflammation (Williams et al., 1994; Benveniste,1997). Moreover, electron microscope studies have demonstrated phagocytosis of myelin by microglia at the site of active lesions (Prineas, 1975; Prineas and Connel, 1978). Activation of microglia by myelin phagocytosis in vitro has also been observed (Williams et al., 1994). These cells may produce inflammatory cytokines (Benveniste,1997), reactive oxygen and nitrogen species (Smith et al., 1999), as well as other costimulatory molecules. Our results confirm Sharief's hypothesis (2002), suggesting that microglia are responsible for the significant increase in CSF levels of CD 137 in multiple sclerosis. Thus, these cells may be involved in interacting with immune peripheral cells and modulating lymphocyte functions in multiple sclerosis. An increase in CD137 immunoreactivity was also observed in microglial cells infected with mycobacteria, suggesting a role of this molecule in CNS tuberculosis.

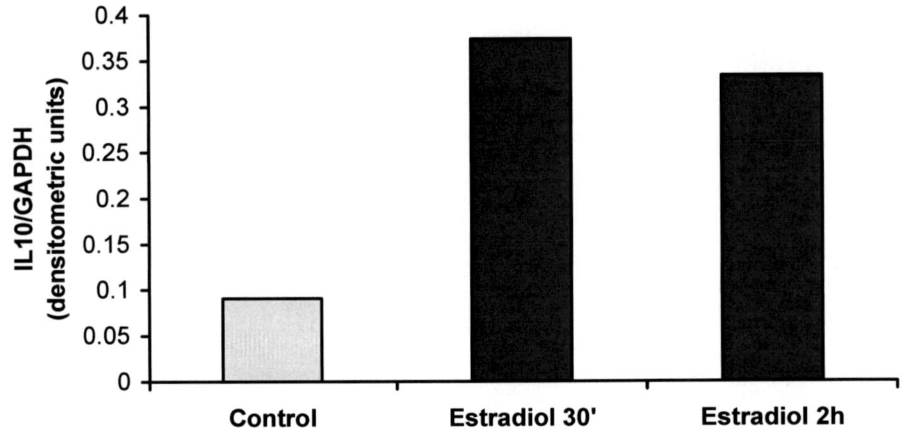

a)

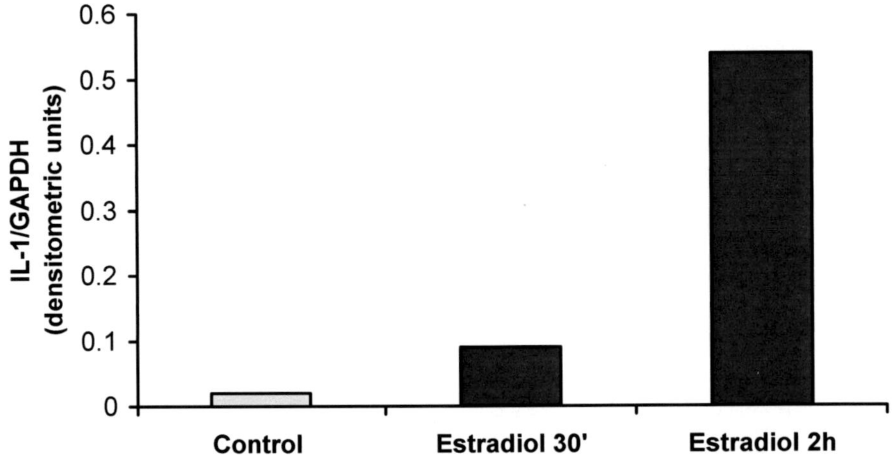

b)

Figure 12: GAPDH normalized results of RT-PCR for IL-1 (a) and IL-10 (b) in microglia treated with 10^{-8} M β-estradiol.

Infection of microglia by mycobacteria determined a modulation of cytokine expression, showing a correlation between virulence and cytokine induction. We observed significant lasting inhibition of IL-1 and IL-10 in cells infected with *M. tuberculosis*. On the other hand, the less pathogenic variant of *M. avium*, SmD, determined an increase in IL-1 expression, whereas infection with SmT induced inhibition of both cytokines, similar to the effect due to *M. tuberculosis*, even if much less severe and short-lived. Interestingly, we observed that the timing and level of TNF-α expression was proportional to the virulence of the mycobacterium type in early phases of infection. Subsequently, the most pathogenic mycobacteria (*M.*

tuberculosis and SmT) induced inhibition of this cytokine expression, which returned to base levels. Thus, our results indicate that the virulence of mycobacteria depends more on cytokine inhibition than induction of their expression. These data agree with observations by Beltan et al., (2000) who found the lowest levels of cytokine expression with pathogenic mycobacterial species, whereas levels approximately 2-8 times higher were observed for non-pathogenic bacteria. It has been demonstrated that a virulence mechanism for several pathogens, including *M. avium*, consists of suppressing TNF-α production (Beuscher et al., 1995; Furney et al., 1992; Shiratsushi et al., 1993).The role of TNF-α in the pathogenesis of tuberculosis has been extensively investigated. Several in vitro and in vivo studies (Denis, 1991; Flynn et al., 1995; Hirsch et al., 1994) have suggested that this cytokine plays a role in controlling intracellular *M. tuberculosis*. Silver et al. (1998) observed that the increase in TNF-α expression in infected monocytes was correlated to mycobacteria virulence. Moreover, TNF-α was found to be present in cerebrospinal fluid during mycobacterial CNS infection by Tsenova et al. (1999), who hypothesized a key role of this cytokine in the pathogenesis and disease progression of CNS tuberculosis. Thus, we may conclude that mycobacterial infection induces immunosuppression, strictly related to the virulence of the bacterium, in brain cells. This may be a mechanism by which non-pathogenic mycobacteria are quickly eliminated, while pathogenic species persist.

Chemokines play an essential role in the immune response to tuberculosis, recruiting peripheral leucocytes into the CNS, a critical step in the development of host defense to pathogens. Several studies have reported the ability of microglia to produce chemokines as a response to inflammatory stimuli, such as bacterial lipopolysaccharide or bacterial and viral infections (for a review see Kielian, 2004). We observed that microglial cells, at basal condition, expressed low levels of two chemokines, IL-8 and MCP-1, whereas MIP and RANTES were undetectable. Mycobacterial infection determined an increase in expression of all chemokines assayed, whose importance corresponded to bacterium virulence. Our results were consistent with the observation by Hu et al. (1999), who found higher levels of RANTES secretion in LPS activated microglia cultures as compared to non-activated cells. The increase in IL-8 expression is consistent with similar results obtained on macrophages infected with different strains of *M. tuberculosis* (Zang et al., 1995).

The induction of chemokine expression may determine other effects in addition to leucocyte recruitment. First of all, chemokines may act in an autocrine fashion, affecting microglial cells which express their receptors. It has been shown that the binding of chemokine receptors induces the production of matrix metalloproteinases in microglia (Cross and Woodroofe, 1999), which may have an important role in the breakdown of the blood-brain barrier necessary for leucocyte infiltration. As suggested by McManus et al.(2000) in HIV infection, chemokine autocrine regulation may produce positive feedback, inducing cells to continue chemokine production, thus perpetuating the inflammatory response. This hypothesis may explain why, unlike results obtained on cytokines, we observed no inhibition in chemokine expression after *M. tuberculosis* infection, except for RANTES, suggesting that microglial cells are not totally inactivated by pathogenic mycobacteria. Moreover, since both neurons and astrocytes express several chemokine receptors on their cell surface (Asensio and Campbell 1999, Glabinski and Ransohoff, 1999), they may also be a target of microglia-produced chemokines.

Our results therefore indicate that microglia response to mycobacterial infection determines modifications in the expression of various factors, modulated according to bacterial virulence.

Moreover, we observed that treatment with estrogens can regulate cytokine expression in microglia cultures. B-estradiol induced an early significant increase in an antinflammatory citokine, IL-10, whereas IL-1 required longer treatment. Increasing evidence supports the neuroprotective role played by estrogens in the CNS after several types of injury, including neurodegeneration. It has been observed that estrogen replacement therapy has beneficial effects on neurodegenerative disorders (Honjo et al., 1995; Tang et al., 1996; Kawas et al., 1997) or inflammatory-related diseases (Horowitz, 1993; Nathan and Chaudhuri, 1997). Vegeto et al. (2001, 2003) showed that microglia express specific receptors through which estrogen prevents a lipopolysaccharide-induced inflammatory response in these cells. We observed an increase in IL-1 expression induced by estrogen. Since we have shown that mycobacterial infection induces the inhibition of cytokine expression in brain cells, estrogen treatment might counteract this process.

In conclusion, microglia respond to various CNS injury producing a range of cytokines and inflammatory molecules. The complete elucidation of the role played by microglia requires further investigation in order to develop new therapeutic strategies to fight or contain such diseases.

References

Aloisi, F. Immune function of microglia. *Glia,* 2001 36,165-179.

Asensio, VC; Campbell, IL. Chemokines in the CNS: plurifunctional mediators in diverse states. *Trends Neurosci,* 1999 22, 504-512.

Bajetto, A; Bonavia, R; Barbero, S; Florio, T; Schettini, G. Chemokines and their receptors in the central nervous system. *Front Neuroendocrinol,* 2001 22, 147-184.

Bajetto, A; Bonavia, R; Barbero, S; Schettini, G. Characterization of chemokines and their receptors in the central nervous system: physiopathological implications. *J Neurochem,* 2002 82, 1311-1329.

Beltan, E; Horgen, L; Rastogi, N; Secretion of cytokines by human macrophages upon infection by pathogenic and non-pathogenic mycobacteria *Microb Pathog,* 2000 28, 313-318.

Benveniste, EN. Inflammatory cytokines within the central nervous system: sources, function, and mechanism of action. *Am J Physiol,* 1992 263, C1-16.

Benveniste, EN. Role of macrophages/microglia in multiple sclerosis and experimental allergic encephalomyelitis. *J Mol Med,* 1997 75, 165-173.

Beuscher, HU; Rodel, F; Forsberg, A; Rollinghoff, M. Bacterial evasion of host immune defense: Yersinia enterocolitica encodes a suppressor for tumor necrosis factor alpha expression. *Infect Immun.* 1995 63, 1270-1277.

Chomczynski, P. A reagent for the single-step simultaneous isolation of RNA, DNA and proteins from cell and tissue samples. *Biotechniques.* 1993 15, 532-537.

Cross, AK; Woodroofe, MN. Chemokine modulation of matrix metalloproteinase and TIMP production in adult rat brain microglia and a human microglial cell line in vitro. *Glia.* 1999 28, 183-189.

Curto, M; Reali, C; Palmieri, G; Scintu, F; Schivo, M.L; Sogos, V; Marcialis, MA; Ennas, MG; Schwarz, H; Pozzi, G; Gremo, F. Inhibition of cytokines expression in human microglia infected by virulent and non-virulent mycobacteria. *Neurochem Int,* 2004 44, 381-392.

De Groot, CJ; Woodroofe, MN. The role of chemokines and chemokine receptors in CNS inflammation. *Prog Brain Res,* 2001 132, 533-544.

Del Rio Ortega, P. Microglia. In Penfield W. editor. *Cytology and cellular pathology of the nervous system. Vol. 2°.* New York: Harbert and Row; 1932; 481-584.

Denis, M. Killing of Mycobacterium tuberculosis within human monocytes: activation by cytokines and calcitriol. *Clin Exp Immunol,* 1991 84, 200-206.

Ferrer, I; Bernet, E; Soriano, E; del Rio, T; Fonseca, M. Naturally occurring cell death in the cerebral cortex of the rat and removal of dead cells by transitory phagocytes. *Neuroscience,* 1990 39, 451-458

Feuerstein, GZ; Wang, X; Barone, FC. The role of cytokines in the neuropathology of stroke and neurotrauma. *Neuroimmunomodulation,* 1998 5, 143-159.

Flynn, JL; Goldstein, MM, Chan, J; Triebold, KJ; Pfeffer, K; Lowenstein, CJ; Schreiber, R; Mak, TW; Bloom, BR. Tumor necrosis factor-alpha is required in the protective immune response against Mycobacterium tuberculosis in mice. *Immunity,* 1995 2, 561-572.

Furney, SK; Skinner, PS; Roberts, AD; Appelberg, R; Orme, IM. Capacity of Mycobacterium avium isolates to grow well or poorly in murine macrophages resides in their ability to induce secretion of tumor necrosis factor. *Infect Immun,* 1992 60, 4410-4413.

Futagawa, T; Akiba, H; Kodama, T; Takeda, K; Hosoda, Y; Yagita, H; Okumura, K. Expression and function of 4-1BB and 4-1BB ligand on murine dendritic cells. *Int Immunol,* 2002 14, 275-286.

Giulian, D; Baker, TJ. Characterization of ameboid microglia isolated from developing mammalian brain. *J Neurosci,* 1986 6, 2163-2178.

Giulian, D. Ameboid microglia as effectors of inflammation in the central nervous system *J Neurosci Res,* 1987 18,155-171.

Glabinski, AR; Ransohoff, RM. Chemokines and chemokine receptors in CNS pathology. *J Neurovirol.* 1999 5, 3-12.

Glabinski, AR; Tani, M; Aras, S; Stoler, MH; Tuohy, VK; Ransohoff, RM. Regulation and function of central nervous system chemokines. *Int J Dev Neurosci,* 1995 13, 153-165.

Gonzalez-Scarano, F; Baltuch, G. Microglia as mediators of inflammatory and degenerative diseases. *Annu Rev Neurosci.* 1999, 22, 219-240.

Gremo, F; Sogos, V; Ennas, MG; Meloni, A; Persichini, T; Colasanti, M; Lauro, GM. Features and functions of human microglia cells. *Adv Exp Med Biol,* 1997 429, 79-97.

Griffin, WS; Stanley, LC; Ling, C; White, L; MacLeod, V; Perrot, LJ; White, CL 3rd; Araoz, C. Brain interleukin 1 and S-100 immunoreactivity are elevated in Down syndrome and Alzheimer disease. *Proc Natl Acad Sci U S A,* 1989 86, 7611-7615.

Halks-Miller, M; Schroeder, ML; Haroutunian, V; Moenning, U; Rossi, M; Achim, C; Purohit, D; Mahmoudi, M; Horuk, R. CCR1 is an early and specific marker of Alzheimer's disease. *Ann Neurol,* 2003 54, 638-646.

Hanisch, UK. Microglia as a source and target of cytokines. *Glia,* 2002 40, 140-155.

Heinisch, IV; Daigle, I; Knopfli, B; Simon, HU. CD137 activation abrogates granulocyte-macrophage colony-stimulating factor-mediated anti-apoptosis in neutrophils. *Eur J Immunol,* 2000 30, 3441-3446.

Hirsch, CS; Ellner, JJ; Russell, DG; Rich, EA. Complement receptor-mediated uptake and tumor necrosis factor-alpha-mediated growth inhibition of Mycobacterium tuberculosis by human alveolar macrophages. *J Immunol,* 1994 152, 743-753.

Hofman, FM. Cytokines in Central Nervous System Disease In Goetz EJ and Spector NH, Editors. *Neuroimmune Networks: Physiology of diseases.* New York: Alan R. Liss 1989. pp 65-71.

Honjo, H; Tanaka, K; Kashiwagi, T; Urabe, M; Okada, H, Hayashi, M; Hayashi, K. Senile dementia-Alzheimer's type and estrogen. *Horm Metab Res,* 1995 27, 204-207.

Horowitz, MC. Cytokines and estrogen in bone: antiosteoporotic effects. *Science,* 1993 260, 626-627.

Hulshof, S; van Haastert, ES; Kuipers, HF; van den Elsen, PJ; De Groot, CJ; van der Valk, P; Ravid, R; Biber, K. CX3CL1 and CX3CR1 expression in human brain tissue: noninflammatory control versus multiple sclerosis. *J Neuropathol Exp Neurol,* 2003 62, 899-907.

Jaranowska, A, Bussolino, F; Sogos, V; Arese, M; Lauro, GM; Gremo, F. Platelet-activating factor production by human fetal microglia. Effect of lipopolysaccharides and tumor necrosis factor-alpha *Mol Chem Neuropathol,* 1995 24, 95-106.

Kawas, C, Resnick, S; Morrison, A; Brookmeyer, R, Corrada, M; Zonderman, A; Bacal, C; Lingle, DD; Metter, E. A prospective study of estrogen replacement therapy and the risk of developing Alzheimer's disease: the Baltimore Longitudinal Study of Aging. *Neurology,* 1997 48, 1517-1521.

Kielian, T. Microglia and chemokines in infectious diseases of the nervous system: views and reviews. *Front Biosci,* 2004 9, 732-750.

Kienzle, G; von Kempis, J. CD137 (ILA/4-1BB), expressed by primary human monocytes, induces monocyte activation and apoptosis of B lymphocytes. *Int Immunol,* 2000 12, 73-82.

Killackey, HP; Glia and the elimination of transient cortical projections. *Trends Neurosci,* 1984 7, 225-226.

Kwon, BS; Tan, KB; Ni, J; Oh, KO; Lee, ZH; Kim, KK; Kim, YJ; Wang, S; Gentz, R; Yu, GL; Harrop, J; Lyn, SD; Silverman, C; Porter, TG; Truneh, A; Young, PR. A newly identified member of the tumor necrosis factor receptor superfamily with a wide tissue distribution and involvement in lymphocyte activation. *J Biol Chem,* 1997 272, 14272-14276.

Kwon, BS; Weissman, SM. cDNA sequences of two inducible T-cell genes *Proc Natl Acad Sci USA,* 1989 86, 1963-1967.

Lauro, GM; Babiloni, D; Buttarelli, FR; Starace, G; Cocchia, D; Ennas, MG; Sogos, V; Gremo, F. Human microglia cultures: a powerful model to study their origin and immunoreactive capacity *Int. J. Devl. Neurosci,* 1995 13, 29-39.

Ling, EA. The origin and nature of microglia. In Fedoroff S, Hertz L, editors. *Advances in cellular neurobiology* Academic Press 1981 pp.33-82.

McManus, CM; Weidenheim, K; Woodman, SE; Nunez, J; Hesselgesser, J; Nath, A; Berman, JW. Chemokine and chemokine-receptor expression in human glial elements: induction by the HIV protein, Tat, and chemokine autoregulation. *Am J Pathol,* 2000 156, 1441-1453.

Minghetti, L; Levi, G. Microglia as effector cells in brain damage and repair: focus on prostanoids and nitric oxide. *Prog Neurobiol.* 1998 54(1):99-125.

Nathan, L; Chaudhuri, G. Estrogen and atherosclerosis. *Annu Rev Pharmacol Toxicol,* 1997 37, 477-515.

Pollok, KE; Kim, YJ; Zhou, Z; Hurtado, J; Kim, KK; Pickard, RT; Kwon, BS. Inducible T cell antigen 4-1BB. Analysis of expression and function *J Immunol,* 1993 150, 771-781.

Powell, EE; Kroon, PA. Measurement of mRNA by quantitative PCR with a nonradioactive label. *J Lipid Res*, 1992 33, 609-14.

Presta, M; Urbinati, C; Dell'era, P; Lauro, GM; Sogos, V; Balaci, L; Ennas, MG; Gremo F. Expression of basic fibroblast growth factor and its receptors in human fetal microglia cells. *Int J Dev Neurosci,* 1995 13, 29-39.

Prineas, JW; Connell F. The fine structure of chronically active multiple sclerosis plaques. *Neurology,* 1978 28, 68-75.

Prineas, JW. Pathology of the early lesion in multiple sclerosis. *Hum. Pathol,* 1975 6, 531-554.

Ragozzino, D. CXC chemokine receptors in the central nervous system: Role in cerebellar neuromodulation and development. *J Neurovirol*, 2002 8, 559-572.

Ransohoff, RM. The chemokine system in neuroinflammation: an update. *J Infect Dis*, 2002 186, 152-156.

Reali, C; Curto, M; Sogos, V; Scintu, F; Pauly, S; Schwarz, H; Gremo, F. Expression of CD137 and its ligand in human neurons, astrocytes, and microglia: modulation by FGF-2. *J Neurosci Res,* 2003 74, 67-73.

Rollins, B. Chemokines. *Blood* 1997 90, 909-928.

Schwarz, H; Valbracht, J; Tuckwell, J; von Kempis, J; Lotz, M. ILA, the human 4-1BB homologue, is inducible in lymphoid and other cell lineages. *Blood,* 1995 85,1043-1052

Sharief, MK. Heightened intrathecal release of soluble CD137 in patients with multiple sclerosis *Eur J Neurol,* 2002 9, 49-54.

Shiratsuchi, H; Toossi, Z; Mettler, MA; Ellner, JJ. Colonial morphotype as a determinant of cytokine expression by human monocytes infected with Mycobacterium avium. *J Immunol,* 1993 150 2945-2954.

Silver, RF; Li, Q; Ellner, JJ. Expression of virulence of Mycobacterium tuberculosis within human monocytes: virulence correlates with intracellular growth and induction of tumor necrosis factor alpha but not with evasion of lymphocyte-dependent monocyte effector functions. *Infect Immun*, 1998 66, 1190-1199.

Smith, KJ; Kapoor, R; Felts, PA. Demyelination: the role of reactive oxygen and nitrogen species. *Brain Pathol,* 1999 9, 69-92

Stanley, LC; Mrak, RE, Woody, RC; Perrot, LJ; Zhang, S; Marshak, DR; Nelson, SJ; Griffin, WS. Glial cytokines as neuropathogenic factors in HIV infection: pathogenic similarities to Alzheimer's disease. *J Neuropathol Exp Neurol,* 1994 53, 231-238.

Streit, WJ; Conde, JR; Harrison, JK. Chemokines and Alzheimer's disease. *Neurobiol Aging,* 2001 22, 909-913.

Tang, MX; Jacobs, D; Stern, Y; Marder, K; Schofield, P; Gurland, B; Andrews, H; Mayeux, R. Effect of oestrogen during menopause on risk and age at onset of Alzheimer's disease. *Lancet,* 1996 348, 429-432.

Vegeto, E; Belcredito, S; Etteri, S; Ghisletti, S; Brusadelli, A; Meda, C; Krust, A; Dupont, S; Ciana, P; Chambon, P; Maggi, A. Estrogen receptor-alpha mediates the brain antiinflammatory activity of estradiol. *Proc Natl Acad Sci USA,* 2003 100, 9614-9619

Vegeto, E; Bonincontro, C; Pollio, G; Sala, A; Viappiani, S; Nardi, F; Brusadelli, A; Viviani, B; Ciana, P; Maggi, A. Estrogen prevents the lipopolysaccharide-induced inflammatory response in microglia. *J Neurosci,* 2001 21, 1809-1818.

Wilcox, RA; Flies, DB; Zhu, G; Johnson, AJ; Tamada, K; Chapoval, AI; Strome, SE; Pease, LR; Chen, L. Provision of antigen and CD137 signaling breaks immunological ignorance, promoting regression of poorly immunogenic tumors *J Clin Invest,* 2002 109, 651-659.

Williams, K; Ulvestad, E; Waage, A; Antel, JP; McLaurin, J. Activation of adult human derived microglia by myelin phagocytosis in vitro. *J Neurosci Res,* 1994 38, 433-443.

Wilms, H; Sievers, J; Dengler, R; Bufler, J; Deuschl, G; Lucius, R. Intrathecal synthesis of monocyte chemoattractant protein-1 (MCP-1) in amyotrophic lateral sclerosis: further evidence for microglial activation in neurodegeneration *J Neuroimmunol,* 2003 144, 139-142.

Zhang, Y; Broser, M; Cohen, H; Bodkin, M; Law, K; Reibman, J; Rom, WN. Enhanced interleukin-8 release and gene expression in macrophages after exposure to Mycobacterium tuberculosis and its components. *J Clin Invest,* 1995 95, 586-92.

In: Focus on Neurochemistry Research
Editor: Robert M. Coleman, pp.139-156
ISBN 1-59454-417-4

Chapter V

Prevention of Ischemic Neuron Death by a Saffron's Carotenoid Pigment Crocin and its Mechanism of Action

***Shinji Soeda*[1], *Takashi Ochiai*[1], *Hiroyuki Tanaka*[2], *Yukihiro Shoyama*[2], *Hiroshi Shimeno*[1]**
[1]Faculty of Pharmaceutical Sciences, Fukuoka University, Fukuoka, Japan
[2]Graduate School of Pharmaceutical Sciences, Kyushu University, Fukuoka, Japan

Abstract

Oxidative stress plays an important role in neuron death associated with various neurodegenerative conditions. Ceramide, the product of sphingomyelin hydrolysis, is reported to be a key mediator of the cell death induced by reactive oxygen species (ROS). Antioxidants that prevent the detrimental consequences of ROS are considered to be a promising approach to neuroprotection. Crocin is a pharmacologically active component of *Crocus sativus L.* (saffron). We describe here the *in vitro* effect of crocin on the oxidative stress-induced death of neuronally differentiated rat pheochromocytoma (PC-12) cells. Depriving the PC-12 cells of serum/glucose causes peroxidation of the cell membrane lipids and a reduction in intracellular superoxide dismutase (SOD) activity. Crocin treatment inhibited the formation of peroxidized lipids, restored SOD activity, and also suppressed the activation of caspase-8 caused by serum/glucose deprivation. These *in vitro* antioxidant effects of crocin were more effective than those of α-tocopherol at the same concentration. Serum/glucose deprivation causes a rapid increase in intracellular ceramide levels, followed by an increase in the activation of c-jun kinase (JNK). The accumulation of ceramide results from the activation of magnesium-dependent neutral sphingomyelinase (nSMase). Serum/glucose deprivation also decreases cellular levels of glutathione (GSH), a potent inhibitor of nSMase. Treating the PC-12 cells with crocin prevented nSMase activation, ceramide formation, and JNK activation. Our studies further demonstrate that crocin can enhance the activities of GSH reductase and γ-glutamylcysteinyl synthase, contributing to a stable GSH supply that blocks the activation of nSMase. We suggest that crocin is a unique and potent antioxidant that combats ischemic stress-induced neuron death by increasing GSH levels, and prevents the activation of the JNK pathway, which is reported to have a role in the signaling cascade downstream ceramide.

Introduction

Crocus sativus L. (saffron) has been used as a spice for flavoring and coloring food preparations and in Chinese traditional medicine as an anodyne and tranquilizer. Recent pharmacological studies have demonstrated that saffron extracts have antitumor effects (Garcia-Olmo *et al.*, 1999; Molnar *et al.*, 2000; Abdullaev, 2002), antinociceptive and anti-inflammatory activities (Hosseinzadeh & Younesi, 2002), and antioxidant properties (Premkumar *et al.*, 2003; Ochiai *et al.*, 2004a). Additionally, behavioral and electrophysiological studies using experimental animals have shown that extracts of saffron had a positive effect on learning and memory (Abe & Saito, 2000). In mice whose learning behavior was damaged by ethanol, an extract of saffron ameliorated the effect of the damage. It also prevented ethanol-induced inhibition of hippocampal long-term potentiation, a form of activity-dependent synaptic plasticity that may underlay learning and memory.

The main constituents of saffron are picrocrocin, the substance causing saffron's bitter taste; safranal, the volatile oil responsible for saffron's characteristic odor; and crocin (crocetin-di-gentiobiose ester), which is a water-soluble carotenoid enriched in saffron (Soeda *et al.*, 2003). Saffron's effect on the central nervous system is attributed to crocin; however, crocetin, the base molecule for crocin does not have this effect (Abe & Saito, 2000; Soeda *et al.*, 2003).

Neuronal cell damage induced by ROS is presumably involved in cerebral ischemia (Cao *et al.*, 1988) and in Alzheimer's disease (AD), Parkinson's disease and others (Behl & Moosmann, 2002). The use of antioxidants, such as α-tocopherol, ascorbic acid, 17β-estradiol and β-carotene, which suppress the effects of ROS is considered a promising approach to neuroprotection. Those constituents of saffron, which have antioxidant properties (Premkumar *et al.*, 2003; Ochiai *et al.*, 2004a), should be candidates for treating neurodegenerative disorders caused by oxidative stress.

Saffron is now used worldwide in folk medicine and is reputed to be useful in treating various human disorders such as heart and blood disorders (Abdullaev, 1993; Rios *et al.*, 1996; Wuthrich *et al.*, 1997; Abdullaev, 2002). However, scientific evidence of saffron's therapeutic potential, safety and the biological properties of each constituent is still lacking. Understanding the regulation of apoptosis in neurons by intracellular pathways is currently a subject of great interest, because of apoptosis role in various neurodegenerative disorders. We have previously shown that saffron's active component, crocin, prevents the death of PC-12 cells when death is induced by tumor necrosis factor (TNF)-α: crocin suppresses the TNF-α-induced expression of latent ICE and Bcl-X_S mRNAs and simultaneously restores the cytokine-induced reduction of Bcl-X_L mRNA expression. The modulating effects of crocin on the expression of Bcl-2 family proteins led to a marked reduction of a TNF-α-induced release of cytochrome c from mitochondria and thereby to inactivation of caspase-3 (Soeda *et al.*, 2001). This review discusses more recent data from our laboratories on the neuroprotective effects of crocin *in vitro*. Special emphasis is placed on its action in PC-12 cells.

Crocin's Neuroprotective Effects

Crocin Prevents the Death of Neuronal Cells by Antioxidant Effects Stronger than those of α-tocopherol

We report here the effects of crocin on PC-12 cells deprived of serum/glucose in comparison with those of α-tocopherol (Ochiai *et al.*, 2004a). Figure 1A and B give the chemical structures of crocin and α-tocopherol, respectively. Depriving the PC-12 cells of serum/glucose caused changes in the morphology and peroxidation of their membrane lipids, and decreased intracellular SOD activity. Figure 2 shows Annexin V staining of the PC-12 cells kept for 3 h in serum/glucose-free culture medium. Although phosphatidylserine (PS) residues are normally present in the inner membrane, oxidative stress moves them to the outer membrane leaflet. PS externalization is an early sign of apoptotic induction. Annexin binds to the negatively charged PS, and the conjugated FITC shows a ring-like stain along the cellular boundary. The cells deprived of serum/glucose show strong ring-like stains (Figure 2B) compared to the control cells (Figure 2A). Figure 2C and D show the effects of 10 μM crocin and α-tocopherol on the serum/glucose-deprived cells. Crocin kept the cell's morphology more intact than α-tocopherol did. In PC-12 cells deprived of serum/glucose for 6 h, the peroxidized membrane lipids increased 1.8-fold over those in the control cells, and SOD activity decreased to 14% of that in the control cells. Crocin significantly decreased the formation of peroxidized membrane lipids and restored the SOD activity, but α-tocopherol did not. The restoration of SOD activity suggests crocin has an important role in the antioxidative effects. Crocin also suppressed the activation of caspase-8 caused by serum/glucose deprivation; the activation was suppressed in a concentration-dependent manner (0.1–10 μM). Crocin did not inhibit caspase-8 activity in the cell lysates, so its inhibitory effect may be caused indirectly by the antioxidant activity.

Vitamin E occurs in nature in eight different forms: α-, β-, γ- and δ-tocopherols and α-, β-, γ- and δ-tocotrienols. It is generally accepted that vitamin E has two major roles: acting as an antioxidant and a stabilizer of cell membranes (Urano, 1993). Vitamin E suppresses the generation of free radicals by terminating the propagation of radical chain reactions (Niki, 1993) and can block the death of PC-12 cells in the presence of anti-SOD antibodies (Troy & Shelanski, 1994). Biochemically, the hydroxyl group in the vitamin E molecule may donate a proton to saturate and detoxify the unpaired electron (Behl & Moosmann, 2002). In AD models, vitamin E prevents the accumulation of oxidative metabolites which induce amyloid

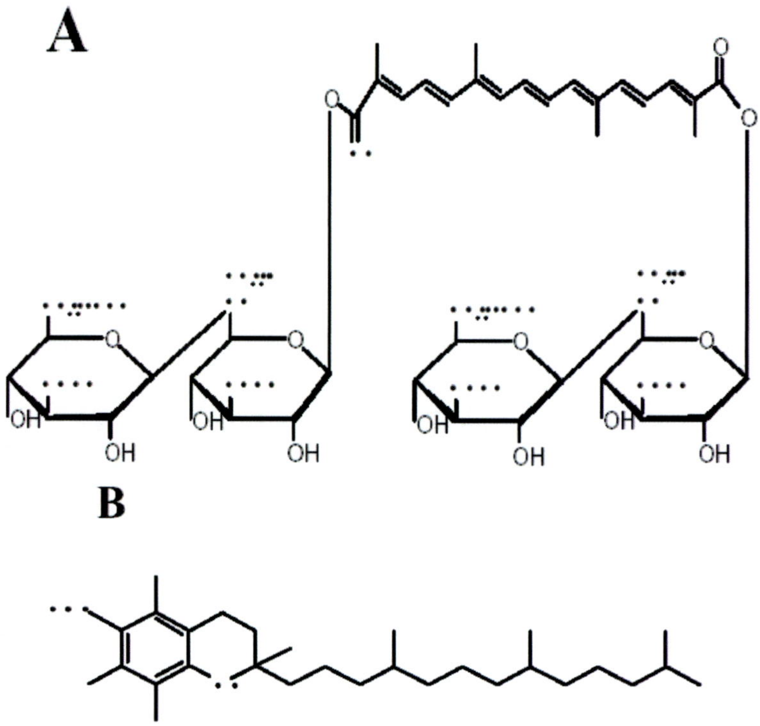

Figure 1. Chemical structures of crocin (A) and α-tocopherol (B).

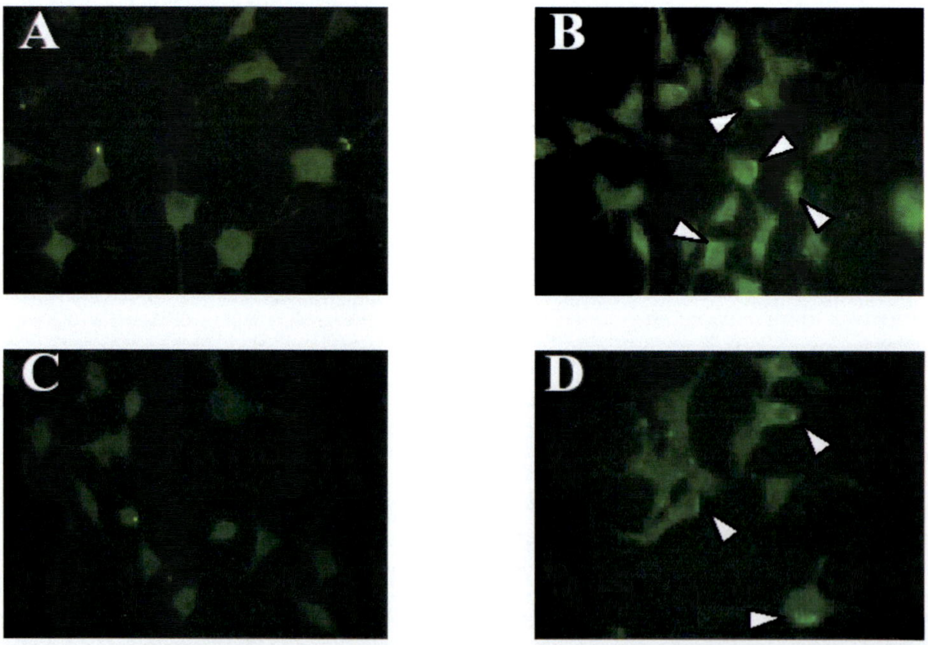

Figure 2. Annexin V staining of PC-12 cells exposed for 3 h in serum/glucose-deprived Dulbecco's modified Eagle's medium (DMEM). (A) control cells in serum/glucose (+) DMEM; (B) cells in (–) DMEM alone; (C) cells in (–) DMEM plus 10 μM crocin; (D) cells in (–) DMEM plus 10 μM α-tocopherol. Arrows also indicate ring-like stains.

β protein toxicity (Behl *et al.*, 1992; Behl *et al.*, 1994; Harris *et al.*, 1995). Based on the protective activity of vitamin E against amyloid β protein toxicity, it has been used in clinical trials for the treatment of AD (Sano *et al.*, 1997). In a multicenter, double-blind, placebo-controlled study on 341 patients with moderately severe AD, a daily dose of approximately 1,350 mg α-tocopherol led to a slight but, nevertheless, significant delay in reaching institutionalization or death. These results suggest that oxidative events play an imortant role in initiating early neurodegenerative processes in AD. Recently, Mishima *et al.* (2003) have demonstrated that α-tocotrienol is more potent as an antioxidant than α-tocopherol in the prevention of cerebral infarction in mice. The key difference in the structures is the three double bonds in α-tocotrienol's lipophilic tail. If antioxidant activity varies directly as the number of double bonds, then crocin, which has seven double bonds, should be an effective antioxidant. Oral administration of saffron's extract at concentrations from 0.1 to 5 g/kg was nontoxic in mice (Abdullaev, 2002). Crocin prevented cerebral infarction in mice (Ochiai *et al.*, unpublished data), as did α-tocotrienol and γ-tocopherol. In summary, crocin is an antioxidant more potent than α-tocopherol against oxidative stress-induced death of neurons *in vitro*. Our data indicate that crocin also combats oxidative stress induced by H_2O_2: approximately 90% of PC-12 cells died by exposing them to 100 μM H_2O_2 for 24 h, while 60% of the cells kept alive by adding 10 μM crocin. These results suggest that crocin has the ability to suppress ROS generated by various oxidative stress.

Figure 3 shows the immunochemically detected movement of crocin in PC-12 cells treated with crocin at intervals from 0 to 60 min. The left and right photographs show the localizations of crocin at 0 and 30 min, respectively. The result suggests that most of the crocin localized in cell membranes as does α-tocopherol but unlike α-tocopherol, crocin is incorporated into the cells in a time-dependent manner (Ochiai *et al.*, 2004a). The exact role of the crocin which is incorporated into the cells remains unknown. However, our previous study suggests that crocin can block cell death induced by intracellular apoptotic stimuli. Daunorubicin, an anthracycline antibiotic, induces apoptosis in cells by activating ceramide synthase (Bose *et al.*, 1995) or sphingomyelinase (SMase) (Jaffrezou *et al.*, 1996). In contrast to TNF-α signaling through the membrane receptors, therefore, daunorubicin is capable of triggering cell death within cells. PC-12 cells treated with daunorubicin for 30 min and incubated in the serum-free medium showed an increased release of mitochondrial cytochrome c and followed by apoptotic cell death. But, crocin inhibited the incorporated daunorubicin-induced release of cytochrome c and cell death (Soeda *et al.*, 2001). This finding provides evidence that crocin is partly incorporated into cells and capable of combating intracellular death signaling.

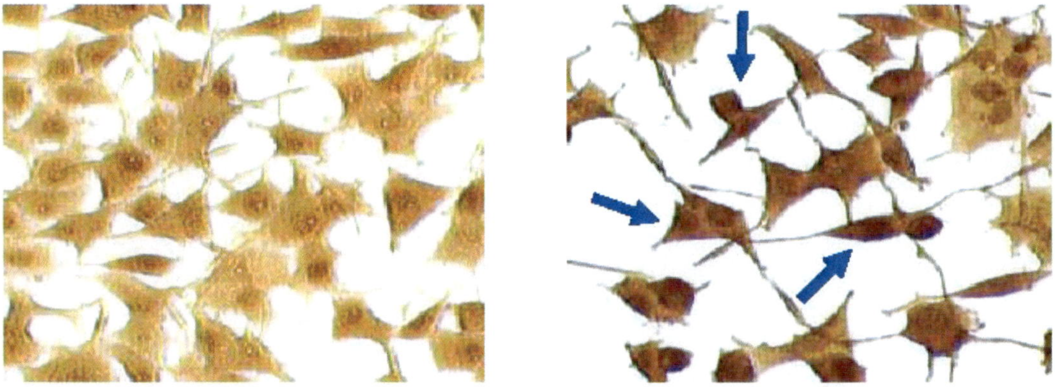

Figure 3. Treatment of PC-12 cells with 10 μM crocin and their immunostaining with crocin antibodies. The left and right photographs indicate the cells at 0 and 30 min, respectively. Arrows also indicate the cells stained strongly with the antibodies.

Crocin Increases Glutathione Synthesis which Prevents Neuron Death through sphIngomyelinase-ceramide Signaling

GSH is a major non-protein antioxidant that protects mammalian cells from oxidative stress; depletion of GSH is thought to precede the onset of apoptosis induced by various agents (Beaver & Waring, 1995; Liu *et al.*, 1998). A drop in GSH levels may activate magnesium-dependent nSMase, thus causing a rapid increase in intracellular ceramide levels resulting in cell death (Liu & Hannun, 1997; Liu *et al.*, 1998; Yoshimura *et al.*, 1999). Ceramide stimulates the stress-activated protein kinase/c-jun kinase (SAPK/JNK) signaling system that precedes induction of apoptosis (Verheij *et al.*, 1996). Willaime-Morawek *et al.* (2003) have further investigated the role of the JNK pathway in ceramide-induced cell death and demonstrated that p38 kinase and JNK/c-Jun act in parallel to induce neuronal apoptosis through ceramide. Several types of SMases which are distinguished by their optimum pH, cellular localization and ion dependence have been identified; these include lysosomal and secreted acidic SMases (aSMase) and cytosolic, magnesium-independent nSMase (Levade & Jaffrezou, 1999; Perry, 1999; Goñi & Alonso, 2002). However, the membrane-bound, magnesium-dependent nSMase may be the SMase most closely linked to the initial production of ceramide induced by depletion of GSH (Liu & Hannun, 1997; Liu *et al.*, 1998; Yoshimura *et al.*, 1999; Lavrentiadou *et al.*, 2001).

GSH is synthesized in the γ-glutamyl cycle, in which γ-glutamylcysteinyl synthase (γ-GCS) is the rate-limiting enzyme (Griffith *et al.*, 1979; Pan & Perez-Polo, 1993). Glutathione peroxidase (GPx) removes H_2O_2 and organic peroxides by converting GSH to the oxidized form (GSSG), whereas glutathione reductase (GR) regenerates GSH in the presence of NADPH. Nerve growth factor (NGF) was shown to prevent the death of PC-12 cells from oxidative stress by increasing the γ-GCS activity (Pan & Perez-Polo, 1993). Recent evidence suggests that interleukin-6 (IL-6) also has the ability to prevent PC-12 cell death by increasing the expression of γ-GCS mRNA, which is followed by an increase in cellular GSH levels (Nakajima *et al.*, 2002). The roles of these neurotrophic factors suggest that chemicals which enable cells to produce enough GSH to prevent oxidative stress may be clinically useful.

In the above section, we showed that crocin acts as a unique and potent antioxidant that combats oxidative stress in PC-12 cells deprived of serum/glucose. Here, we describe the effect of crocin on the nSMase-ceramide signalimg pathway in the serum/glucose-deprived cells (Ochiai *et al.* 2004b). Crocin enhances the activities of GR and γ-GCS, contributing to a stable GSH supply that blocks the activation of nSMase. Increased GSH levels also prevent the activation of JNK pathway, which has a role of the signaling cascade downstream ceramide for neuronal cell death.

Serum (Oppenheim, 1991; Batistatou & Green, 1991) or NGF (Mesner *et al.*, 1992; Pittman *et al.*, 1993) deprivation is reported to induce the death of PC-12 cells. A recent study by Colombaioni *et al.* (2002) found that serum deprivation increases intracellular ceramide levels in undifferentiated HN9.10e cells and induces their apoptosis. These findings suggest that ceramide levels in PC-12 cells may increase in the absence of serum/glucose. In our studies, PC-12 cells kept in serum/glucose-free medium for 3 h showed a 3.5-fold increase in ceramide levels above the basal level, while the addition of 10 μM crocin caused a significant decrease in ceramide levels. Fumonisin B_1 in concentrations from 10–30 μM inhibits *de novo* ceramide synthesis in cells (Wang *et al.*, 1991; Merrill *et al.*, 1993). This ceramide synthase inhibitor had no significant effect on the ceramide levels, suggesting that the accumulation of ceramide following a 3-h exposure to serum/glucose-free conditions is not caused by an enhancement of *de novo* synthesis. During stress-induced cell death in U937 cells and BAE cells, the sphingomyelin pathway and SAPK/JNK signaling systems are coordinated (Verheij *et al.*, 1996). Figure 4A and B show the effect of crocin on JNK phosphorylation (activation). Deprivation of serum/glucose increases JNK phosphorylation to approximately 3.7 times that in the control cells. The presence of 10 μM crocin prevented the increase in phosphorylated JNK. These results suggest that crocin inhibits ceramide accumulation in PC-12 cells, and thereby contributes to the inhibition of ceramide-induced activation of JNK that leads to cell death. The ceramide that was formed early (within 3 h) may be the product of sphingomyelin hydrolysis by magnesium-dependent nSMase which occurs during hypoxic PC-12 cell death (Yoshimura *et al.*, 1998, 1999).

To confirm the origin of the accumulated ceramide, we measured the nSMase activity present on the membrane of PC-12 cells, using 2-hexadecanoylamino-4-nitrophenylphosphorylcholine as a substrate. The nSMase activity in the cells exposed to serum/glucose-free conditions peaked at 1 h after the start of starvation and decreased to a level close to the control cell's at 3 h. On the other hand, under these conditions, the activity of aSMase in the cells was unaffected for at least 3 h. The presence of 0.1–10 μM crocin in the culture medium suppressed the enhancement of nSMase activities in a dose-dependent manner: however, direct addition of crocin to the reaction mixture had no inhibitory effect on nSMase activity (Ochiai *et al.*, 2004b). The addition of 1 or 10 mM GSH, which is reported to be a physiological inhibitor of magnesium-dependent nSMase (Liu & Hannun, 1997; Liu *et al.*, 1998; Yoshimura *et al.*, 1999), inhibited the enzyme activity. In cells, nSMase is inactive in the presence of physiological concentrations (1–20 mM) of GSH, but aSMase is not (Liu & Hannun, 1997). Therefore, our results suggest that the nSMase activity in the

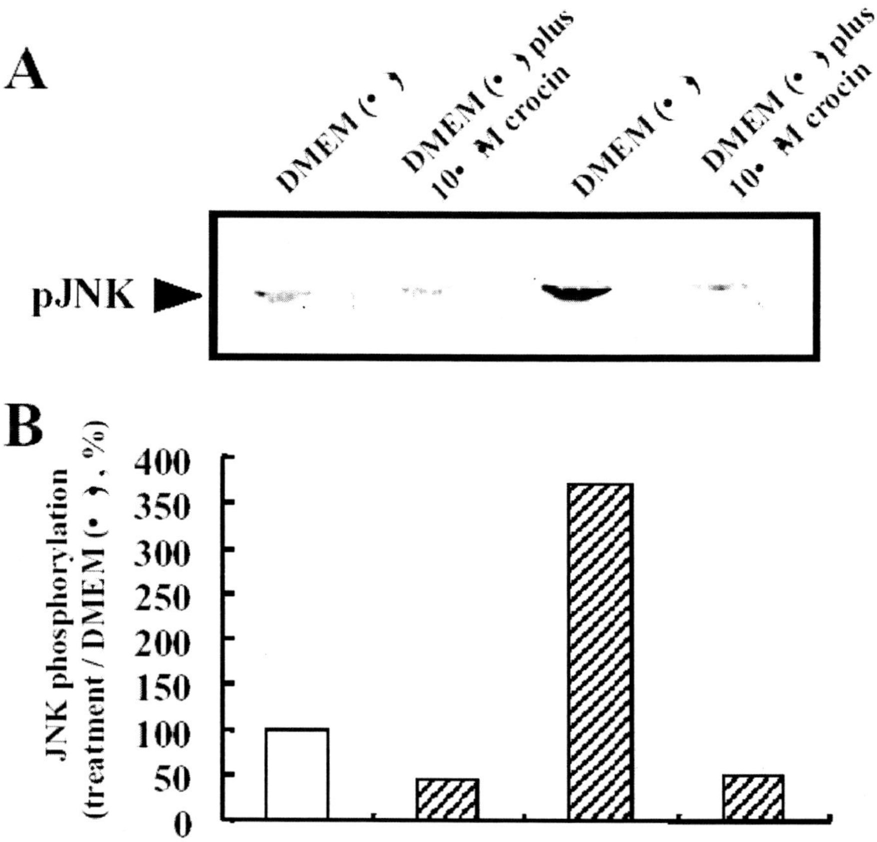

Figure 4. The effect of crocin on the phosphorylation of JNK in serum/glucose-deprived PC-12 cells. The phosphorylated JNK was made visible by Western blotting (A) and the bands were quantified using an NIH Imager (B). JNK in PC-12 cells incubated in serum/glucose (+) DMEM was used as the control.

reaction medium is caused by magnesium-dependent nSMase present in plasma membranes and that the observed nSMase inhibition by crocin is not caused by its direct action to the enzyme. We hypothesize that crocin might prevent the activation of nSMase in serum/glucose-derived PC-12 cells by a GSH-dependent mechanism of inhibition. As shown in Figure 5, the GSH levels in PC-12 cells exposed for 3 h to serum/glucose-free medium decreased by half compared with levels in the control cells. However, the addition of crocin to the medium increased the intracellular GSH levels in a dose-dependent manner and kept levels higher than normal for 3 h. We mathematically converted the total amount of GSH in the presence of 10 μM crocin at 1 h (15.4 ± 0.01 μmol/mg protein) into a final concentration of 9 mM GSH in the cells. The quantity of GSH was enough to inactivate the nSMase. We next investigated how crocin increased the GSH levels. The GR activities in serum/glucose-deprived PC-12 cells decreased in a time-dependent manner, whereas the co-presence of 10 μM crocin enhanced the GR activity each hour (approximately, four-fold elevation at 6 h). However, crocin had no significant effect on the GPx activity in the cells. This suggests that crocin increases GSH levels by promoting the conversion of GSSG to GSH.

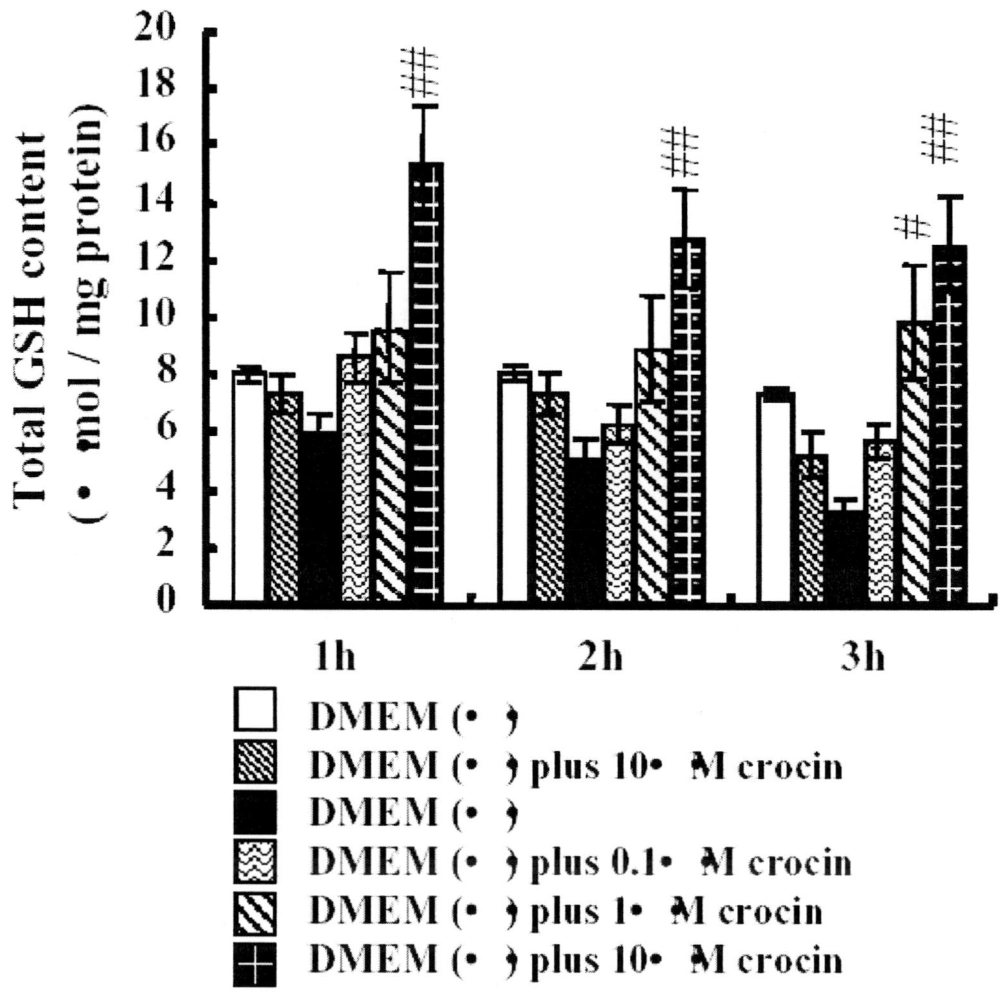

Figure 5. The effect of crocin on the intracellular GSH in serum/glucose-deprived PC-12 cells. Each bar represents the mean ± S.D. of three experiments. $^{\#}P< 0.05$, $^{\#\#}P< 0.01$, compared to (–) DMEM alone.

GSH synthesis is regulated by the rate limiting enzyme γ-GCS. This enzyme is thought to be regulated by several mechanisms. In mouse endothelial cells, TNF-α or IL-1β-induced increase in γ-GCS activity is associated with an increase in the mRNA expression (Urata *et al.*, 1996). In PC-12 cells, IL-6 stimulates the expression of γ-GCS mRNA and increases the enzyme activity, which leads to increased GSH levels (Nakajima *et al.*, 2002). In contrast, Pan & Perez-Polo (1993) reported that NGF had no ability to increase the activity of γ-GCS at the transcription level; rather it extended the half-life of γ-GCS mRNA. Figure 6 shows the effect of crocin on the mRNA expression and activity of γ-GCS in PC-12 cells exposed to serum/glucose-rich or -free medium for 6 h. Crocin (10 μM) doubled the γ-GCS mRNA expression in PC-12 cells in serum/glucose-free medium, while it had no effect on the mRNA levels in the control cells (Figure 6A and B). As shown in Figure 6C, the crocin-induced increase in γ-GCS mRNA expression is reflected in an increase in the enzyme activity of the

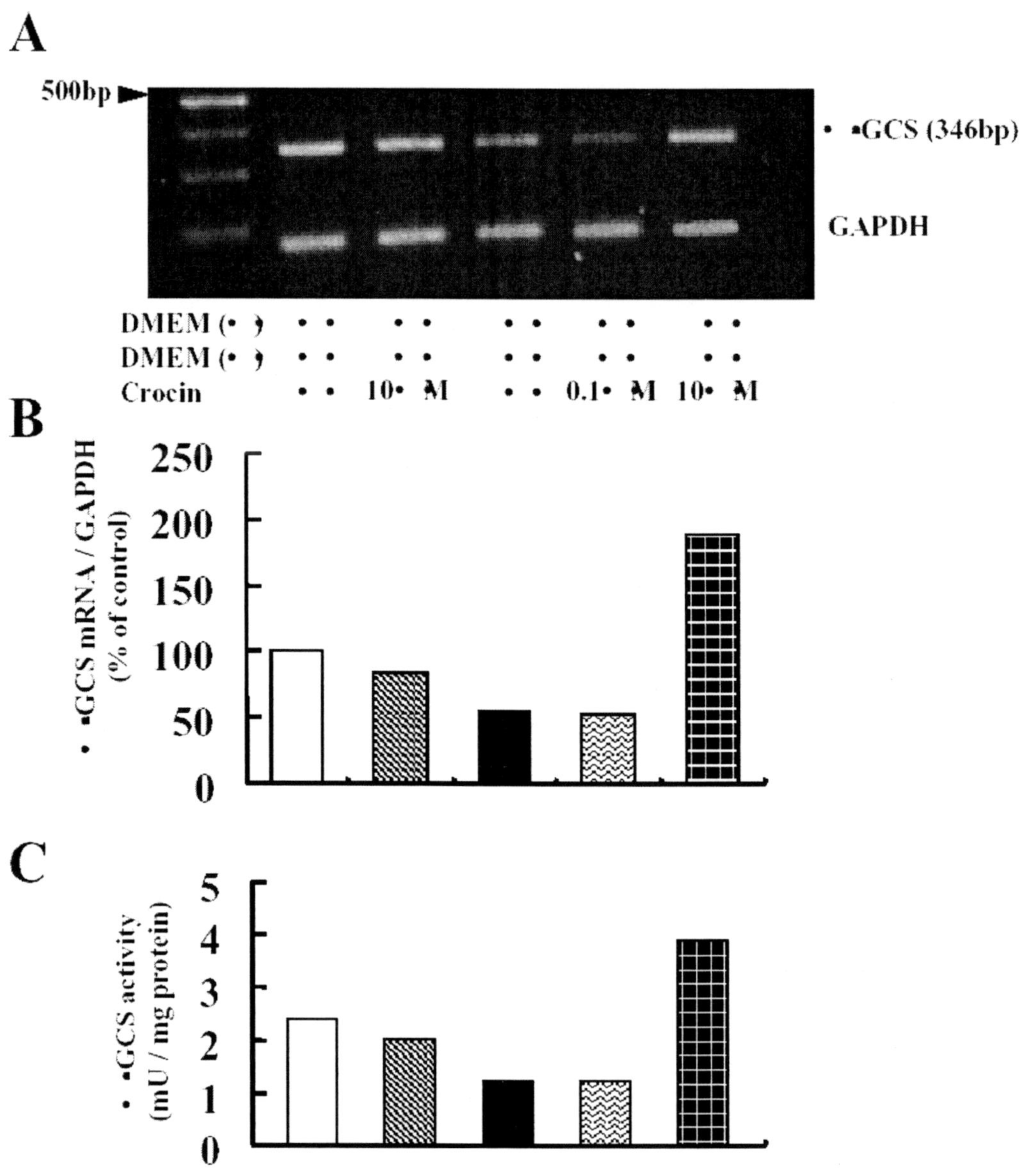

Figure 6. The effect of crocin on the mRNA expression (A and B) and activity (C) of γ-GCS in serum/glucose-deprived PC-12 cells. γ-GCS and GAPDH mRNA levels in the cells were evaluated by RT-PCR. The bands were quantified using an NIH Imager. The enzyme activity was measured as described by Ray et al. (1999).

cells. These results suggest that crocin can increase intracellular GSH levels by increasing the activities of both GR and γ-GCS. GSH levels play a key role in the crocin's survival-promoting effects. Serum/glucose-deprived PC-12 cells were treated with crocin and buthionine sulfoximine for 6 h. This chemical is a widely used inhibitor of GSH synthesis (Griffith *et al.*, 1979). The data show that the inhibitor can reverse crocin's anti-apoptotic effect (Ochiai *et al.*, 2004b). In conclusion, crocin can prevent oxidative stress-mediated depletion of GSH in cells by promoting both GSH biosynthesis and conversion of GSSG to GSH. The GSH-dependent pathway involves activation of nSMase and appears at least to be responsible for the earlier ceramide response to induction of apoptosis, as evidenced by the presented crocin's action.

The Role of Sphingomyelinase-ceramide Signaling in Neuron Death Caused by Ischemic Stress

We have investigated the effect of crocin on an infarcted area caused by occlusion of the middle cerebral artery (MCA) in ddY mice. Crocin (10 mg/kg i.v. each time) was administrated immediately before and 3 h after the MCA occlusion. The administration of crocin reduced the infarct volume significantly (Ochiai *et al.*, unpublished data). The result suggests that crocin can prevent neuron death caused by ischemic stress: however, whether it prevented the neuronal cell death by GSH-dependent inhibition of nSMase activity followed by ceramide formation remains unknown. Recently, we have designed and synthesized sphingomyelin difluoromethylene analogues (SMAs) which inhibit nSMase in bovine brain microsomes with IC_{50} values of 3.3–377 μM (Yokomatsu *et al.*, 2001, 2003). As shown in Figure 7, administration of SMA-7 (10 mg/kg i.v.) with IC_{50} =3.3 reduced the size of the cerebral infarcts significantly, compared to the control mice (Soeda *et al.*, 2004). These results suggest that inhibiting or stopping nSMase activity is a key strategy to prevent neuronal cell death caused by cerebral ischemia and that crocin's prevention of developing cerebral ischemia may be largely due to its inhibiting nSMase.

In contrast, Yu *et al.* (2000) have reported that transient focal cerebral ischemia in mice induces an increase in aSMase activity. In mice lacking aSMase, ischemia-induced brain tissue damage decreased, and the behavioral outcome improved. The authors suggest that inhibitors of aSMase may prove beneficial in stroke patients. However, the central nervous system involves predominantly membrane-bound nSMases (Bernardo *et al.*, 2000), and the genes producing these enzymes have recently been cloned (Tomiuk *et al.*, 1998; Hofmann *et al.*, 2000). The exact roles of both aSMase and nSMase during ischemic injury remain an open issue. It is conceivable that stress signaling may be initiated by nSMase-catalyzed hydrolysis of the plasma membrane sphingomyelin and later achieved by the activation of lysosomal/endosomal aSMase that leads to cell-death. Interestingly, our synthesized SMA-7 has the same inhibitory effect on both aSMase and nSMase in the brain (Yokomatsu *et al.*, 2001). The ability of SMA-7 to prevent cerebral ischemia induced by MCA occlusion in mice may have been caused by its ability to inhibit both nSMase and aSMase at the same level of effectiveness. In our previous experiments using PC-12 cells, however, the activity of aSMase in the cells was unaffected by serum/glucose deprivation for at least 3 h. Further studies are needed to clarify the roles of both aSMase and nSMase in ischemic neuron death and the effect of crocin on the ceramide signaling induced specifically by activation of aSMase.

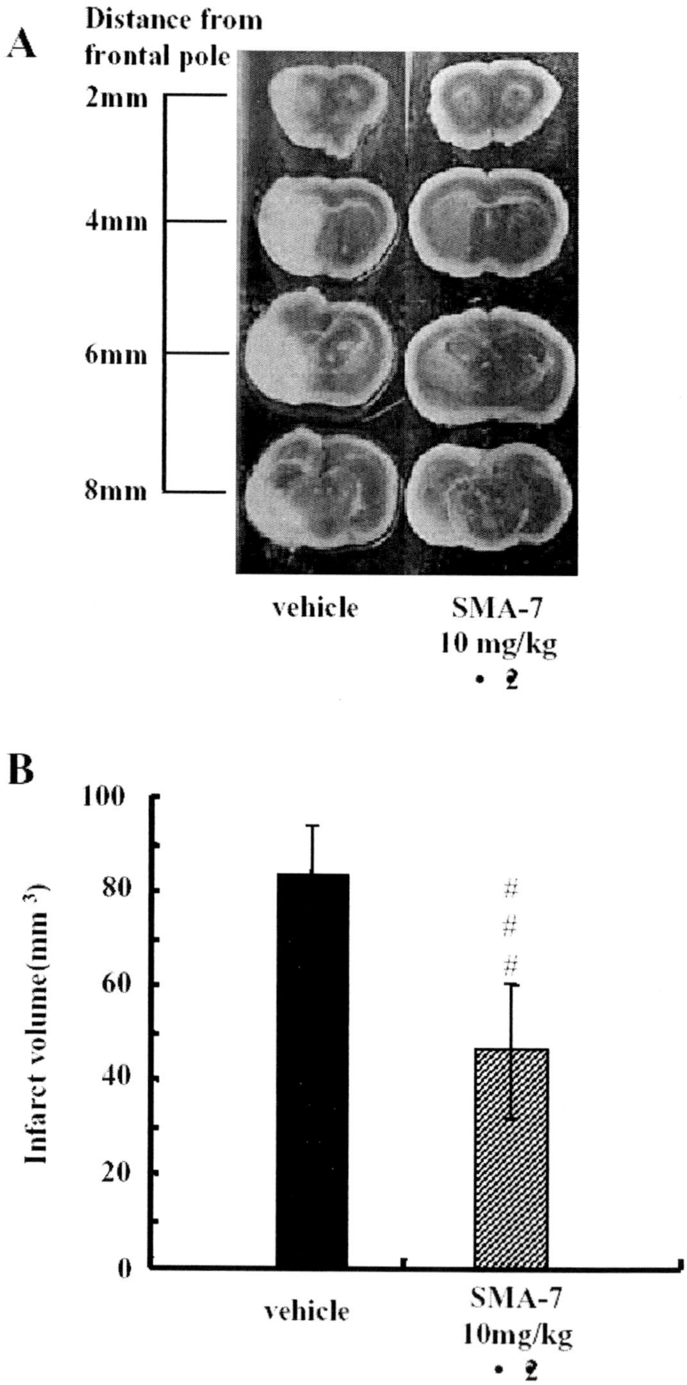

Figure 7. The effect of SMA-7 on the infarct volume induced by MCA occlusion in ddY mice. SMA-7 was injected i.v. immediately before and 3 h after the MCA occlusion. Slices were immediately stained with 2% 2,3,5-triphenyltetrazolium chloride. (A) Representative infarcted area caused in each group of mice. (B) Each bar represents the mean ± S.E. of five to seven mouse brains. ###$P< 0.001$, compared to the control, vehicle alone.

Conclusions and Future Perspectives

In folk medicine, as well as in the modern pharmacy, saffron has been reported to be useful in the treatment of many human diseases including insomnia and paralysis (Abdullaev et al., 1993; Rios et al., 1996; Wuthrich et al., 1997). We found that the saffron's major component crocin has antioxidant effects stronger than those of α-tocopherol and reduces neuron death by increasing GSH synthesis. To our knowledge, this is a novel pharmacological activity found in this medicinal plant. It is well established that oxidative modifications of cellular molecules are related to neurodegenerative disorders. With respect to AD, two factors are believed to contribute to neuronal dysfunction and degeneration in AD: one is increased oxidative stress and another is increased production of neurotoxic forms of amyloid β-peptide (Mattson, 1997). Although the pathogenesis of AD is tightly linked to amyloid β-peptide deposition and oxidative stress, it remains unclear how these factors result in neuronal dysfunction and death. Recently, Cutler *et al.* (2004) have demonstrated that treating neurons with α-tocopherol or an inhibitor of sphingomyelin synthesis prevents accumulation of ceramide and cholesterol and protects them against death induced by amyloid β-peptide. The finding suggests a sequence of events in the pathogenesis of AD: amyloid β-peptide induces membrane-associated oxidative stress, resulting in perturbed ceramide and cholesterol metabolism which in turn triggers a neurodegenerative cascade. In fact, the risk of AD is affected by inheritance of different isoforms of apolipoprotein E, and changes in cholesterol metabolism can affect the production of amyloid β-peptide (Puglielli *et al.*, 2001; Runz *et al.*, 2002). Drugs that lower cholesterol levels may reduce the risk of AD (Fassbender *et al.*, 2001). On the other hand, Lee *et al.* (2004) have demonstrated that amyloid β-peptide induces oligodendrocyte apoptosis *via* an oxidative mechanism. The molecular mechanisms are: activation of nSMase, but not aSMase by amyloid β-peptide and followed by production of ceramide that induces apoptosis, because the inhibition of nSMase by 3-*O*-methyl-sphingomyelin or by gene knockdown using antisense oligonucleotides attenuates the cell death. Interestingly, GSH precursors such as N-acetylcysteine inhibit the activation of nSMase by amyloid β-peptide and prevent oligodendrocyte death. Thus, crocin, which prevents the activation of nSMase-ceramide signaling pathway by increasing GSH synthesis and also has antioxidant properties more potent than those of α-tocopherol, may be an attractive drug candidate for retarding the amyloid β-peptide-induced neurodegenerative process.

α-Tocopherol is still one of the most potent antioxidants for AD prevention and therapy. However, the great pharmacological disadvantage of this antioxidant is its very limited passage through the blood-brain barrier (Vatassery, 1998). Blood-brain barrier permeability is one of the major hurdles in developing neuroprotective antioxidants (Gilgun-Sherki *et al.*, 2001). In our animal experiments, administration of crocin reduced significantly the infarct volume induced by MCA occlusion, suggesting that the chemical passes through the blood-brain barrier in mice and prevents neuron death caused by ischemic stress. Nevertheless, the activity of the natural product crocin may still be the starting point for a future drug design.

Acknowledgements

This work was supported in part by funds from the Central Research Institute of Fukuoka University and from the Advanced Materials Institute of Fukuoka University. Our thanks to Mr. Steven Sabotta for reading the manuscript.

References

Abdullaev, F.I. (1993). Biological effects of saffron. *BioFactors*, *4*: 83–86.

Abdullaev, F.I. (2002). Cancer chemopreventive and tumoricidal properties of saffron (Crocus sativus L.). *Exp. Biol. Med.*, *227*: 20–25.

Abe, K. & Saito, H. (2000). Effects of saffron extract and its constituent crocin on learning behaviour and long-term potentiation. *Phytother. Res.*, *14*: 149–152.

Batistatou, A. & Green, L.A. (1991). Aurintricarboxylic acid rescues PC12 cells and sympathetic neurons from cell death caused by nerve growth factor deprivation: correlation with suppression of endonuclease activity. *J. Cell Biol.*, *115*: 461–471.

Beaver, J.P. & Waring, P. (1995). A decrease in intracellular glutathione concentration precedes the onset of apoptosis in murine thymocytes. *Eur. J. Cell. Biol.*, *68*: 47–54.

Behl, C., Davies, J.B., Cole, G.M. & Schubert, D. (1992). Vitamin E protects nerve cells from amyloid β protein toxicity. *Biochem. Biophys. Res. Commun.*, *186*: 944–950.

Behl, C., Davies, JB., Lesley, R. & Schubert, D. (1994). Hydrogen peroxide mediates amyloid β protein toxicity. *Cell*, *77*: 817–827.

Bernardo, K., Krut, O., Wiegmann, K., Kreder, D., Micheli, M., Schäfer, R., Sickman, A., Schmidt, W.E., Schroder, J.M., Meyer, H.E., Sandhoff, K. & Krone, M. (2000). Purification and characterization of a magnesium-dependent neutral sphingomyelinase from bovine brain. *J. Biol. Chem.*, *275*: 7641–7647.

Behl, C. & Moosmann, B. (2002). Antioxidant neuroprotection in Alzheimer's disease as preventive and therapeutic approach. *Free Radic. Biol. Med.*, *33*: 182–191.

Bose, R., Verheji, M., Haimovitz-Friedman, A., Scotto, K., Fuks, Z. & Kolesnick, R. (1995). Ceramide synthase mediates daunorubicin-induced apoptosis: an alternative mechanism for generating death signals. *Cell*, *82*: 405–414.

Cao, W., Carney, J.M., Duchon, A., Floyd, R.A. & Chevion, M. (1988). Oxygen free radical involvement in ischemia and reperfusion injury to brain. *Neurosci. Lett.*, *88*: 233–238.

Colombaioni, L., Frago, L.M., Varela-Nieto, I., Pesi, R. & Garcia-Gil, M. (2002). Serum deprivation increases ceramide levels and induces apoptosis in undifferentiated HN9.10e cells. *Neurochem. Int.*, *40*: 327–336.

Cutler, R.G., Kelly, J., Storie, K., Pedersen, W.A., Tammara, A., Hatanpaa, K., Troncoso, J.C. & Mattson, M.P. (2004). Involvement of oxidative stress-induced abnormalities in ceramide and cholesterol metabolism in brain aging and Alzheimer's disease. *Proc. Natl. Acad. Sci. USA*, *101*: 2070–2075.

Fassbender, K., Simons, M., Bergmann, C., Stroick, M., Lutjohann, D., Keller, P., Runz, H., Kuhl, S., Bertsch, T., von Bergmann, K., Hennerici, M., Beyreuther, K. & Hartmann, T.

(2001). Simvastatin strongly reduces levels of Alzheimer's disease β-amyloid peptides Aβ 42 and Aβ 40 in vitro and in vivo. *Proc. Natl. Acad. Sci. USA*, *98*: 5856–5861.

Garcia-Olmo, D.C., Riese, H.H., Escribano, J., Ontanon, J. Fernandez, J.A., Atienzar, M. & Garcia-Olmo, D. (1999). Effects of long-term treatment of colon adenocarcinoma with crocin, a carotenoid from saffron (Crocus sativus L.): an experimental study in the rat. *Nutr. Cancer*, *35*: 120–126.

Gilgun-Sherki, Y., Melamed, E. & Offen, D. (2001). Oxidative stress-induced neurodegenerative diseases: the need for antioxidants that penetrate the blood-brain barrier. *Neuropharmacology*, *40*: 959–975.

Goñi, F.M. & Alonso, A. (2002). Sphingomyelinases: enzymology and membrane activity. *FEBS Lett.*, *531*: 38–46.

Griffith, O.W., Anderson, M.E. & Meister, A. (1979). Inhibition of glutathione biosynthesis by prothionine sulfoximine (*S-n*-propyl homocysteine sulfoximine), a selective inhibitor of γ-glutamylcysteine synthase. *J. Biol. Chem.*, *254*: 1205–1210.

Harris, M.E., Hensley, K., Butterfield, A., Leedle, R.A. & Carney, J.M. (1995). Direct evidence of oxidative injury produced by the Alzheimer's β-amiloid peptide (1–40) in cultured hippocampal neurons. *Exp. Neurol.*, *131*: 193–202.

Hofmann, K., Tomiuk, S., Wolff, G. & Stoffel, W. (2000). Cloning and characterization of the mammalian brain-specific, Mg^{2+}-dependent neutral sphingomyelinase. *Proc. Natl. Acad. Sci. USA*, *97*: 5895–5900.

Hosseinzadeh, H. & Younesi, H.M. (2002). Antinociceptive and anti-inflammatory effects of Crocus sativus L. stigma and petal extracts in mice. *BMC Pharmacol.*, *2*: 1–8.

Jaffrezou, J.P., Levade, T., Bettaieb, A., Andrieu, N., Bezombes, C., Maestre, N., Vermeetsch, S., Rousse, A. & Laurent, G. (1996). Daunorubicin-induced apoptosis: triggering of ceramide generation through sphingomyelin hydrolysis. *EMBO J.*, *15*: 2417–2424.

Lavrentiadou, S.N., Chan, C., Kawcak, T.-N., Ravid, T., Tsaba, A., van der Vliet, A., Rasooly, R. & Goldkorn, T. (2001). Ceramide-mediated apoptosis in lung epithelial cells is regulated by glutathione. *Am. J. Respir. Cell. Mol. Biol.*, *25*: 676–684.

Lee, J.-T., Xu, J., Lee, J.-M., Ku, G., Han, X., Yang, D.-I., Chen, S. & Hsu, C.Y. (2004). Amyloid β-peptide induces oligodendrocyte death by activating the neutral sphingomyelinase-ceramide pathway. J. Cell Biol., *164*: 123–131.

Levade, T. & Jaffrezou, J.P. (1999). Signalling sphingomyelinases: which, where, how and why? *Biochim. Biophys. Acta*, *1438*: 1–17.

Liu, B., Andrieu-Abadie, N., Levade, T., Zhang, P., Obeid, L.M. & Hannun, Y.A. (1998). Glutathione regulation of neutral sphingomyelinase in tumor necrosis factor-α-induced cell death. *J. Biol. Chem.*, *273*: 11313–11320.

Liu, B. & Hannun, Y.A. (1997). Inhibition of the neutral magnesium-dependent sphingomyelinase by glutathione. *J. Biol. Chem.*, *272*: 16281–16287.

Mattson, M.P. (1997). Celluar actions of β-amyloid precursor protein and its soluble and fibrillogenic derivatives. Physiol. Rev., *77*: 1081–1132.

Merrill Jr., A.H., van Echten, G., Wang, E. & Sandhoff, K. (1993). Fumonisin B1 inhibits sphingosine (sphinganine) N-acyltransferase and de novo sphingolipid biosynthesis in cultured neurons in situ. *J. Biol. Chem.*, *268*: 27299–27306.

Mesner, P.W., Winter, T.R. & Green, S.H. (1992). Nerve growth factor withdrawal-induced cell death in neuronal PC12 cells resembles that in sympathetic neurons. *J. Cell Biol.*, *119*: 1669–1680.

Mishima, K., Tanaka, T., Pu, F., Egashira, N., Iwasaki, K., Hidaka, R., Matsunaga K., Takada, J., Karube, Y. & Fujiwara, M. (2003). Vitamin E isoforms α-tocotrienol and γ-tocopherol prevent cerebral infarction in mice. *Neurosci. Lett.*, *337*: 56–60.

Molnar, J., Szabo, D., Pusztai, R., Mucsi, I., Berek, L., Ocsovszki, I., Kawata, E. & Shoyama, Y. (2000). Membrane associated antitumor effects of crocine-, ginsenoside and cannabinoid derivatives. *Anticancer Res.*, *20*: 861–867.

Nakajima, A., Yamada, K., Zou, L.-B., Yan, Y., Mizuno, M. & Nabeshima, T. (2002). Interleukin-6 protects PC12 cells from 4-hydroxynonenal-induced cytotoxicity by increasing intracellular glutathione levels. *Free Radic. Biol. Med.*, *32*: 1324–1332.

Niki, E. (1993). Function of vitamin E as antioxidant in the membranes. In: *Vitamin E-Its Usefulness in Health and in Curing Disease*, Mino, M., Nakamura, H., Diplock, A.T. & Kayden, H.J. (eds.), pp. 23–30. Japan Sci. Soc. Press, Tokyo.

Ochiai, T., Ohno, S., Soeda, S., Tanaka, H., Shoyama, Y. & Shimeno, H. (2004a). Crocin prevents the death of rat pheochromocytoma (PC-12) cells by its antioxidant effects stronger than those of α-tocopherol. *Neurosci. Lett.*, *362*: 61–64.

Ochiai, T., Soeda, S., Ohno, S., Tanaka, H., Shoyama, Y. & Shimeno, H. (2004b). Crocin prevents the death of PC-12 cells through sphingomyelinase-ceramide signaling by increasing glutathione synthesis. *Neurochem. Int.*, *44*: 321–330.

Oppenheim, R.W. (1991). Cell death during development of the nervous system. *Ann. Rev. Neurosci.*, *14*: 453–501.

Pan, Z. & Perez-Polo, R. (1993). Role of nerve growth factor in oxidant homeostasis: glutathione metabolism. *J. Neurochem.*, *61*: 1713–1721.

Perry, D.K. (1999). Ceramide and apoptosis. *Biochem. Soc. Trans.*, *27*: 399–404.

Pittman, R.N., Wang, S., DiBenedetto, A.J. & Mills, J.C. (1993). A system for characterizing cellular and molecular events in programmed neuronal cell death. *J. Neurosci.*, *13*: 3669–3680.

Premkumar, K., Abraham, S.K., Santhiya, S.T. & Ramesh, A. (2003). Protective effects of saffron (Crocus sativus L.) on genotoxins-induced oxidative stress in Swiss albino mice. *Phytother. Res.*, *17*: 614–617.

Puglielli, L., Konopka, G., Pack-Chung, E., Ingano, L.A., Berezovska, O., Hyman, B.T., Chang, T.Y., Tanzi, R.E. & Kovacs, D.M. (2001). Acyl-coenzyme A: cholesterol acyltransferase modulates the generation of the amyloid β-peptide. *Nat. Cell Biol.*, *3*: 905–912.

Ray, S., Misso, N.L.A., Lenzo, J.C., Robinson, C. & Thompson, P.J. (1999). γ-Glutamylcysteine synthase activity in human lung epithelial (A549) cells: factors influencing its measurement. *Free Radic. Biol. Med. 27*: 1346–1356.

Rios, J.L., Recio, M.C., Giner, R.M. & Mañez, S. (1996). An update review of saffron and its active compounds. *Phytother. Res.*, *10*: 189–193.

Runz, H., Rietdorf, J., Tomic, I., de Bernard, M., Beyreuther, K., Peooerkok, R. & Hartmann, T. (2002). Inhibition of intracellular cholesterol transport alters presenilin localization and amyloid precursor protein processing in neuronal cells. *J. Neurosci.*, *22*: 1679–1689.

Sano, M., Ernesto, C., Thomas, R.G., Klauber, M.R., Schafer, K., Grundman, M., Woodbury, P., Growdon, J., Cotman, C.W., Pfeiffer, E., Schneider, L.S. & Thal, L.J. (1997). A controlled trial of selegiline, alpha-tocopherol, or both as treatment for Alzheimer's disease. *N. Engl. J. Med., 336*: 1216–1222.

Soeda, S., Ochiai, T., Paopong, L., Tanaka, H., Shoyama, Y. & Shimeno, H. (2001). Crocin suppresses tumor necrosis factor-α-induced cell death of neuronally differentiated PC-12 cells. *Life Sci., 69*: 2887–2898.

Soeda, S., Ochiai, T., Shimeno, H., Saito, H., Abe, K., Sugiura, M., Tanaka, H., Taura, F., Morimoto, S. & Shoyama, Y. (2003). Promising pharmacological actions of crocin in Crocus sativus L. on the central nervous system. In: *Studies in Natural Products Chemistry*, Atta-ur-Rahman (ed.), vol. 28, pp. 313–329. Elsevier Science B.V.

Soeda, S., Tsuji, Y., Ochiai, T., Mishima, K., Iwasaki, K., Fujiwara, M., Yokomatsu, T., Murano, T., Shibuya, S. & Shimeno, H. (2004). Inhibition of sphingomyelinase activity helps to prevent neuron death caused by ischemic stress. *Neurochem. Int., 45*: 619–626.

Tomiuk, S., Hofmann, K., Nix, M., Zumbabsen, M. & Stoffel, W. (1998). Cloned mammalian neutral sphingomyelinase: Functions in sphingolipid signaling? *Proc. Natl. Acad. Sci. USA, 95*: 3638–3643.

Troy, C.M. & Schelanski, M.L. (1994). Down-regulation of copper/zinc superoxide dismutase causes apoptotic death of PC 12 neuronal cells. *Proc. Natl. Acad. Sci. U.S.A., 91*: 6384–6387.

Urano, S. (1993). Membrane stabilization by vitamin E. In: *Vitamin E-Its Usefulness in Health and in Curing Disease*, Mino, M., Nakamura, H., Diplock, A.T. & Kayden, H.J. (eds.), pp. 41–50. Japan Sci. Soc. Press, Tokyo.

Urata, Y., Yamamoto, H., Goto, S., Tsushima, H., Akazawa, S., Yamashita, S., Nagataki, S. & Kondo, T. (1996). Long exposure to high glucose concentration impairs the responsive expression of γ-glutamylcysteine synthase by interleukin-1β and tumor necrosis factor-α in mouse endothelial cells. *J. Biol. Chem., 271*: 15146–15152.

Vatassery, G.T., Fahn, S. & Kuskowski, M.A. (1998). α-Tocopherol in CSF of subjects taking high-dose vitamin E in the Datatop study. Parkinson study group. *Neurology, 50*: 1900–1902.

Verheij, M., Bose, R., Lin, X.H., Yao, B., Jarvis, W.D., Grant, S., Birrer, M.J., Szabo, E., Zon, L.I., Kyriakis, J.M., Haimovits-Friedman, A., Fuks, Z. & Kolesnick, R.N. (1996). Requirement for ceramide-initiated SAPK/JNK signalling in stress-induced apoptosis. *Nature, 380*: 75–79.

Wang, E., Norred, W.P., Bacon, C.W., Riley, R.T. & Merrill Jr., A.H. (1991). Inhibition of sphingolipid biosynthesis by fumonisins: implications for diseases associated with Fusarium moniforme. *J. Biol. Chem., 266*: 14486–14490.

Willaime-Morawek, S., Brami-Cherrier, K., Mariani, J., Caboche, J. & Brugg, B. (2003). c-Jun N-terminal kinases/c-Jun and p38 pathways cooperate in ceramide-induced neuronal apoptosis. *Neuroscience, 119*: 387–397.

Wuthrich, B., Schmid-Grendelmeyer, P. & Lunberg, M. (1997). Anaphylaxis to saffron. *Allergy, 52*: 476–477.

Yokomatsu, T., Murano, T., Akiyama, T., Koizumi, J., Shibuya, S., Tsuji, Y., Soeda, S. & Shimeno, H. (2003). Synthesis of non-competitive inhibitors of sphingomyelinase with siginificant activity. *Bioorg. Med. Chem. Lett.*, *13*: 229–236.

Yokomatsu, T., Takechi, H., Akiyama, T., Shibuya, S., Kominato, T., Soeda, S. & Shimeno, H. (2001). Synthesis and evaluation of a difluoromethylene analogue of sphingomyeline as an inhibitor of sphingomyelinase. *Bioorg. Med. Chem. Lett.*, *11*: 1277–1280.

Yoshimura, S., Banno, Y., Nakashima, S., Takenaka, K., Sakai, H., Nishimura, Y., Sakai, N., Shimizu, S., Eguchi, Y., Tsujimoto, Y. & Nozawa, Y. (1998). Ceramide formation leads to caspase-3 activation during hypoxic PC-12 cell death. *J. Biol. Chem.*, *273*: 6921–6927.

Yoshimura, S., Banno, Y., Nakashima, S., Hayashi, K., Yamakawa, H., Sawada, M., Sakai, N. & Nozawa, Y. (1999). Inhibition of neutral sphingomyelinase activation and ceramide formation by glutathione in hypoxic PC12 cell death. *J. Neurochem.*, *73*: 476–477.

Yu, Z.F., Nikolova-Karakashian, M., Zhou, D., Cheng, G., Schuchman, E.H. & Mattson, M.P. (2000). Pivotal role of acidic sphingomyelinase in cerebral ischemia-induced ceramide and cytokine production, and neuronal apoptosis. *J. Mol. Neurosci.*, *15*: 85–97.

In: Focus on Neurochemistry Research
Editor: Robert M. Coleman, pp.157-195
ISBN 1-59454-417-4

Chapter VI

Neurochemistry of Epileptic Seizures: The Role of f-Actin, Intercellular Glutamate, and Glutamate Ionotropic Receptors Location

Germán Sierra-Paredes,* Araceli Vázquez-López, Teresa Oreiro-García, Alejandra Núñez-Rodríguez and Germán Sierra-Marcuño

Neuroscience Division, Department of Biochemistry and Molecular Biology, School of Medicine, University of Santiago, San Francisco, Santiago de Compostela, Spain

Abstract

The molecular basis for developing epilepsy remains ill defined. It is hypothesized that increased excitatory synaptic activity may activate the N-methyl-D-aspartate (NMDA) receptor- Ca^{2+} transduction pathway, which induces long-lasting plasticity changes leading to recurrent epileptiform discharges. In epileptic patients and several experimental models of epileptogenesis, the integrity of dendritic spines and the molecular organization of the postsynaptic density, both essential for normal neurotransmission and controlled neuronal excitability, are thought to be dramatically altered. In order to determine *in vivo* if these effects are due to the disruption of f-actin filaments in the dendritic spines and the subsequent changes in receptor location, we have perfused the hippocampus of conscious freely-moving rats with the F-actin depolimeryzing agent latrunculin-A and the actin filaments stabilizer jasplakinolide. Single perfusions of latrunculin-A and jasplakinolide through microdialysis probes

* Address correspondence to Dr. G. Sierra-Paredes at Department of Biochemistry and Molecular Biology, School of Medicine, University of Santiago, San Francisco 1, 15782 Santiago de Compostela, Spain. Telephone: +43-981-582382. Fax: +43-981-587801. E-mail: bngersp@usc.es

decrease and increase picrotoxin seizure threshold respectively. Rats treated with latrunculin A become permanently susceptible to increased glutamate and glicine concentrations. Furthermore, repeated perfusions of both latrunculin-A and jasplakinolide induce acute epileptic seizures and a variable number of spontaneous seizures during a period of six months after the treatment. In those animals a long-term increase in neuronal excitability as measured by picrotoxin seizure threshold is observed. We have also investigated the effect of protein kinase and protein phosphatese inhibitors on experimental epileptic seizures. We found that ascomycin (100μM), a specific calcineurin inhibitor, prevents picrotoxin-induced seizures. H-9 dihydrochloride (100μM), an inhibitor of several protein kinases, also shows antiepileptic effect.

Our results show that *in vivo* neurochemical research methods such as chronic microdialysis are a good tool to explore the dynamic molecular changes associated with neuronal plasticity and epileptogenesis, complementing *in vitro* studies. They also suggest that actin disruption, and the subsequent loss of dendritic spines followed by dramatic changes in AMPA and NMDA receptor location, is not just a consequence but might be one of the molecular basis of epileptic seizures and epileptogenesis. Finally, we propose a new, complementary hypohesis for epileptogenesis, and a new experimental model in rats to study the biochemical changes that may lead to chronic seizures and the chronic effect of potential antiepileptic drugs.

Introduction

The neurochemical events underlying the hyperexcitability of neurons and neuronal networks which may lead to epileptic seizures are still under debate. Despite of the significant progress in epilepsy research during the last decade, even in individuals known to be at a high risk for developing epilepsy there is currently no effective method of preventing the development of the disease. Current therapies are entirely symptomatic. Understanding the molecular mechanisms of epileptogenesis, as well as the conditions for seizure start and arrest, may provide new molecular targets for antiepileptic drugs (Lees et al., 2000).

There is a considerable volume of scientific data indicating that the pathogenesis of epileptic seizures may result from alterations of the synaptic function and several intrinsic properties of neurons. Although development of circuitry with recurrent excitatory synapses is emerging as a common theme in many experimental models of epilepsy, it seems probable that the intrinsic properties of the neurons within a network will have a powerful influence on its excitability (McNamara, 1999, Scharfman, 2002).

Much basic research on epileptogenesis has focused in the role of intracellular Ca^{2+} concentration mediated by N-methyl-D-aspartate receptor (NMDAR) receptor activation. In several animal models, prolonged activation of the NMDAR-Ca^{2+} transduction pathway induces long-lasting plasticity changes in hippocampal neurons causing increased excitability leading to the occurrence of recurrent epileptiform discharges (Stasheff et al., 1989; Dingledine et al., 1990; DeLorenzo et al., 1998).

Ca^{2+} influx through NMDARs causes a depolymerization of F-actin (Bonfoco et al., 1996; Shorte, 1997) and inhibits its interaction with NMDARs through the competitive inhibition of α-actinin binding by Ca^{2+}-calmodulin (Wyszynski et al., 1997; Zhang et al.,

1998). Although this induces a Ca^{2+}-dependent inactivation of NMDA currents (Rosenmund and Westbrook, 1993; Krup et al., 1999) which may be responsible for seizure arrest (Sierra-Paredes et al., 1999), it is commonly accepted that NMDAR-mediated Ca^{2+} influx is the trigger for long-term potentiation and other neurochemical events increasing neuronal excitability (DeLorenzo et al., 1998).

The postsynaptic density is a highly dynamic structure which is constantly reorganized in an activity-dependent manner. The molecular organization of the postsynaptic density is thought to be essential for the fidelity and precission of signaling events. Clustering and immobilization of neurotransmitter receptors and ion channels is maintained by an intrincate system of protein-protein interactions. It has been shown (Wyneken et al., 2001) that intense synaptic activity associated with seizures modifies the protein composition of postsynaptic densities, and has profound consequencies on the function of several receptors present in them.

The integrity of dendritic spines seems also to be important for neuronal excitability. NMDARs and other glutamate receptor subtypes are clustered in dendritic spines (Craig et al., 1994; Kornau et al., 1995; Rao and Craig, 1997; O'Brien et al., 1998), which serve as integrative units in synaptic circuitry and participate in synaptic plasticity (for review, see Harris and Kater, 1994; Yuste and Denk, 1995; Matus 2000). The accumulation of glutamate receptor clusters in spines is governed by excitatory synaptic activity (Rao and Craig, 1997; O'Brien et al., 1998), and F-actin, a cytoskeletal protein that is concentrated in dendritic spines (Matus et al., 1982; Kaech et al., 1997), provides the main structural basis for their cytoskeletal organization (Halpain 2000; Matus 2000). The shape of a spine depends largely on the degree of actin polymerization. Dendritic spines are highly enriched with actin filaments (Fifkova and Delay,1982; Matus et al.,1982), which are thought to be involved in spine plasticity (Matus, 2000; Rao and Craig,2000).

The localization of NMDARs at synaptic sites is achieved through interactions between their intracellular domains, cytoskeletal elements (Zhang et al., 1998; Allison et al., 1998; Ehlers et al., 1998) and other cytoplasmically located submembrane proteins in the postsynaptic density (Ponting et al., 1997), many of them scaffolding proteins (Smythies, 2002). F-actin may be responsible for targeting NMDARs to synaptic sites, because treatment with actin-depolymerizing agents selectively reduces the numbers of synaptic NMDAR clusters without affecting nonsynaptic clusters (Allison et al., 1998; Allison et al., 2000; Sattler et al., 2000).

Furthermore, cultured hippocampal neurons exposed to NMDA, L-glutamate, AMPA, and ionomycin exhibit a rapid and extensive loss of dendritic spines (Halpain et al., 1998). The close relationship between NMDA receptor signaling and dendrite development is strenghten by recent research (Aizawa et al., 2004) showing that the lasting effects of neuronal activity on dendrite development involve calciun-dependent gene expression via CREST, a calcium responsive transactivator.

When neurons from both hippocampus and neocortex are examined from patients with chronic focal epilepsy, they often show dramatic dendritic abnormalities (Sheibel and Sheibel, 1973; Sheibel et al., 1994; Isokawa and Levesque, 1991; Müller et al., 1993; Multani et al., 1994; Belichenko and Dahlstrom, 1995). Dendritic spine loss has been repeatedly reported and is suggested to be more severe with increasing duration of a seizure disorder

(Multani et al., 1994). A decrease in dendritic branching is commonly observed. Also, redistribution of glutamate receptors and glutamate receptor subunits has been reported (de Lanerolle et al., 1998). However, it is difficult to know if these alterations are a cause or a consequence of epileptic seizures. The sequence of biochemical events leading to increase neuronal excitability is still unknown. There are multiple sites by which Ca^{2+} may enter the neuron, and Ca^{2+} can activate numerous specific enzyme systems or trigger specific Ca^{2+} transduction pathways (Bading et al., 1993; Ghosh and Greenberg, 1995). Also, other stimuli apart from Ca^{2+} influx through NMDARs may induce alterations in dendritic spines and cytoskeletal proteins (Halpain, 2000).

The Importance of Ion Channels and Receptors Localization

Much recent research has focused in the role of ion channels and receptors localization and mobility. Misonou et al. (2004), report activity-dependent changes in the localization and biophysical properties of Kv2.1 K^{+} channels in the kainate model of continuous seizures in rats. After seizures, there is an in vivo loss of Kv2.1 clustering in pyramidal neurons, as well as a marked dephosphorilation of Kv2.1. In rat hippocampal pyramidal neurons, glutamate stimulation rapidly causes dephosphorilation of Kv2.1 and channel traslocation from original clusters to a more uniform localization, accompanied by a shift in the voltage-dependent activation. Groc et al. (2004), using a single-molecule tracking technique, found that changes in neuronal altivity modifies AMPA receptor mobility, whereas PKC activation modified both AMPA and NMDA receptors mobility. These results suggest an important link between excitatory neurotransmission, channel location, and the intrinsec excitability of pyramidal neurons, and also that receptor and ion channel diffussion to different cell domains might be involved in synaptic plasticity.

Glutamate receptors are mainly concentrated in the postsynaptic complex of central synapses. This implies a highly organized and stable postsynaptic membrane with tightly anchored receptors. The identification of many anchorin proteins such as rapsyn, gephyrin and PSD-95, and the association of postsynaptic density proteins into a molecular scaffold has strenghten this view (Smythies, 2002). Nevertheless, recent evidence suggests that AMPA and NMDA receptors at synapses are highly dynamic (Liu and Cull-Candy, 2000; Tovar and Westbrook, 2002). It is becoming evident that scaffolding proteins have much more dynamic properties than simply act as innert scaffolds for other molecules (Smythies, 2002). Both the rapid incorporation of AMPA and NMDA receptors into synapses and their removal by endocytosis are activity dependent, suggesting that this exchange may underlie several forms of synaptic plasticity. Tovar and Westbrook (2002) demonstrated that NMDA receptors move laterally between synaptic and extrasynaptic pools, providing an alternate mechanism for altering synaptic strenght on the time course of minutes. Synaptic and extrasynaptic NMDA receptors have been show to play a different role in excititoxicity (Sattler et al., 2000; Sinor et al., 2000). NMDA receptors expressed in a single neuron can be differentially regulated based on subcellular localization. Li et al. (2002) proposed that distinct regulation of synaptic versus extrasynaptic NMDA receptors provides a mechanism for receptor adaptation in response to a variety of stimuli. Furthermore, Hardingham et al. (2002) provided evidence that a key determinant of the nature of NMDA receptor signaling is the location of the activated receptor. They reported that synaptic and extrasynaptic NMDA receptors have opposite effects on CREB (cAMP response element binding protein) function, gene regulation and neuron survival. Calcium entry through synaptic NMDA receptors induced CREB activity and brain-derived neurotrophic factor gene expression as strongly as did stimulation of L-type calcium channels. In contrast, calcium entry through extrasynaptic NMDA receptors, triggered by bath glutamate exposure or hypoxic/ischemic conditions, activated a general and dominant CREB shut-off pathway that blocked induction of BDNF expression. Synaptic NMDA receptors have anti-apoptotic activity, whereas stimulation of extrasynaptic NMDA receptors caused loss of mitochondrial membrane potential and cell death.

The previous data suggest a sequence of events which may be involved in the long-term response of a postsynaptic neuron: increased excitatory activity in the presynaptic neuron, calcium influx, phosphatases and/or protein kinase activation, phosphorilation/dephosphorilation of receptors, ion channels and other proteins, cytoskeletal changes, receptor and ion channels mobilization, and, finally, long term changes in postsynaptic excitability. If this reflects the physiological sequence leading to biochemical and morphological changes involved in the normal function of neuronal networks, it is possible that biochemical and/or morphological alterations in any of these steps might produce pathological consequences. For instance, deregulated mobility caused by alterations in receptor subunit composition, protein kinases/phosphatases or cytoskeletal proteins is likely to be involved in hyperexcitability leading to epileptic seizures.

¿How might hippocampal neurons distinguish between synaptic and extrasynaptic NMDA receptor signals? Synaptic and extrasynaptic receptors may be composed of different subunits (Riccio and Ginty, 2002; Tovar and Westbrook, 1999), each potentially endowed with a unique signaling capacity. Lu et al. (2001) have shown that activation of synaptic NMDA receptors promotes LTP, whereas activation of extrasynaptic receptors supports LTD. Recent research (Liu et al., 2004) indicates that changes in neural excitability are dependent on both stimulation frequency and NMDA receptor subunit composition.

Glutamate in the Intercellular Space

The other important question to be adressed is the origin of intercellular glutamate which may bind to extrasynaptic receptors.

In vivo microdialysis technique has been used to quantify, in a defined brain region, a presumed increase in extracellular glutamate release involved in the onset of epileptic seizures in several animal models and in humans (Rogawski, 1995; Chapman, 1997). The relevance of these studies lies in the assumption that excessive exocytosis of glutamate at the pre-synaptic level may escape from the uptake in the synaptic cleft and diffuse to the neuronal extracellular milieu. However, experimental data suggest that no link exists between seizure activity and the increase in extracellular glutamate concentrations (Miece et al., 1996; Obrenovitch et al., 1996; Sierra-Paredes et al., 1999, 2000).

Evidence supporting the role of AMPA and NMDA receptors in the development and expression of epileptic seizures has been documented (Rogawski, 1995; Chapman, 1997). On these grounds, the modifications of extracellular glutamate and aspartate levels may be related to ionotropic glutamate receptors activity or it may depend on the interplay between both AMPA and NMDA receptors. On the other hand, there is a little doubt that both neurons and glia release substances in the brain extracellular microenvironment (Martin, 1992). Some transmitter amino acids (glutamate, aspartate, glycine and taurine) are present at high concentrations in astrocytes (Levy and Patrizio, 1992), which have the enzymatic machinery for their synthesis (Hertz, 1982). Release of neuroactive amino acids from glia can be induced by receptor stimulation (Lehmann and Hansson, 1988; Barnes, 1991; Martin, 1992) or other stimuli that evoke Ca^{2+} oscillations (Fellin et al., 2004). It is also well known that astroglia expresses ion channels and membrane receptors for most of the known neurotransmitters and neuromodulators (Murphy and Pearce, 1987; Seifert and Steinhäuser, 1995; Farb et al., 1995; Steinhäuser and Gallo, 1996). These data suggest that astrocytes play some role in the regulation of the concentration of amino acids, ions and other neuroactive substances in the extracellular space. It is possible, also, that glutamate in the neuronal microenvironment may be released by axon terminals mismatched from the receptor localization. Furthermore, the active role of astrocytes in synaptic plasticity (Yang et al., 2003) and neuronal synchronization (Fellin et al., 2004) has been recently demostrated

On these grounds, the decrease of glutamate and aspartate in the brain extracellular microenvironment related to seizures (Sierra-Paredes et al., 1998) may be the result of either a reduced release or an increased reuptake by neurons or glia.

Recent research suggest that the brain extracellular environment appears to be a dynamic entity which maintains a steady intracellular milieu with regard to ions and small molecules, critical for the neuronal function, under the control of neurons and glial cells (Nicholson, 1995).

While we have no clear picture of the role of neuronal and non-neuronal cells in keeping the levels of neuroactive amino acids in the extracellular milieu in vivo, these problems have been approached in vitro using neuronal and astroglial cultures, and also brain slice preparations (Lehmann and Hanson, 1988; Nakanishi, 1992; Steinhauser et al., 1994).

However, it is difficult to interpret the data obtained in experiments in vivo through in vitro research because the tools for this comparison have not yet been developed. Occasionally, some of the results from in vivo experiments are contrary to those obtained in brain slice preparations (Sierra-Paredes et al., 1999). An interesting approach to this problem was that used by Rothstein et al (1996), which compared the levels of extracellular glutamate between knockout animals for neuronal or glial glutamate transporters.

Bearing in mind all these problems, the analysis of our results is simplified if we take also into account the concept of non-synaptic transmission (Fuxe and Agnati, 1991; Bach-y-Rita, 1991a; Vizzi and Kiss, 1998) as a complementary hypothesis to the glutamate action on synaptic transmission.

At the synaptic level, there is much experimental data to support the role of glutamate and glutamate receptors in the genesis of epileptiform discharges in the hippocampus (Rogawski, 1995; Wheal et al., 1991).

The excitatory post-synaptic potentials mediated by glutamate receptors may be increased by recurrent stimulation through a mecanism of glutamate exocytosis positive feedback (Herrero et al., 1992, a, b) which is mediated by post-synaptically generated arachidonic acid (Lynch et al., 1989) and which induces an enhancement in the amount of glutamate release by synaptic vesicles. The effects of the increased release of glutamate is also potentiated by blocking the GABAergic inhibition with picrotoxin. Both NMDA and non-NMDA receptors participate in this enhanced synaptic stimulation (Andreasen et al., 1989)

The epileptiform synaptic potentials in the hippocampus are clearly dependent on the vesicular glutamate pool. However, it is difficult to understand the role of extracellular glutamate and aspartate concentrations based only on its synaptic effect.

Astroglial cells encapsulate synapses, and astroglial glutamate carriers have sufficient capacity to remove all glutamate released from the presynaptic site (Hertz, 1979). Glial cells respond to synaptically released glutamate by activation of electrogenic transporters, which generate a current that is directly proportional to the amount of glutamate released (Diamond et al., 1998) The high capacity and affinity glutamate uptake by astrocytes in the synaptic region maintains synaptic transmission possible for long periods of time with a high signal-to-noise ratio (Rönnback and Hansson, 1997).

The need for astroglial cells working at the synaptic region as effective system for clearance of the presynaptically released glutamate may be clearly understood taking into account the role of the glutamate-glutamine cycle in both the presynaptic region and the glial cells (Schousboe, 1981). Upon release into the synaptic cleft, the glutamate diffuses away, and the glial glutamate transporters internalize the extracellular glutamate. Inside the glial cells, glutamate is metabolized into glutamine, which is transported out of the glial cells and into the synaptic terminal for subsequent resynthesis of glutamate. Since glutamine does not act on the glutamate receptor, this release does not interfere with synaptic transmission (Schousboe et al., 1977). Also, using knockout rats (Rothstein et al., 1996), it was shown that extracellular glutamate levels were unmodified in the animals which had lost neuronal glutamate transport, while the animals which had lost the glutamate transport in glia showed an increase in extracellular glutamate concentrations, suggesting a major role for glial glutamate transporters in the clearance of extracellular glutamate.

This special role of the astrocytes at the synaptic region may be explained by the astroglial heterogeneity (Hansson, 1990). We also need to take into account the morphological (DeRobertis, 1965) and the pharmacological (Curtis and Eccles, 1958) evidence of synaptic barriers, and also the fact that during the period in which picrotoxin induced only inter-ictal discharges without concomitant seizures, no modifications in the extracellular glutamate and aspartate were observed (Sierra-Paredes et al., 1998).

Taken together, all these data suggest that the epileptiform synaptic potential is generated by the glutamate synaptic pool without the participation of the extracellular glutamate pool.

It also explains why the increase in the extracellular glutamate concentration does not produce electrophysiological changes indicative of excessive excitation (Obrenovitch et al.,

1996), and appears insufficient to induce neuronal degeneration (Massieu et al., 1995). The increase in extracellular glutamate observed during seizures was related to neuronal damage (Liu et al., 1997; Peña and Tapia, 1999).

A different picture is observed on the role of glutamate and aspartate in non-synaptic transmission.

Fagni et al. (1983) suggested that the L-glutamate added to the bath stimulated different glutamate receptors to those activated by the endogenous neurotransmitter, and postulated the existence of extrasynaptic receptors.

Evidence of extrasynaptic glutamate receptors has been reported (Aoki et al., 1994; Kullman et al., 1996), and a comparison of some properties of synaptic and non-synaptic NMDA receptors has been published (Clark et al., 1997, Hardingham et al., 2002). Some data suggest that the extracellular glutamate from astrocytic origins may act on neuronal and glial NMDA non-synaptic receptors (Farb et al., 1995; Seifert and Steinhäser, 1995, Fellin et al., 2004).

The decrease in extracellular glutamate and aspartate levels related to seizures is not prevented by THA (Sierra-Paredes et al., 1999), a potent inhibitor of glutamate and aspartate reuptake. This fact may have two explanations. The first is that recurrent paroxysmal depolarization induced by glutamate at the synaptic level may block the glutamate release by glial cells. The ability of the glial cells to change the extracellular environment has been postulated as mechanism which simultaneously modulates the excitability of many neurons (Olsson et al., 1997). Thus, the decrease in extracellular glutamate and aspartate may be a way to reduce the neuronal cluster excitability.

It has been stated that seizures are always the consequence of an abnormal synchronizations of millions of neuronal discharges (Traub and Wong, 1982). A second possibility would be that the paroxismal depolarization generated at the synapses induces a massive binding of extracellular glutamate and aspartate at non-synaptic glutamate receptors and it is this which induces abnormal neuronal synchronization. Repolarization or hyperpolarization of the mass of neurons by diazepam stops the seizures and produces a rapid return to basal glutamate and aspartate levels (Sierra-Paredes et al., 1997a). Furthermore, these mechanisms explain the effect of MK-801 preventing the seizures.

The role of non-synaptic receptors in mass sustained convulsive activity has been postulated (Bach-y-Rita, 1991 b). An increase in glutamate binding in surgically removed brain tissue from patients with intractable seizures has also been reported (Hosford et al., 1991). Recently, Fellin et al. (2004) have shown that neuronal synchrony can be achieved by activation of extrasynaptic NMDA receptors with astrocytic glutamate.

Taking into account the decrease of extracellular glutamate observed before the start of the seizures (Sierra-Paredes et al., 1999), it is possible that both processes may work in a temporal sequence: firstly, an increase in synaptic activity might increase extracellular glutamate released by astrocytes and activate extrasynaptic NMDA receptors on both neurons and glia. Extracellular glutamate then binds massively to extrasynaptic NMDA receptors to induce an abnormal synchrony, and later, a decrease of glial release may decrease neuronal activity in order to stop the seizures.

Protein Phosphatases and Epileptic Seizures

Modulation of intracellular calcium levels is important in a number of normal neuronal processes including regulation of gene expression, development of learning and memory, and neurotransmitter synthesis and release (Churn 1995; Chittajallu et al. 1998). However, alterations in calcium-regulated systems and loss of calcium homeostasis have also been implicated in many pathological conditions, such as ischemia (Choi 1987; Parsons et al. 1997, 1999), traumatic brain injury (Rzigalinski et al. 1998), and status epilepticus (Pal et al. 1999; Parsons et al. 2000). Specifically, influx of calcium through the NMDA subtype of glutamate receptor is suspected to be important for physiological changes occurring after status epilepticus (Rice and DeLorenzo 1998; Pal et al. 1999). NMDA-linked increases in intracellular calcium affect a number of calcium-controlled cellular mechanisms and enzymes including calcium/calmodulin-dependent kinase II (Churn et al. 1995; Kochan et al. 2000), calpain I (del Cerro et al. 1994), and calcineurin (Montoro et al. 1993).

Calcineurin is a calcium/calmodulin-stimulated phosphatase highly enriched in neural tissue (Pallen and Wang 1985). Calcineurin-mediated dephosphorylation is an important modulatory factor in several cellular processes, including development of learning and memory (Riedel 1999), regulation of long-term potentiation (LTP) and long-term depression (LTD) (Groth et al., 2003), and induction of apoptosis (Springer et al. 2000). Additionally, calcineurin may depress the activity of the $GABA_A$ receptor (Huang and Dillon 1998). These characteristics make calcineurin an intriguing enzyme for study in epileptic seizures.

Kurz et al., (2001) have reported a significant increase in calcineurin activity in the rat pilocarpine model of status epilepticus, occurring through a NMDA-dependent mechanism.

Protein Kinases and Epileptic Seizures

Regulation of NMDA (Westphal et al., 1999; Lan et al., 2001) and AMPA receptors by cAMP-dependent protein kinase (PKA) is essential for synaptic transmission. Ca^{2+} influx through NMDA receptors is closely related with protein kinase activity. PKC phosphorilates the NR1 C-terminal domain upregulating Ca^{2+}-calmodulin dependent inactivation, but it can also indirectly enhance NMDA evoked currents (Lu et al., 2000). PKA phosphorilation of the AMPA receptor subunits GluR4 and GluR1 directly controlls the synaptic incorporation of AMPA receptors in the rat hippocampus (Esteban et al., 2003), indicating that PKA phosphorilation of AMPA receptor subunits contributes to diverse mechanisms underlying synaptic plasticity. Lan et al (2001) reported that PKC increases NMDA channel opening rate and delivers new NMDA channels to the plasma membrane through regulated exocytosis. It was shown that increased excitability in neurons underlying epilepsies would be maintained by abnormalities in protein phosphorylation systems (Walaas and Greengard, 1991), however, most reserch about the role of PKA in epileptic seizures has been done exclusively on seizure-prone genetic models in mice and rats. Tehrani and Barnes (1995) showed that an abnormal elevation of protein kinase A activity in tottering mouse brain contributes to an impairment of $GABA_A$ receptor function, and suggested that the resulting loss of inhibition could play a role in induction of the seizures. Recently, Yechikhov et al. (2001) compared the functioning of the cAMP-dependent system of protein phosphorylation in homogenates of neocortex and hippocampus in three animal groups: genetically prone to audiogenic seizures (GPAS) rats, GPAS rats exposed to daily repeated audiogenic seizures and nonepileptic Wistar ones. They found significant differences in phosphorylation of 270, 58, 54 and 42 kDa proteins in neocortex and hippocampus of GPAS rats in comparison with Wistar ones. Daily repeated seizures induced modifications of phosphorylation of these proteins in the hippocampus.

Purpose

Due to methodological reasons, many of the molecular events related to receptor and synaptic dinamics have been exclusively studied on *in vitro* preparations such as neuronal cultures, synaptosomes or hippocampal slices. However, it has been pointed out that most in vitro paradigms do not work in vivo preparations (Baudry and Davis, 1997).

To adress these questions *in vivo*, we have used a new experimental design on a whole-animal model in which partial seizures can be elicited repeatedly on different days without changes in threshold or seizure patterns. Picrotoxin seizure thresholds remain constant in the same animal in repeated experiments for time periods as long as six months, thus providing a good model to study possible modifications in neuronal excitability (Sierra-Paredes et al., 1996). In vivo microdialysis has been widely used for measuring extracellular neurotransmitter concentrations as a direct or indirect method to evaluate changes in neuronal transmitter release. However, some neurotransmitters found in the brain extracellular fluid do not only have a neuronal origin, but they are released and reuptaked by glial cells. Thus, measuring the brain extracellular fluid neurotransmitter concentrations does not allow us know the origin of the analysed transmitters, but it allows us to study de brain internal milieu and how it is modified by drugs.

In order to investigate the neurochemical mechanisms of epileptic seizures and to find new pharmacological targets for potential antiepileptic drugs, we have been using in vivo microdialysis as a “two way road” which allows us to study, in the same animal, aspects related to the extracellular and intracellular regulation of neurotransmitter receptors and reuptake systems.

We have been working on the hypothesis that extracellular amino acids, such as glutamate, aspartate and glycine, may play a role in modulating neuronal excitability, probably by binding to extrasynaptic NMDA receptors. Hippocampal microperfusion in freely moving rats has been demonstrated to be a good method to investigate the in vivo effect of substances with intracellular or extracellular action on neurons (Sierra-Paredes et al., 1999, 2000a, 2001). Using the microdialysis method with a new experimental design on a new chronic whole-animal model (Sierra-Paredes and Sierra-Marcuño, 1996a; Sierra-Paredes and Sierra-Marcuño, 1996b) we have shown that during the interictal discharges, no changes in the extracellular levels of glutamate, aspartate, γ-amminobutyric acid (GABA), glycine or taurine occur. However, during ictal discharges a significant decrease in the levels of extracellular glutamate and aspartate was found (Sierra-Paredes et al., 1998). These data suggest that modifications in extracellular glutamate and aspartate may be related to seizures rather than to paroxysmal activity, supporting the neurophysiological differences between ictal and non-ictal activity (Matsumoto and Ajmone-Marsans, 1964 a, b; Wyler, 1982).

We have found that microdialysis combined with electrophysiological recording techniques allows to study in vivo some aspects of the receptor regulation which had been studied only *in vitro*.

In order to test if the integrity of the dendritic spine cytoskeleton affects directly to neuronal excitability, we have tested, for the first time, the effects of the actin-depolymerizing agent latrunculin-A and actin filaments stabilizer jasplakinolide in the hippocampus of freely moving rats.

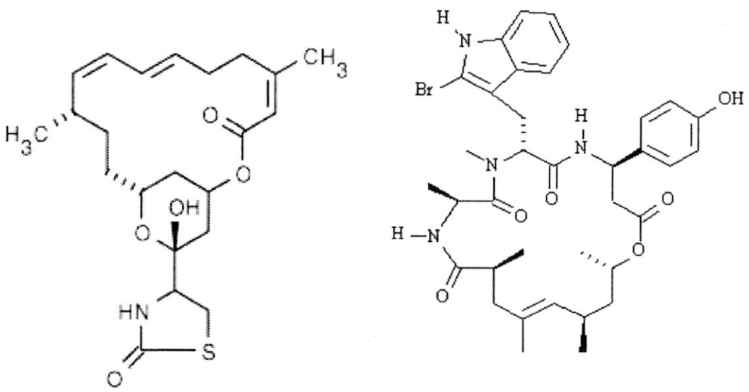

Jasplakinolide

Latrunculin A Jasplakinolide

Figure 1. Chemical structure of Latrunculin A and Jasplakinolide.

Latrunculin-A (Spector et al., 1983), affects actin polymerization by the formation of a 1:1 molar complex with G-actin, causing net actin depolymerization (Spector et al., 1989). Latrunculin-A was recently used (Allison et al., 1998; Halpain et al., 1998; Sattler et al., 2000) to study the role of F-actin in anchoring NMDARs to synaptic sites. Jasplakinolide stabilizes actin filaments in vitro and induces in vivo polymerization of actin Bubb et al., 2000). We have also studied the effect of increasing glutamate and glycine concentrations on rats previously treated with latrunculin A.

Latrunculin-A, picrotoxin, jasplakinolide as well as amino-acids were dialyzed through the probe to avoid possible dynamic effects imposed by the blood-brain barrier on certain systemic administered drugs (Aguilar-Veiga et al., 1991).

Methods

Drugs and Solutions

Latrunculin-A and jasplakinolide which were purchased from Molecular Probes (Eugene, Oregon, USA). Stock solutions of latrunculin-A and jasplakinolide were prepared in ethanol (100 μg/μl) and kept at -20°C until used. Solutions in Ringer's fluid appropiate for microperfusion containing a maximum of 1:18750 ethanol were prepared weekly at concentrations of 2, 4 and 6 μg/ml. Control Ringer's fluid and picrotoxin solutions containing 1:18750 ethanol were prepared weekly. Ascomycin and H-9 dihydrochloride were obtained from Tocris (UK).

Animals and Surgical Procedure

Twenty adult male Sprague-Dawley rats, initially weighing 250-300 g were used. They were housed in groups of three under controlled environmental conditions (ambient temperature 21±1.8 °C, humidity 50-60%, 12:12 h light/dark cycle) with free access to food and water except during testing. Rats were obtained from the animalary of the University of Santiago. All experiments were performed in a laboratory with controlled environmental conditions and at the same time in the morning to avoid circadian variations. All efforts were made to minimize animal suffering, and our chronic animal protocols were designed to reduce the number of animals used (Sierra-Paredes et al., 1996). Animal care followed Spanish legislation on Protection of Animals Used in Experimental and Other Scientific Purposes, in agreement with the European Union regulations (O.J. of E.C. L358/1 18/12/1986)

The rats were anaesthetized with pentobarbital (40 mg/kg) and placed in a stereotaxic instrument (D. Kopf, Tujunga, CA, USA). Under aseptic conditions, 2 stainless steel microscrews to be used as electrodes for EEG recording were positioned in the skull above the frontal and occipital areas of each hemisphere; one screw, used as a reference electrode, was anchored in the mid-line, 7-9 mm rostrally to the coronal suture. The intracerebral guide for the microdialysis probe (CMA/12, CMA/Microdialysis AB, Stockholm, Sweden) was implanted vertically into the ventral hippocampus. Stereotaxis coordinates derived from the atlas of Paxinos and Watson (1997) were 5 mm posterior, 4.8 mm lateral and 4 mm ventral for the tip of the cannula relative to bregma and dural surface.

Wires from the microscrews were soldered to a miniature plug (Cannon MD1-9SL1, ITT Cannon, Santa Ana, USA) and fixed firmly to the skull with dental cement. After surgery, the rats were housed in individual cages and received antibiotic therapy for 4-5 days.

The experiments were carried out on conscious, freely moving rats 10 days after surgery. From the fourth day the animals were placed daily for three hours in the experimental unit for habituation. Bipolar cortical EEGs were recorded on magnetic tapes using a holter-EEG system (Oxford-Medilog 9200, Oxford, U.K.), and also with a Minihuit electroencephalograph (Alvar Electronic, Paris, France).

Seizure Thresholds

In our experiments, seizure threshold was defined as the lowest picrotoxin concentration which produced a specific pattern of EEG and/or behavioural seizures after a 5 minute perfusion through the rat hippocampus. Only one picrotoxin dose was perfused in each experimental session. The lowest picrotoxin concentration used was 100 μ*M*, and the dose was slowly increased (+25μ*M* each step) in each animal in successive experimental sessions at 7 day intervals until an EEG-behavioral seizure was induced. This seizure was defined as the threshold seizure. Each animal was randomly perfused 4 times with the same picrotoxin concentrations which had induced seizures to ensure that these doses would produce the same type of seizure on different days. Seizure types and rest periods between experimental sessions were described previously in detail (Sierra-Paredes et al., 1996).

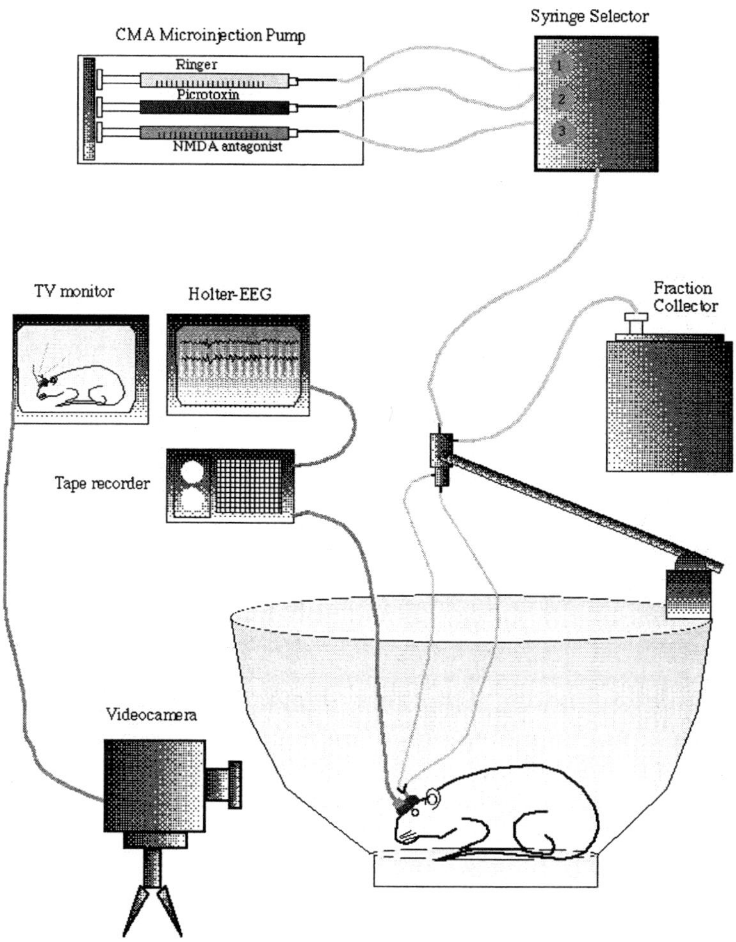

Figure 2. Schematic diagram showing the microdialysis and recording systems

Microdialysis

We used CMA/120 system for freely moving animals (CMA/Microdialysis AB, Stockholm, Sweden) and CMA/12 microdialysis probes with 4 mm of membrane length. The probe was connected via polyethylene tubing to a syringe selector (CMA/111), and to 1 ml syringes mounted on a microinjection pump (CMA/100). Before starting each experiment, the probe was perfused with ethanol and distilled water. After checking the integrity of the probe under a light microscope, it was perfused with a sterile Ringer's solution (NaCl 147 m*M*, KCl 4.0 m*M*, CaCl2 2.4 m*M*) for 10 min, tested routinely for in vitro recovery before every experiment and then introduced into the rat hippocampus through the chronically implanted intracerebral guide.

For the control experiments, Ringer's solution was perfused at a constant flow rate of 2 μl/min throughout the experiment.

Latrunculin-A and Jasplakinolide Experiments

We used latrunculin-A for a selective actin depolymerization at the dendrite level (Allison et al., 1998), and jasplakinolide for its two possible actions: stabilization of actin filaments (Halpain et al., 1998), and disruption of actin filaments inducing polymerization of monomeric actin in amorphous masses (Bubb et al., 2000.)

For the latrunculin-A experiments, rats were divided into four groups: In the first group, after the picrotoxin seizure threshold was established, latrunculin-A was perfused for two hours, and then a subthreshold dose of picrotoxin was administrated. In the second group, in order to test the acute effect of latrunculin-A alone, it was dissolved in Ringer and perfused daily for 8 hours/day with continuous EEG monitoring for five consecutive days. In the third group, in order to test the chronic effect of a single latrunculin-A perfusion, the picrotoxin threshold was established, and, after a week of rest, latrunculin-A was perfused for 8 hours with continuous EEG monitoring. In the fourth group, latrunculin-A was perfused weekly for 8 hours for a maximum period of 8 weeks. The picrotoxin threshold was checked every month and latrunculin-A concentrations of 2, 4 and 6 μg/ml were used. Frequent control EEGs were recorded in all the animals during all the experimental period (at least twice a week), and they were directly observed and videotaped in search for spontaneous seizures.

The same protocol was repeated in order to test the effect of jasplakinolide.

Increased Glutamate and Glycine Concentrations

In order to test the effect of glutamate and glycine on animals pretreated with latrunculin A, each was dissolved in Ringer's solution at concentration 200 μM and 1mM(Sierra-Paredes et al., 2001a). The microdialysis probe recovery was 16.8 % ± 2.5% for glutamate, and 18,2% ± 3.1% for glycine as measured by HPLC and fluorescence derivatization af amino acids (Sierra-Paredes et al., 1999). This means estimated maximal intercelular concentrations of 168 ± 25 μM for glutamate y 182 ±31 μM for glycine. Each solution was perfused continuously throughout the experiment in all the animals on different days, following the same protocol for Ringer's solution and picrotoxin administration in the control experiments.

PKA and Calcineurin Inhibitors

Ascomycin and H-9 dihydrochloride were dissolved in Ringer at a 100 µM concentration and perfused continuously throughout the experiment in all the animals on different days, following the same protocol for Ringer's solution and picrotoxin administration in the control experiments. H-9 dihydrochloride is a cell-permeable inhibitor of protein kinase A (Ki = 1.9 □M), protein kinase C (Ki = 18 □M), and protein kinase G (Ki = 870 nM) (Hidaka et al., 1984). Ascomycin is a FK 506 analog with high affinity to FKBP which inhibits calcineurin phosphatase in the nM range (Meadows, 1993).

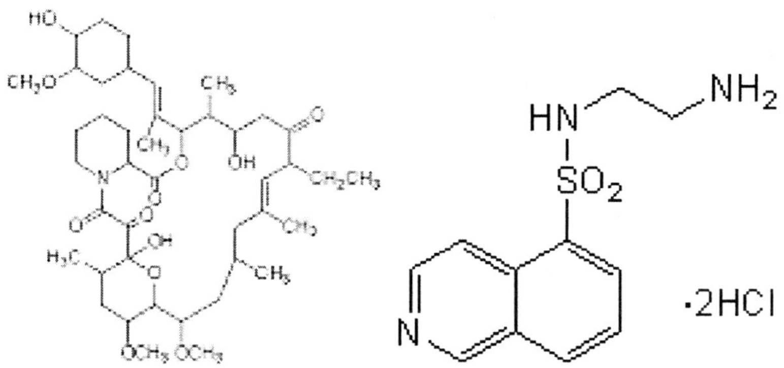

Ascomycin H-9 dihydrochloride

Figure 3. Chemical structure of ascomycin and H-9 dihydrochloride

EEG Recording

EEG records were analyzed using the Medilog 9200 software, version 7.2. Spike and wave discharges duration, seizure duration, and seizure latencies were evaluated after picrotoxin administration. Statistical significance of the difference in number of seizures, seizure duration, and seizure onset times was determined by Student's paired t test.

At the end of the experiments rats were anaesthetized with Nembutal and killed by decapitation. A probe was introduced and perfused with Sudan black to help locate the position of the probe. Then, the brain was removed and placed in 4% phosphate buffered formaldehyde solution. A week later 50 µm coronal sections were cut and stained with cresyl violet, and the position of the probe was checked under light microscopy.

Results

In vivo Alterations of Actin Dynamics Induces Epileptic Seizures

Latrunculin-A induces a short-term increase on neuronal excitability in the rat hippocampus which leads to epileptic seizures, and a permanent effect as measured by picrotoxin seizure threshold (PCTS threshold). The degree of hyperexcitability is not dose-dependent at the concentrations studied (2, 4 and 6 μg/ml), but related to the perfusion duration. In animals not treated with picrotoxin, no modifications were detected in basal EEG during wakefulness and sleep periods during a single 2 hour perfusion of latrunculin-A. However, latrunculin-induced seizures were observed after 6 hours of continuous perfusion in 20% of the animals studied (Figure 4a). When latrunculin-A was perfused for several consecutive days (8 hours a day, see Methods), seizures were observed in 80% of the animals studied (Figure 4a, c). The severity, duration and onset time of the seizures showed considerable variation among animals, but 90%±6% of the seizures were observed during the second and third days of consecutive perfusion (Figure 4a). No seizures were observed after the fourth day of perfusion (Figure 4a). Weekly perfusions of latrunculin-A induced seizures during perfusion after the 4th week. As in the consecutive perfusion experiments, the onset of the first seizures showed considerable variation, but all the animals studied presented several seizures between weeks 4 and 6 (Figure 5 a, c). Those animals also show an abnormal EEG response to handling and tapping with slow waves and a significant increase in sleep periods as compared to control EEGs in the same animals (Figure 5 e). Other sensory stimuli, such as auditory random stimuli did not elicit any EEG response. A few sporadic spontaneous seizures (0.4±0.3 seizures/month) were observed during controls in the three months following the last latrunculin-A perfusion (Fig 7a). The effect observed here seems to be receptor activation-independent and related solely to actin depolymerization. In all the animals in which PCTS threshold was tested, a significant decrease up to 56%±7% of the basal PCTS theshold was observed after a 2 hours perfusion (Figure 6). The chronic experiments showed that the decrease in PCTS threshold is permanent and progressive for a period of three months in all the animals studied (Figure 7b). Seizures induced by latrunculin A can be prevented by the use of several antiepileptic drugs (unpublished data).

Repeated and consecutive jasplakinolide perfusions (see Methods) into the hippocampus of living rats induces an increase on neuronal excitability in the rat hippocampus which leads to epileptic seizures in 30% of the animals studied, a permanent effect as measured by picrotoxin seizure threshold (PCTS threshold) in 80% of the rats, and sporadic spontaneous seizures in 60%. The degree of hyperexcitability is also related to the perfusion duration and is not dose-dependent at the concentrations studied (2, 4 and 6 μg/ml). In animals not treated with picrotoxin, no modifications were detected in basal EEG during wakefulness and sleep periods during a single jasplakinolide perfusion. Jasplakinolide-induced seizures were observed after three consecutive days in 30% of the animals studied (Figure 1a,d). In all the animals in which PCTS threshold was tested, a significant increase up to 33% of the basal PCTS theshold was observed after a 2 hour perfusion (Figure 3). However, chronic experiments showed that a permanent decrease in PCTS threshold is detected one week after jasplakinolide administrations, lasting for a period of three months in 80% of the animals studied (Figure 4b). Weekly perfusions of jasplakinolide induced seizures during perfusion after the 3rd week (Figure 2a, d), and 80% of the animals studied presented several sporadic seizures between weeks 3 and 12 (1,7±1,3 seizures/month, Fig 4a). The severity, duration and frequency of the seizures showed considerable variation among animals, ranging from rapidly recurrent seizures during 24 hours to short (48±21 seconds) partial seizures. During the three

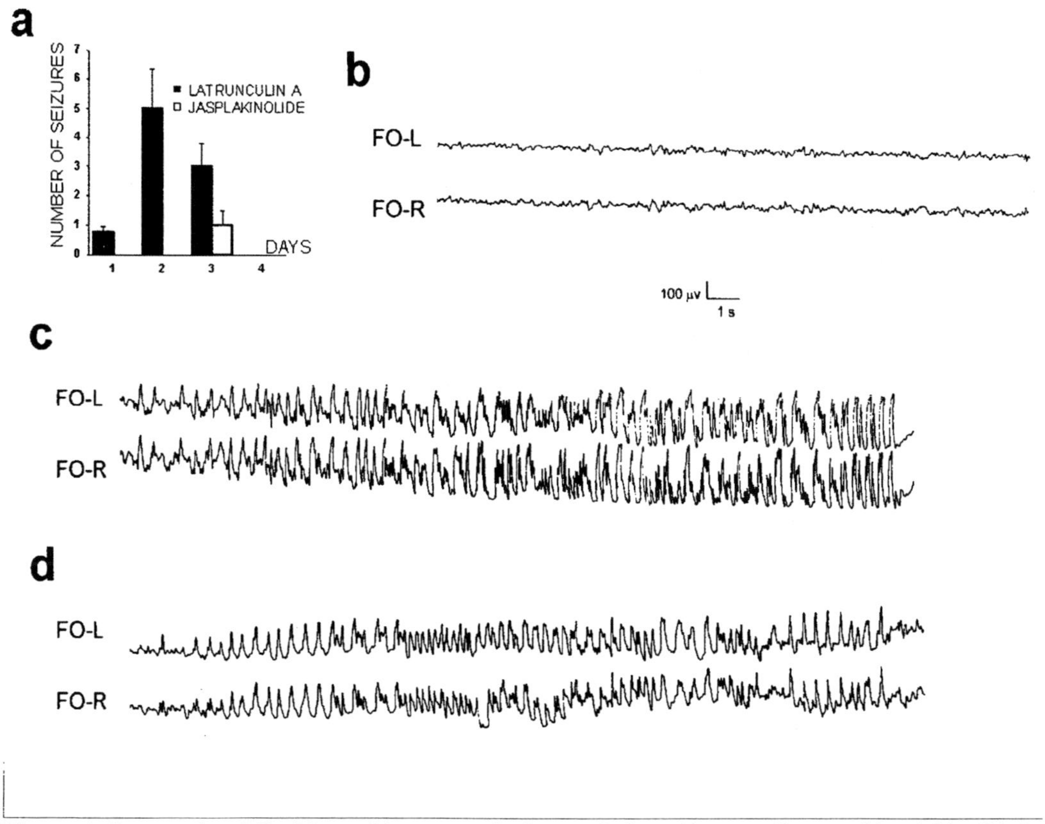

Figure 4. Consecutive daily 8 hours hippocampal microperfusion of latrunculin-A and jasplakinolide in the hippocampus of conscious rats induces epileptiform seizures. (a) Average number of seizures induced by latrunculin-A (n=10) and jasplakinolide (n=10) from days 1-4 of consecutive perfusion. (b) Control EEG recording (FOR fronto-occipital right; FOL fronto-occipital left). (c) EEG seizure induced by latrunculin-A (4 µg/ml), correlated with rearing, alimentary automatism and forelimb clonus. (d) EEG seizure induced by jasplakinolide (4 µg/ml), with a similar behavioural outcome.

months following the last jasplakinolide administration, spontaneous focal discharges and abnormal EEG responses to handling, tapping and whisk stimulation were observed in the 80% of the animals (Figure 2f), and also a significant increase in sleep periods as compared to control EEGs in the same animals (data not shown).

Previous studies in vitro and cell cultures showed that AMPA and NMDA receptors are anchored in the dendritic spines by actin cytoskeleton (Allison et al., 1998; Halpain, 2000). Several findings support the use of latrunculin-A to perturb NMDA and AMPA receptor clusters (Allison et al., 1998; Sattler et al., 2000), but do not establish a functional significance for this effect. Because the actin-depolymerizing agents failed to show any effect on macroscopic whole-cell NMDA currents (Sattler et al., 2000), it was supposed that although NMDAR localization might be rearranged, function may be grossly unaffected

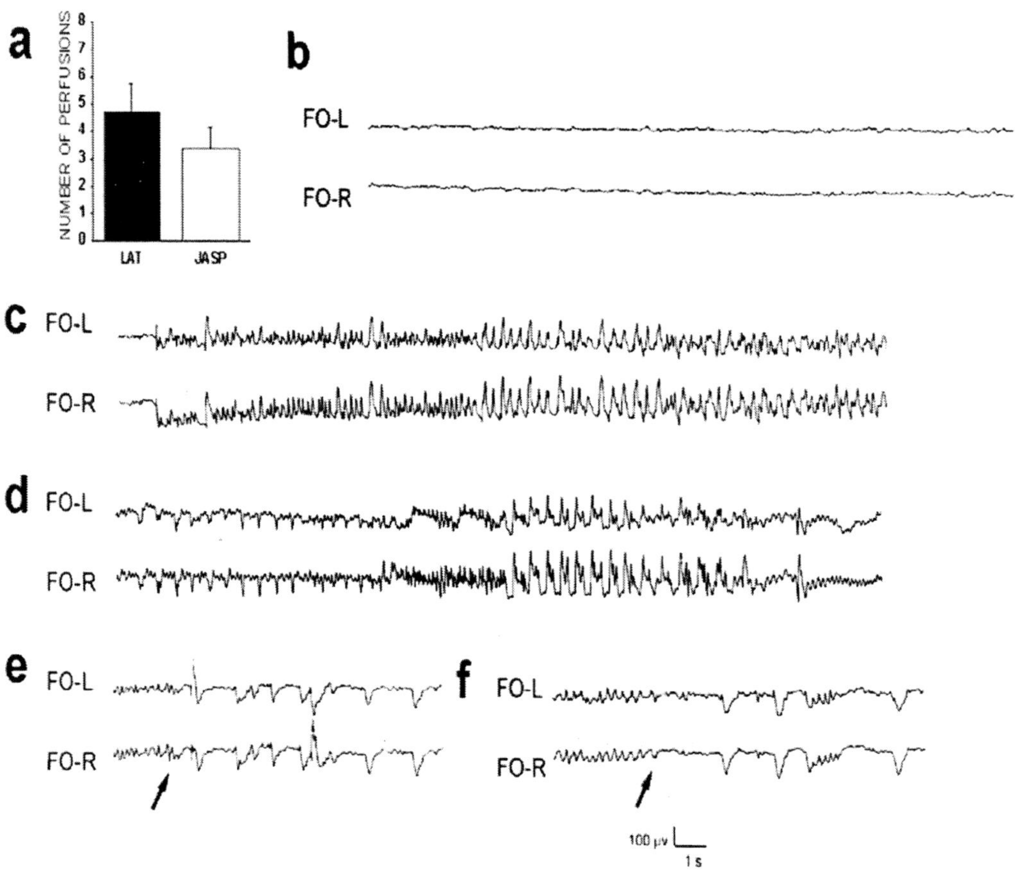

Figure 5. Consecutive weekly 8 hours hippocampal microperfusion of latrunculin-A and jasplakinolide in the hippocampus of freely moving rats induces epileptiform seizures with a variable frequency. (a) Summary of the average number of weekly perfusions of latrunculin -A(n=10) and jasplakinolide (n=10) to induce seizures in rats. (b) Control EEG recording (FOR fronto-occipital right; FOL fronto-occipital left). (c) EEG seizure induced by latrunculin-A (4 µg/ml). (d) EEG seizure induced by jasplakinolide (4 µg/ml). (e) EEG response to handling (this pattern appears 10-15 sec after stimulation) in a rat treated with latrunculin-A. (f) EEG response 5 seconds after whisk stimulation in a rat treated with jasplakinolide.

(Allison et al., 2000). Furthermore, direct evidence of the in vivo effect of actin cytoskeleton disruption was lacking. We show here, for the first time, that actin disrupting agents can be perfused in conscious, freely moving animals in order to investigate the in vivo actin dynamics, which through depolymerization or abnormal polymerization, modifies neuronal excitability and leads to epileptic seizures.

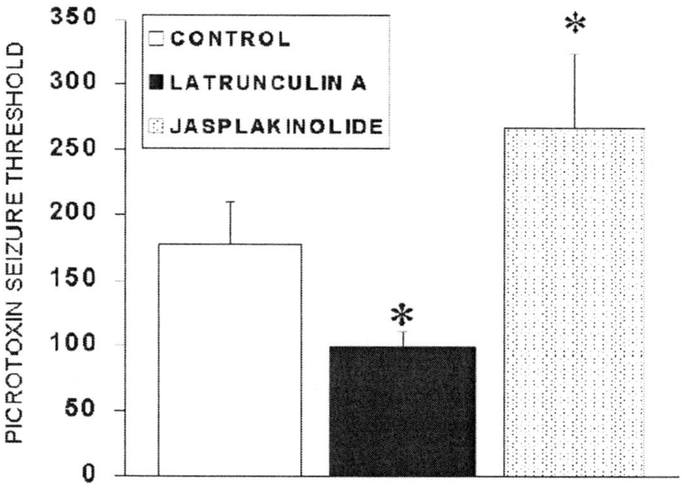

Figure 6. A single 2 hour perfusion of a 4 mg/ml latrunculin-A solution decreases significantly (*p< 0.01) the picrotoxin seizure threshold in freely moving rats (n=8). A 2 hour jasplakinolide perfusion (4 μg/ml), however, sigificantly increases (*p< 0.01) picrotoxin seizure threshold (n=8).

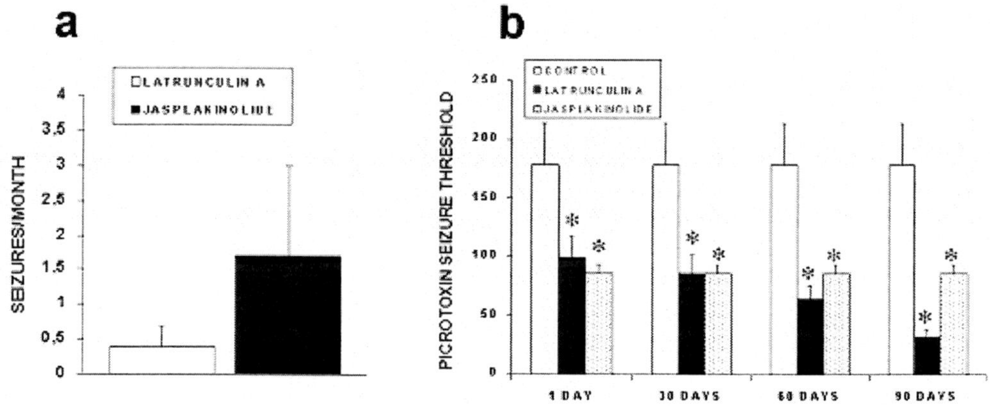

Figure 7. Repeated latrunculin-A and jasplakinolide administrations (a minimum of 3 perfusions, 4 μg/ml, daily or weekly) produce a long-lasting effect in neuronal excitability. (a) Average number of seizures per month observed the three months following weekly latrunculin-A (n=10) and jasplakinolide (n=10) administration. (b) Repeated latrunculin-A perfusions induce a permanent and progressive decrease in picrotoxin seizure threshold (*p< 0.01) which lasts for three months after the last latrunculin-A administration (n= 10). Repeated jasplakinolide perfusions produce a significant (*p< 0.01) decrease which lasts unmodified for three months (n=10).

Much experimental evidence shows that an increase in glutamate excitatory action modifies the excitability of postsynaptic neurons via the actin cytoskeleton. Alterations in other actin filaments related proteins, such as acidic calponin (Ferhat et al., 2003), have been linked to experimental epileptic seizures. It is well known that excitatory synaptic activity modulates the distribution of AMPA and NMDA receptors in the postsynaptic sites of hippocampal neurons. The mechanism for receptor redistribution involves calcium-mediated actin depolymerization (Bonfoco et al., 1996; Halpain et al., 1998; Lin et al., 2000) and is related to rapid dendritic spine plasticity (Halpain, 2000; Fischer et al., 1998; Shi et al., 1999). This receptor redistribution has been shown to participate in mechanisms such as LTP and LTD (Carroll et al., 1999; Lin et al., 2000), as well as in pathological processes leading to epileptogenesis (de Lanerolle rt al., 1998). Most of these studies have centered on the increased or decreased number of postsynaptic receptors; however, receptor location may be also an important factor in neuronal excitability, and may be changed by alterations in the actin cytoskeleton (Sattler et al., 2000). Our results show that this effect can be induced in living animals by perfusing latrunculin-A and jasplakinolide in the hippocampus. Alterations in receptor density or distribution on dendritic segments would be expected to significantly modify the effectiveness of synaptic transmission. Furthermore, alterations in the distribution of recurrent excitatory synapses on dendrites could lead to an enhanced ability of these synapses to produce action potentials and, in turn, promote the reverberation of recurrent excitation in networks of mutually excitatory pyramidal cells.

Spontaneous seizure activity was never present in the first month after latrunculin A or jasplkinolide administration. This "latency" period is consistent with the time needed for plasticity changes in the hippocampus such as sprouting. It is not completely clear when sprouting occurs after seizures, although in models involving a period of status epilepticus (i.e.,kainic acid or pilocarpine), it appears that sprouting begins sometime in the first 1 to 2 weeks,and is most robust after many weeks (Scharfman, 2002). Eventhough latrunculin A and jasplakinolide do not induce status epilepticus, this latter condition is characterized by massive actin disruption in dendritic spines (Isokawa 2000; Ferhat et al., 2003).

A differential action of actin disrupting agents among excitatory pyramidal cells and GABAergic interneurons cannot be ruled out. F-actin is necessary for the proper synaptic localization of AMPA receptors in GABA cells (Allison et al., 1998), and the receptor relocation and complete recovery may be more difficult in GABA cells than in the dendritic spines of pyramidal cells, maintaining a permanent lack of recurrent inhibition.

We observed a diferent effect between latrunculin-A and jasplakinolide in the acute administration. Depolymerization of actin by latrunculin-A has been reported to transiently promote neurotransmitter release by a mechanism independent of extracellular Ca^{2+}, and this effect is antagonized by jasplakinolide (Morales et al., 2000). However, this presynaptic effect does not explain the chronic effect of both latrunculin-A and jasplakinolide. In cell cultures, a 24 h treatment with latrunculin-A is required for maximum effect (Allison et al., 1998), and we obtain the maximum effect after three perfusions of 8 hours each. For jasplakinolide, it has been shown that it can disrupt actin filaments and induce polymerization of monomeric actin in amorphous masses (Bubb et al., 2000), which may explain the similar effect when both compunds are perfused for several consecutive days. The permanent decrease in seizure threshold is more likely to be related to morphological changes in the number or shape of dendritic spines, or a permanent reorganization in the location of glutamate receptors or other proteins within the post-synaptic density which are highly dependent on F-actin for their localization, such as CaMKII, spectrin, myosin V, α-adducin, neurabin, neurabinII/spinophilin, cortactin, and many others (Allison et al., 2000). The major function of those actin-associated components proteins of the post-synaptic membrane appears to be in signal transduction and modification of the microfilament arrays in response to synaptic activation, events thought to mediate long-term synaptic plasticity.

In the rat pilocarpine model, dendrites of dentate granule cells displayed a generalized spine loss immediately after the status epilepticus induced by pilocarpine injection. This spine loss was transient and was followed by a recovery in spine density that started 3 days after status epilepticus and reached a plateau level 15 days later (Isokawa, 2000). The subsequent recovery of dendritic spine density probably reflects the formation of new spines on preexisting granule cell dendrites and/or development of spines on outgrowing dendrites of newly formed granule cells subsequent to epilepsy-induced neurogenesis (Parent et al.,1997). Using this model, Ferhat et al. (2003) showed that an increase in acidic calponin is observed during this period of recovery of spine density, and they suggest that acidic calponin is involved in the formation of new dendritic spines. Those new spines might be involved in the formation of aberrant synapses with new mossy fiber terminals. Actin filament disruption induced by latrunculin A or jasplakinolide migh trigger molecular mechanisms for important remodeling of dendritic spine shape through the action of acidic calponin and/or synaptopodin (Roth et al., 2001).

The F-actin cytoskeleton is vulnerable to disruption by elevated intracellular calcium, a condition observed in several neuropathologies such as epilepsy, neurotrauma, and other degenerative neurological diseases. The results reported here support the idea that NMDARs and voltage-dependent Ca^{2+} channels remain suitable targets for future antiepileptic drugs, but they also support the search for other intracellular pharmacological targets such as calcineurin (Halpain et al., 1998), calpain (Sierra-Paredes et al., 1999), PKC, CaMKII (MacDonald et al., 1996) and several other proteins related to actin dynamics (Allison et al., 2000; Roth et al., 2001; Ferhat et al. 2003). Furthermore, the seizures induced by latrunculin-A or jasplakinolide present distinctive characteristics, such as low frequency and a permanent threshold decrease, that are completely different from the chemical convulsants widely used today, and could provide a new alternative experimental model of epilepsy for testing antiepileptic drugs.

Increased Intercellular Glutamate and Glycine Concentrations are Epileptogenic in Rats Pretereated with Latrunculin A

We have previously reported the effect of high concentrations of glutamate and glycine on picrotoxin-induced seizure thresholds on freely moving rats (Sierra-Paredes et al., 2000). Our results showed that high extracellular glutamate and glycine levels during prolonged microperfusion do not lead to excitatory neuronal damage or paroxismal activity in the living rat brain.

We found, however, that elevating glutamate and glycine concentrations in the hippocampus significantly increases neuronal excitability, as reflected by a significant reduction in picrotoxin seizure threshold (Sierra-Paredes et al., 2001). Glutamate and glycine acting alone are not enough to induce epileptiform seizures, but their continuous action may activate extrasynaptic receptors which facilitate seizure onset when GABAergic inhibition is impaired. The fact that glutamate and glycine produce almost the same effect indicates that their action is likely to be mediated by NMDA receptors. This is also confirmed by the fact that 5,7 DKA, a competitive inhibitor of the glycine binding site in the NMDA receptor, completely prevents the effect of both glutamate and glycine (Sierra-Paredes et al., 2001).

When rats are pretreated with latrunculin A, both glycine and glutamate induce seizures when 1mM concentrations are perfused into the hippocampus (table 1, fig 8). This effect can be observed from one to six months after latrunculin A administration. However, rats pretreated with picrotoxin do not show any seizures or EEG discharges when treated with the same glutamate or glycine concentrations (table 1). This means that latrunculin A, not seizure activity, induces permanent changes in neuronal excitability.

Table 1. Probability of Observing a Seizure in Controls and Rats Pretreated with Picrotoxin or Latrunculin A during Microperfusion of Vehicle, 1mM Glutamate and 1mM Glycine

Treatment	Vehicle	1mM Glu	1mM Gly
None	0%	0%	0%
Picrotoxin	0%	0%	0%
Latrunculin A	5,1%	78%	86%

As discussed above, latrunculin A induces long-term changes in neuronal excitability leading to the onset of spontaneous seizures and a permanent decrease in picrotoxin seizure threshold. One hypothesis which might explain the action of latrunculin A is the displacement of synaptic receptors to extrasynaptic locations, where they may be easily activated by intercellular glutamate. Our results are compatible with two usual findings in the brain of epileptic patients: morphological alterations in dendritic spines, and the increase in the concentrations of glutamate, glycine and other amino acids.

These results, taken together with previous data about the lack of effect of increased extracellular glutamate in normal rats (Sierra-Paredes et al., 2000 a, 2001 a; Obrenovitch et al., 1997 a,b) and the evidences of dendrite isolation from intercellular space (Ventura and Harris, 1999), support the proposal that exogenous glutamate never reaches postsynaptic space, and its action is only produced in special conditions by activating extrasynaptic receptors (Sinor et al., 2000; Sierra-Paredes et al., 1999).

Several studies suggest that NMDA receptors compatimentalization and distribution, as well as the NMDA receptor interaction with other proteins, are essential for excitotoxic effects (Sattler et al., 1998, 1999; Hardingham et al., 2002). It is possible that relationships among NMDA receptors and the intracellular processes which lead to increased excitability or cellular death may be different in extrasinaptic locations (Sinor et al., 2000; Hardingham et al., 2002). This suggests that glutamate activity may be different with the increase in the number of extrasynaptic receptors.

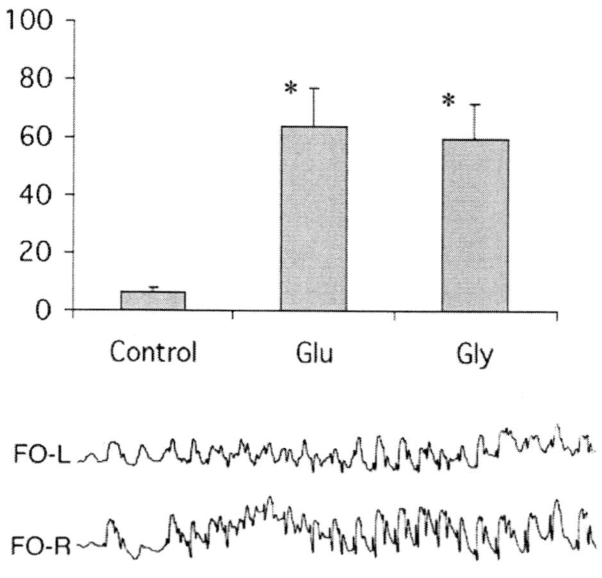

Figure 8. Probablility of recording a seizure in rats pretreated with latrunculin A when perfusing Ringer solution (control), 1mM glutamate (Glu) and 1mM glycine (Gly). Data are mean ± s.d. from 12 rats. EEG activity recorded in a rat perfused with 1mM glutamate.

The evidence suggesting that AMPA and NMDA receptors at synapses are highly dynamic and able to move laterally between synaptic and extrasynaptic pools (Tovar and Westbrook, 2000; Carroll et al., 1999; Hayashi et al., 2000; Liu and Cull-Candy, 2000), taken together with the in vitro studies about the action of latrunculin A indicating that actin filaments depolymerization induces AMPA and NMDA receptors displacement to extrasynaptic locations (Allison et al., 1998; Allison et al., 2000; Sattler et al., 2000), indicate that the epileptogenic action of intercellular glutamate and glycine occurs only when the structure or localization of glutamate ionotropic receptors is altered.

Our results do not support the idea that signals from extrasynaptic NMDA receptors necessarily trigger cell death pathways (Hardingham et al., 2002), but it should be taken into account the differences in the intercelular signals reaching a neuron in a living brain comparing to those reaching a cultured neuron. For instance, the culture used by Hardingham et al. was not enriched with astrocytes, which provide living brain neurons with many antiapoptotic signals (Rothstein et al., 1996).

On the other hand, recent evidencias about a role of astrocityc glutamete in mediating neuronal synchrony (Fellin et al., 2004), strongly support the role of extrasynaptic NMDA receptors in epileptic neuronal synchronization.

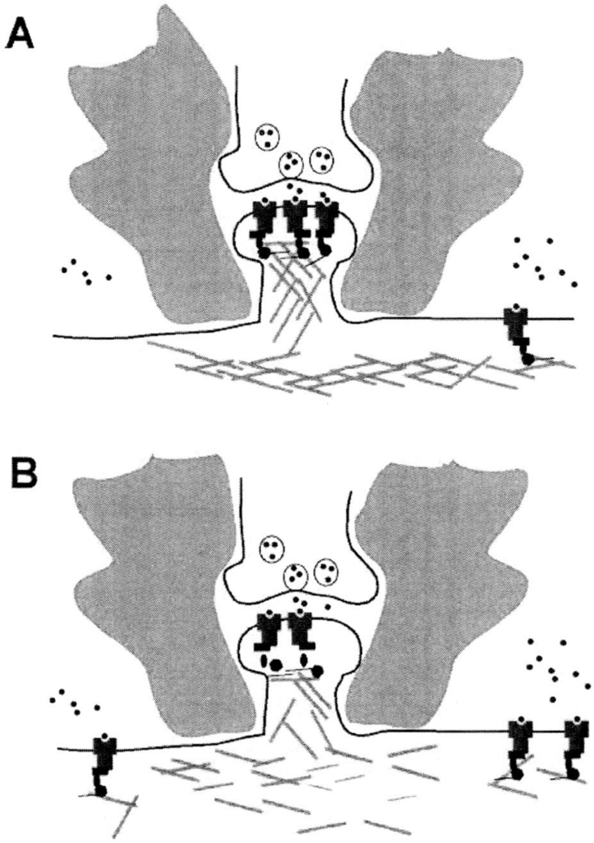

Figure 9. Schematic diagram of the proposed effects of actin depolymerization.
A: Under normal conditions, most NMDA receptors (in red) are located in the dendritic spine. B: after actin depolymerization with latrunculin A, some NMDA receptors are moved to extrasynaptic locations, where they can be activated by extracellular glutamate (black dots).

Altough most papers (Allison et al., 1998; Tovar and Westbrook, 2002) refer the effect of latrunculin A as "mobilization" of synaptic receptors to extrasynaptic locations, it is not easy to demonstrate that receptors found im extrasynaptic locations are the same that appeared previously in the postsynaptic space. Is seems probable that actin filaments depolimerization might lead to receptor internalization and proteolysis (Zhou et al., 2000), while new receptors, probably with a different subunit composition, would be carried to the cell membrane without the precise location control which happens under physiological conditions. Those receptors may be activated more easily by the glutamate perfused through microdialysis probes producing biochemical modifications leading to epileptic seizures. Actin filament depolymerization by latrunculin A migh also alter protein anchoring to NMDA receptors, modifying the intracellular signaling induced by both synaptic and extrasynaptic receptors.

The proconvulsant effect of amino acids is higher in those rats which presented seizures during latrunculin A microperfusion. It is possible that seizures potentiate the effect of latrunculin A, because during seizures, massive calcium entry through NMDA receptors produces by itself actin filament depolymerization.

The higher effect of glycine as comapred with glutamate is intriguing. It may be due to the fact that we have used the same concentration for both amino acids, while physiological intercellular levels of glycine are lower than glutamate levels (Sierra-Paredes et al., 1998). Furthermore, both glial and vascular transport mechanisms are much more active for glutamate (Hertz, 1979; O'Kane et al., 1999).

Our results, taken together with previous research discussed above, favor the hypothesis that extrasynaptic glutamate receptors might be involved in diverse pathological consequences linked to this neurotransmitter. An interesting conclusion is that selective inhibition of extrasynaptic NMDA receptors might have interesting therapeutic consequencies in the treatment of epilepsy, as well as in prevention of epileptogenesis in patientes with brain lesions, cerebral isquemia or stroke. During these injuries, it seems probable that the excess of extracellular glutamate of any source (including blood-brain barrier alterations), migh activate extrasynaptic NMDA receptors (Riccio and Ginty, 2002) contributing to epileptogenesis.

Generalized inhibition of NMDA receptors is highly problematic, because receptor funcion is required for normal synaptic transmission. The use of NMDA inhibitors, both competitive and non-competitive, with strong antiepileptic effect in experimental models of epilepsy, has not shown to be useful in the clinical trials performed to these moment (Rogawsky, 1995; Vizzi and Kiss, 1998). One of the future proposals in the search for new antiepileptic drugs might involve the development of selective methods for blocking extrasynaptic NMDA receptors without inactivating synaptic receptors.

Understanding the cellular and molecular changes that take place during latrunculin A-induced seizures is essential to understanding the physiological mechanisms that cause this activity.

PKA and Calcineurin Inhibitors Show Antiepileptic Effects

We have found that ascomycin (100 μM), a calcineurin inhibitor, has a good antiepileptic affect against picrotoxin seizures when perfused into the rat hippocampus (figs 10, 11). During picrotoxin seizures, significant changes in calcineurin activity were not found, however, a significant increase in calcineurin activity was observed during latrunculin A induced seizures (unpublished data). These differences might be due to the biochemical changes leading to the permanent changes produced by latrunculin A, and also to the biochemical consequencies of repeated seizures. Kurtz et al. (2001) demonstrated a significant increase in calcineurin activity in cortical and hippocampal homogenates during status epilepticus induced by pilocarpine.

Calcineurin is a calcium/calmodulin-stimulated enzyme (Klee et al. 1998) and would be stimulated by increased intracellular calcium concentrations. An increase in intracellular free calcium has been found during and after status epilepticus (Pal et al. 1999). The increased intracellular calcium associated with status epilepticus could be responsible for activating calcineurin above its normal physiological level. Status epilepticus induces a loss of function of the endoplasmic reticulum Mg 21 /Ca 21 ATPase (Parsons et al. 2000). This enzyme sequesters calcium ions into the microsomes of the smooth endoplasmic reticulum, providing a high affinity mechanism for regulating intracellular calcium concentration (Carafoli 1987; Miller 1991). After status epilepticus, ATPase-mediated uptake of calcium into the microsomes is less efficient (Parsons et al. 2000), which could potentially result in higher than normal resting calcium concentrations inside the cell did not affect the status epilepticus induced increase in calcineurin dephosphorylation.

Another explanation of the increase in calcineurin activity during status epilepticus and latrunculin seizures might be related with the activity of the calcium-stimulated neutral protease, calpain I. Calpain cleaves the catalytic calcineurin A subunit, removing the calmodulin binding domain and creating a highly active, 43 kDa, calcium-independent form of the enzyme (Kurtz et al., 2001; Tallant et al. 1988; Wang et al. 1989). Some researchers also propose that calpain-mediated proteolysis removes the autoinhibitory domain present in the A subunit (Wang et al. 1989). Calpain may be activated by the increased intracellular calcium concentrations present during status epilepticus, activating calcineurin. We have previously shown that calpain is involved in seizure arrest (Sierra-Paredes et al., 1999), so during status epilepticus calpain might be overactive, being responsible for the increase in calcineurin activity.

Calcineurin activity might also be involved in the biochemical changes leading to picrotoxin-induced epileptic seizures, but changes could bo too small to be easily detected. The results of one kindling study already suggest that calcineurin may have an epileptogenic role (Moia et al. 1994). Neuronal mechanisms which regulate calcineurin dephosphorylation under normal conditions could be highly altered as a result of impaired GABAergic activity. One important calcineurin-mediated mechanism is modulation of the $GABA_A$ receptor. GABA receptors are the primary receptor responsible for the fast inhibitory response in neuronal tissue (Macdonald and Olsen 1994), and play a major role in preventing the neuronal hyperexcitability associated with epilepsy. Several recent studies have demonstrated an inhibitory modulation of $GABA_A$ receptor function by calcineurin (Chen and Wong 1995; Amico et al. 1998; Lu et al. 2000). Picrotoxin binding to $GABA_A$ receptors may result in increased dephosphorylation, producing a net disinhibition of cellular excitability.

Increased excitability in neurons can be maintained by abnormalities in protein phosphorilation systems (Yechikhov et al., 2001). It was shown that PKC modulates receptor trafficking and gating (Lan et al., 2001) underlying synaptic plasticity (Esteban et al., 2003), and it is possible that in our latrunculin model receptor mobilization would be a direct effect of actin filaments depolymerization associated to changes in the phosphorilation mechanisms. In vivo PKA inhibition showed a slight antiepileptic effect on picrotoxin induced seizures (figures 10, 11), supporting the idea that normal phosphorilation and dephosphorilation are involved in seizure activity. Picrotoxin and latruculin-induced seizures are followed by a non significant increase in the activity of several protein kinases in the rat hippocampus (unpublished data). It seems possible that some long-term epileptogenic changes involve modifications in protein kinase and/or protein phosphatase activity, and the chronic biochemical or morphological changes induced by direct actin depolymerization might require the activation of those mechanisms. It is also possible that the location of enzimes would be critical for their action, modifying neuronal response (Hardingham et al., 2003). Further research is required in order to elucidate which biochemical and/or morphological alterations are epileptogenic in the latrunculin A/jasplakinolide model.

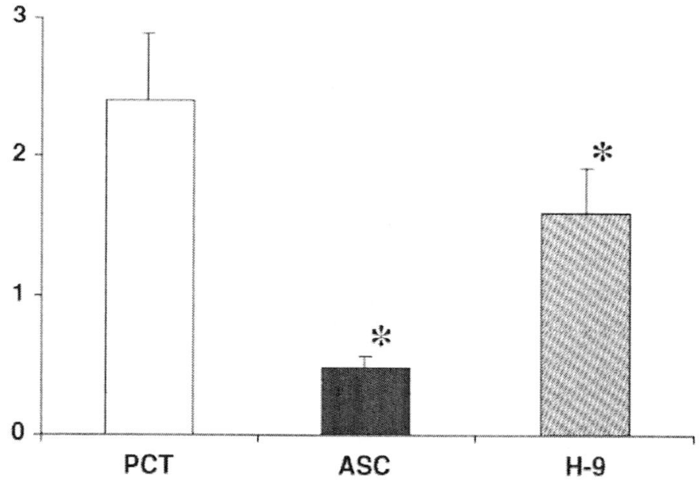

Figure 10. Ascomycin (100μM) and H-9 dihydrochloride (100μM) significantly reduce the number of seizures (mean ± s.d. from 8 rats) when perfused in the hippocampus of freely moving rats.

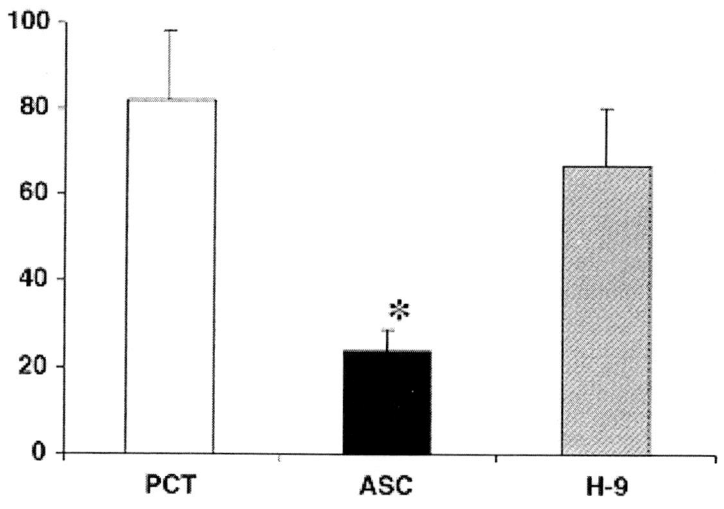

Figure 11. Ascomycin (100μM) significantly the mean seizure duration in picrotoxin treated rats. H-9 dihydrochloride (100μM) effect was not significant when compared with control rats.

Conclusions

Our results, showing that epileptic seizures may be caused by the direct disruption of the actin cytoskeleton, indicate that the biochemical study of actin-dependent receptor anchoring, related transduction pathways and many other PSD proteins seems to be an important approach to the neuropathology and neuropharmacology of epilepsy. Is seems crear now that neurotransmitter receptors and synapses are not fixed structures, but are subjected to continuous changes and replacements. Receptor and enzyme location seems to be a decisive feature for the transmission of neural messages among neurons, and it is very probable that cahanges in receptor location might be involved in learning associated plasticity changes as well as in some brain diseases. One important question remains to be answered: How and when these molecular dynamics produce long term changes in neuronal circuits, and why sometimes the changes do not seem to be easily reversible?. In the case of epileptogenesis, cell death and/or excessive axon prunning is one possible answer. However, further research is required to define the mechanisms of the permanent neuronal alterations induced by in vivo actin cytoskeleton disruption.

These results, taken together with previous research discussed above, favor the hypothesis that extrasynaptic NMDA and AMPA glutamate receptors might be involved in diverse pathological consequences linked to this neurotransmitter, including epileptic seizures. Altered biochemical conditions which may lead to an increase in extrasynaptic receptor number or activity can be related with epileptogenic processes. An interesting conclusion is that selective inhibition of extrasynaptic NMDA receptors and/or intracellular molecules involved in directing receptor location might have interesting therapeutic consequencies in the treatment of epilepsy, as well as in prevention of epileptogenesis in patientes with brain lesions, cerebral isquemia or stroke.

Acknowledgements

The authors gratefully acknowledge the work of the laboratory of Neuroanatomy (Director: Prof. J.L. Labandeira) who performed the histological study and Prof. J.L. Otero-Cepeda (Department of Statistics) for statistical analysis.

This work was supported by grants XUGA PGIDT00PXI20807PR and XUGA PGIDIT03PXIB20803PR from the Consellería de Educación e Ordenación Universitaria, Xunta de Galicia, Galicia, Spain.

References

Aguilar-Veiga, E. Sierra-Paredes, G., Galán-Valiente, J., Soto-Otero, R., Méndez-Alvarez, E. and Sierra-Marcuño, G. (1991) Correlation between ethosuximide brain levels measured by high performance liquid chromatography and its antiepileptic potential. *Res. Com. Chem, Pathol. Pharmacol. 71,* 351-364.

Aizawa, H., Hu, S.C., Bobb, K., Balakrishnan, K., Ince, G., Gurevich, I., Cowan, M. And Ghosh, A. (2004) Dendrite development regulated by CREST, a calcium-regulated transcriptional activator. *Science 303*: 197-202.

Allison, D.W., Gelfand, V.I., Spector, I. and Craig, A.M. (1998) Role of actin in anchoring postsynaptic receptors in cultured hippocampal neurons: differential attachment of NMDA versus AMPA receptors. *J Neurosci. 18,* 2423–2436.

Allison, D.W., Chervin, A.S., Gelfand, V.I. and Craig, A.M. (2000) Postsynaptic scaffolds of excitatory and inhibitory synapses in hippocampal neurons: maintenance of core components independent of actin filaments and microtubules. *J. Neurosci. 20,* 4545-4554.

Amico, C., Cupello, A., Fossati, C. and Robello, M. (1998) Involvement of phosphatase activities in the run-down of GABAA receptor function in rat cerebellar granule cells in culture. *Neuroscience 84*, 529-535.

Andreasen, M., Lambert, J.D.C. and Jensen, M.S., 1989. Effects of new non-N-methyl-D-aspartate antagonists on synaptic transmission in the *in vitro* rat hippocampus. *J. Physiol (Lond.)* 414, 317-336.

Aoki, C., Venkatesan, C., Go, C.G., Mong, J.A. and Dawson, T.M., 1994. Cellular and subcellular localization of NMDA-R1 subunit immunoreactivity in the visual cortex of adult and neonatal rats. *J. Neurosci.* 14, 5202-5222.

Bading, H., Ginty, D.D. and Greenberg, M.E. (1993) Regulation of gene expression in hippocampal neurons by distinct calcium signaling pathways. *Science 260,* 181-186.

Bach-y-Rita, P., 1991 Thoughts on the role of volumen transmission in normal and abnormal mass sustained functions. In: Fuxe, K. and Agnati, L.F. (Eds) *Volumen Transmission in the Brain* . New York: pp: Raven Press 489-496.

Bach-y-Rita, P., 1993 Nonsynaptic diffusion neurotransmission (NDN) in the brain. *Neurochem. Int.,* 23, 297-318.

Barnes, B.A., 1991. New roles for glia. *J. Neurosci.,* 11, 3685-3694.

Belichenko, P.V. and Dahlstrom, A. (1995) Studies on the 3-dimensional architecture of dendritic spines and varicosities in human cortex by confocal laser scanning microscopy and Lucifer yellow microinjections. *J. Neurosci. Meth. 57,* 55-61.

Bonfoco, E., Leist, M., Zhivotovsky, B., Orrenius, S., Lipton, S.A., and Nicotera, P. (1996) Cytoskeletal breakdown and apoptosis elicited by NO donors in cerebellar granule cells require NMDA receptor activation. *J. Neurochem. 67,* 2484-2493.

Bubb, M.R. Spector, I., Beyer, B.B. and Fosen, K.M. (2000) Effects of jasplakinolide on the kinetics of actin polymerization. *J. Biol. Chem. 275,* 5163-5170.

Carafoli, E. (1987) Intracellular calcium homeostasis. *Annu. Rev. Biochem. 56*, 395-433.

Carroll, R.C., Lissin, D.V., von Zastrow, M., Nicoll, R.A. and Malenka, R.C. (1999) Rapid redistribution of glutamate receptors contributes to long-term depression in hippocampal cultures. *Nat. Neurosci. 2,* 454-460.

Chapman A.G., 1997. Microdialysis. In: Engel Jr, J. and Pedley, T.A. (Eds.) *Epilepsy: A Comprehensive Textbook.* Lippincott-Raven Publishers, Philadelphia, pp. 1067-1072.

Chen, Q.X. and Wong, R.K.S. (1995) Suppression of GABAA receptor responses by NMDA application in hippocampal neurons acutely isolated from the adult guinea-pig. *J. Physiol. (Lond.) 482*, 353-362.

Chittajallu, R., Alford, S. and Collingridge, G.L. (1998) Ca^{2+} and synaptic plasticity. *Cell Calcium 24*, 377-385.

Choi, D.W. (1987) Ionic dependence of glutamate neurotoxicity. *J. Neurosci. 7*, 369-379.

_____ . (1988) Calcium-mediated neurotoxicity: relationship to specific channel types and role in ischemic damage. *Trends Neurosci. 11*, 465-467.

Churn, S.B., Limbrick, D., Sombati, S. and De-Lorenzo, R.J. (1995) Excitotoxic activation of the NMDA receptor results in inhibition of calcium/calmodulin kinase II activity in cultured hippocampal neurons. *J. Neurosci. 15*, 3200-3214.

Clark, B.A., Farrant, M. and Cull-Candy, S.G., 1997. A direct comparison of the single-channel properties of synaptic and extrasynaptic NMDA receptors. *J. Neurosci.* 17, 107-116.

Conti, F. and Weinberg, R.J., 1999. Shaping excitation at glutamatergic synapses. *Trends Neurosci.,* 22, 451-458.

Craig, A.M., Blackstone, C.D., Huganir, R.L. and Banker, G. (1994) Selective clustering of glutamate and γ-aminobutyric acid receptors opposite terminals releasing the corresponding neurotransmitters. *Proc. Natl. Acad. Sci. USA 91,* 12373-12377.

Curtis, D. R. and Eccles, R. M., 1958. The effect of diffusional barriers upon the pharmacology of cells within the central nervous system. *J. Physiol (Lond.)* 141, 446-463.

de Lanerolle, N.C., Eid, T., von Campe, G., Kovacs, I., Spencer, D.D., and Brines, M. (1998) Glutamate receptor subunits GluR1 and GluR2/3 distribution shows reorganization in the human epileptogenic hippocampus. *Eur. J. Neurosci. 10,* 1687-1703.

Del-Cerro, S., Arai, A., Kessler, M., Bahr, B.A., Vanderklish, P., Rivera, S. and Lynch, G. (1994) Stimulation of NMDA receptors activates calpain in cultured hippocampal slices. *Neurosci. Lett. 167*, 149-152.

DeLorenzo, R.J., Pal, S. and Sombati, S. (1998) Prolonged activation of the N-methyl-D-aspartate receptor-Ca^{2+} transduction pathway causes spontaneous recurrent epileptiform discharges in hippocampal neurons in culture. *Proc. Natl. Acad. Sci. USA 95,* 14482-14487.

De Robertis, E., 1965. Some new electron microscopical contribution to the biology of neuroglia. In: De Robertis, E. and Carrera, R. (Eds) *Progress in Brain Research, vol 15: Biology of Neuroglia* New York: Elsevier pp: 1-11.

Diamond, J.S., Bergles, D.E., and Jahr, C.E. , 1998. Glutamate release monitored with astrocyte transporter currents during LTP. *Neuron*, 21, 425-433.

Dichter, M. and Ayala, G.F., 1987. Cellular mechanisms of epilepsy: a status report. *Science*, 237, 157-164.

Dingledine, R., McBain, C.J. and McNamara, J.O. (1990) Excitatory amino acids in epilepsy. *Trends Pharmacol. Sci. 11,* 334-338.

Ehlers, M.D., Fung, E.T., O'Brien, R.J. and Huganir, R.L. (1998) Splice variant-specific interaction of the NMDA receptor subunit NR1 with neuronal intermediate filaments. *J Neurosci 18,* 720-730.

Esteban, J.A., Shi, S.H., Wilson, C., Nuriya, M., Huganir, R.L. and Malinow, R. (2003) PKA phosphorylation of AMPA receptor subunits control synaptic trafficking underlying plasticity. *Nature Neurosci. 6*, 2, 136-143.

Fagni, L., Baudry, M and Lynch, G., 1983. Clasification and properties of excitatory amino acid receptors in the hippocampus I. Electrophysiological studies of an apparent desensititazion and interactions with drugs which block transmission. *J. Neurosci.*, 3, 1538-1546.

Farb, C.R., Aoki, T. and LeDoux, J.E., (1995). Differential localization of NMDA and AMPA receptor subunits in the lateral and basal nuclei of amygdala: a light and electron microscope study. *J. Comp. Neurol.*, 362, 86-108.

Fellin, T., Pascual, O., Gobbo, S., Pozzan, T., Haydon, P.G. and Carmignoto, G. (2004) Neuronal synchrony mediated by astrocytic glutamate through activation of extrasynaptic NMDA receptors. *Neuron, 43*:729-743.

Fifkova, E. And Delay, R.J. (1982) Cytoplasmic actin in neuronal processes as a possible mediator of synaptic plasticity. J. Cell. Biol. 95:345-350.

Fischer, M., Kaech, S., Knutti, D. and Mattus, A. (1998) Rapid actin-based plasticity in dendritic spines. *Neuron 20,* 847-854.

Fuxe, K. and Agnati, L.F., 1991. Two principal modes of electrochemical communications in the brain: volume versus wiring transmission. *Adv. Neurosci.* 1, 1-9.

Ghosh, A. and Greenberg, M.E. (1995) Calcium signaling in neurons: molecular mechanisms and cellular consequences. *Science 268,* 239-247.

Gloor, P. and Fariello, R.G., 1988. Generalized epilepsy: some of its cellular mechanisms differ from those of focal epilepsy. *Trends Neurosci.,* 11, 63-68.

Groc, L., Heine, M., Cognet, L., Brickley, K., Stephenson, F. A., Lounis, B. and Choquet, D. (2004) Differential activity-dependent regulation of the lateral mobilities of AMPA and NMDA receptors. *Nature Neurosci. 7*, 695-696.

Groth, R.D., Dunbar, R.L. and Mermelstein, P.G. (2003) Calcineurin regulation of neuronal plasticity. *Biochem. Biophys. Res. Com. 311,*1159-1171.

Halpain, S., Hipolito, A. and Saffer, L. (1998) Regulation of F-actin stability in dendritic spines by glutamate receptors and calcineurin. *J. Neurosci. 18,* 9835-9844.

Halpain, S. (2000) Actin and the agile spine: how and why do dendritic spines dance? *Trends Neurosci. 23,* 141-146.

Hardingham, G.E., Fukunaga, Y. and Bading, H. (2002) Extrasynaptic NMDARs oppose synaptic NMDARs by triggering CREB shut-off and cell death pathways. *Nature Neurosci. 5*:5, 405-414.

Harris, K.M. and Kater, S.B. (1994) Dendritic spines: cellular specializations imparting both stability and flexibility to synaptic function. *Annu. Rev. Neurosci. 17,* 341-371.

Herkenham, M., 1991. Mismatches between neurotransmitter and receptor localizations: Implications for endocrine functions in the brain. In: Fuxe, K. and Agnati, L.F. (Eds) *Advances in Neuroscience, Vol 1: Volume Transmission in the Brain. Novel Mechanisms for Neural Transmission.* New York, Raven Press, 1991, pp: 63-87.

Herrero, I., Miras-Portugal, M.T. and Sanchez-Prieto, J. 1992a. Positive feedback of glutamate exocytosis by metabotropic presynaptic receptor stimulation. *Nature.* 360, 163-166.

_____ . 1992b Activation of protein kinase C by phorbol esters and arachidonic acid required for the optimal potentiation of glutamate exocytosis. *J. Neurochem.* 59, 1574-1577.

Hertz, I. 1979. Functional interactions between neurons and astrocytes I. Turnover and metabolism of putative amino acids transmitters. *Prog. Neurobiol.* 13, 277-323.

_____ . 1982. Astrocytes. In: Lajtha, A. (Ed.) *Handbook of Neurochemistry*. New York: Plenum Press, pp: 319-355.

Hidaka, H., Inagaki, M., Kawamoto, S. and Saaki, Y. (1984) Isoquinolinesulfonamides: novel and potent inhibitors of cyclic nucleotide dependent protein kinase and protein kinase C. *Biochemistry 23*, 50365041.

Hosford, D.A., Crain, B.J., Cao, Z., Bonhaus, D.W., Friedman, A.H., Okazaki, M.M., Nadler, J.V. and McNamara, J.O., 1991. Increased AMPA-sensitive quisqualate receptor binding and reduced NMDA receptor binding in epileptic human hippocampus. *J. Neurosci.*, 11, 428-434.

Huang, R. and Dillon, G.H. (1998) Maintenance of recombinant type A □-aminobutyric acid receptor function: role of tyrosine phosphatases and calcineurina. *J. Pharmacol. Exp. Ther. 286*, 243-255.

Isokawa, M. and Levesque, M.F. (1991) Increased NMDA responses and dendritic degeneration in human epileptic hippocampal neurons in slices. *Neurosci. Lett. 132,* 212-216.

Isokawa, M., Levesque, M.F., Fried I. and Engel J. Jr. (2000) Glutamate Currents in Morphologically Identified Human Dentate Granule Cells in Temporal Lobe Epilepsy. *J Neurophysiol 77,* 3355-3369.

Kaech, S., Fischer, M., Doll, T. and Matus, A. (1997) Isoform specificity in the relationship of actin to dendritic spines. *J. Neurosci. 17,* 9565-9572.

Klee, C.B., Ren, H. and Wang, X. (1998) Regulation of the calmodulin-stimulated protein phosphatase, calcineurin. *J. Biol. Chem. 273*, 13367-13370.

Kochan, L.D., Churn, S.B., Omojokun, O., Rice, A. and De-Lorenzo, R.J. (2000) Status epilépticus results in an N-metyl-D-aspartate receptor-dependent inhibition of Ca^{2+}/calmodulin-dependent kinase II activity in the rat. *Neuroscience 95*, 735-743.

Kornau, H.C., Schenker, L.T., Kennedy, M.B. and Seeburg, P.H. (1995) Domain interaction between NMDA receptor subunits and the postsynaptic density protein PSD-95. *Science 269,* 1737-1740.

Krupp, J.J., Vissel, B., Thomas, C.G., Heinemann, S.F. and Westbrook, G.L. (1999) Interactions of calmodulin and alpha-actinin with the NR1 subunit modulate Ca^{2+}-dependent inactivation of NMDA receptors. *J. Neurosci. 19,* 1165-1178.

Kullman, D.M., Erdemli, G. and Asztély, F., 1996. LTP of AMPA and NMDA receptor-mediated signals: Evidence for presynaptic expression and extrasynaptic glutamate spill-over. *Neuron,* 17, 461-474.

Kurz, J.E., Sheets D., Travis-Parsons, J., Rana, A., Delorenzo, R.J. and Churn, S.B. (2001) A significant increase in both basal and maximal calcineurin activity in the rat pilocarpine model of status epilepticus. *J. Neurochem. 78*, 304-315.

Lan, J., Skeberdis, V.A., Jover, T., Grooms, S.Y., Lin, Y., Araneda, R.C., Zheng, X., Bennett, M.V.L. and Zukin, S. (2001) Protein kinase C modulates NMDA receptor trafficking and gating. *Nature Neurosci. 4*:4, 382-390.

Lees, G.V., and Jones, E.G. (2000) Epilepsy. *Neurobiol. Dis. 7*, 549-551.

Lehmann, A. and Hansson, E., 1988. Kainate-induced stimulation of amino acids release from primary astroglial cultures of the rat hippocampus. *Neurochem. Int.*, 13, 557-561.

Levi, G. and Patrizio, M., 1992. Astrocytes heterogeneity: endogenous amino acid levels and release evoked by non-N-methyl-aspartate receptor agonists and by potassium-induced swelling in type-1 and type-2 astrcytes. *J. Neurochem.,* 58, 1943-1952.

Li, B., Chen, N., Luo, T., Otsu, Y., Murphy, T.H. and Raymond, L.A. (2002) Differential regulation of synaptic and extra-synaptic NMDA receptors. *Nature Neurosci. 5:*9, 833-834.

Lin, J.W., Ju, W., Foster, K., Lee, S.H., Ahmadian, G., Wyszynski, M., Wang, Y.T., and Sheng, M. (2000) Distinct molecular mechanisms and divergent endocytotic pathways of AMPA receptor internalization. *Nat. Neurosci. 3,* 1282-1290.

Liu, S.J. and Cull-Candy, S.G. (2000) Activity at calcium-permeable AMPA receptors induces a switch in receptor subtype. *Nature 405*, 454-458.

Liu, L., Wong, T.P. Pozza, M.F., Lingenhoehl, K. Wang, Y., Sheng, M., Auberson, Y. P. and Wang, T.Y. (2004) Role of NMDA receptor subtypes in governing the direction of hippocampal synaptic plasticity. *Science 304*, 1021-1024.

Liu, Z., Stafstrom, C.E., Sarkisian, M.R., Yang, Y., Hori, A., Tandon, P. and Holmes, G.L., 1997. Seizure-induced glutamate release in mature and immature animals: in vivo microdialysis. *Neuroreport*, 8, 2019-2023.

Löscher, W. and Hönack, D., 1994. Effects of non-NMDA antagonist NBQX and the 2,3-benzodiazepine GYKI 52466 on different seizure types in mice: comparison with diazepam and flumazenil. *Br. J. Pharmacol.,* 113, 1349-1357.

Lynch, M.A., Errington, M.L. and Bliss, T.V.P., 1989. Nordihydroguaiaretic acid blocks the synaptic component of long-term potentiation and the associated increase in release of glutamate and arachidonate: an in vivo study in the dentate gyrus in the rat. *Neuroscience* 30, 693-701.

Lu W, Man H, Ju W, Trimble WS, MacDonald JF, Wang YT. (2001) Activation of synaptic NMDA receptors induces membrane insertion of new AMPA receptors and LTP in cultured hippocampal neurons. Neuron. 29(1):243-54.

MacDonald, J.F., Browning, M.D. and Wang, L.Y. (1996) Long-term enhancement of excitability and the regulation of glutamate receptors by protein kinases. In *Progressive Nature of Epileptogenesis* (U. Heinnemann, J. Engel Jr., G. Avanzini, B.S., Meldrum, A. Mouritzen-Dam, and C. Wasterlain, Eds) pp.275-282. Elsevier, Amsterdam.

Macdonald, R.L. and Olsen, R.W. (1994) $GABA_A$ receptor channels. *Annu. Rev. Neurosci. 17*, 569-602.

Matus, A., Ackermann, M., Pehling. G., Byers, H.R. and Fujiwara, K. (1982) High actin concentrations in brain dendritic spines and postsynaptic densities. *Proc. Natl. Acad. Sci. USA 79,* 7590 –7594.

Matus, A., Brinkhaus, H. And Wagner, U. (2000) Actin dynamics in dendritic spines: A form of regulated plasticity at excitatory synapses. *Hippocampus 10*: 555-560.

Matus, A. (2000) Actin-based plasticity in dendritic spines. *Science 290,* 754-758.

McNamara, J.O. (1999) Emerging insights into the genesis of epilepsy. *Nature 399,* A15-A22.

Martin, D.L., 1992. Synthesis and release of neuroactive substances by glial cells. *Glia* 5, 81-94.

Massieu, L., Morales-Villagrán, A. and Tapia, R., 1995 Accumulation of extracellular glutamate by inhibition of its uptake is not sufficient for inducing neuronal damage: An in vivo microdialysis study. *J. Neurochem.* 64, 2262-2272.

Matsumoto, H. and Ajmone-Marsan, C. 1964a. Cortical cellular phenomena in experimental epilepsy: Inter-ictal manifestations. *Exp. Neurol* 9, 286-304.

_____ . 1964b. Cortical cellular phenomena in experimental epilepsy: Ictal manifestations. *Exp. Neurol* 9, 305-326.

Meadows R.P., Nettesheim D.G., Xu R.X., Olejniczak E.T., Petros A.M., Holzman T.F., Severin J., Gubbins E., Smith H. and Fesik S.W. (1993) Three-dimensional structure of the FK506 binding protein/ascomycin complex in solution by heteronuclear three- and four-dimensional NMR. *Biochemistry. 32*(3):754-765.

Millan, M.H., Obrenovitch, T.P., Sarna, G.S., Lok, S.Y., Symon, L. and Meldrum, B.S., 1991 Changes in rat brain extracellular glutamate concentration during seizures: an in vivo dialysis study with on line enzimatic detection. *Epilepsy Res. 9,* 86-91.

Miece, M., Boutelle, M.G., and Fillenz, M., 1996. The source of physiologically stimulated glutamate efflux from the striatum of conscious rats. *J. Physiol. (Lond.)* 497, 745-751.

Miller, R.J. (1991) The control of neuronal Ca^{2+} homeostasis. *Annu. Rev. Biochem. 56*, 395-433.

Misonou, H., Mohapatra, D.P., Park, E.W., Leung, V., Zhen, D., Misonou, K., Anderson, A.E. and Trimmer, J.S. (2004) Regulation of ion channel localization and phosphorylation by neuronal activity. *Nature Neurosci. 7*, 711-718.

Moia, L., Matsui, H., de Barrose, G.A.M., Tomizawa, K., Miyamoto, K., Kuwata, Y., Tokuda, M., Itano, T. and Hatase, O. (1994) Immunosupressants and calcineurin inhibitors cyclosporin A and FK506, reversibly inhibit epileptogenesis in amigdaloid kindled rat. *Brain Res. 648*, 337-341.

Montoro, R.J., Diaz-Nido, J., Avila, J. and Lopez-Barneo, J. (1993) N-metyl-D-aspartate stimulates the dephosphorylation of the microtubule-associated protein 2 and potentiates excitatory synaptic pathways in the rat hippocampus. *Neuroscience 54*, 859-871.

Morales, M., Colicos, M.A. and Goda, Y. (2000) Actin-dependent regulation of neurotransmitter release at central synapses. *Neuron 27,* 539-550.

Müller M, Gähwiler BH, Rietschin L and Thompson SM (1993) Reversible loss of dendritic spines and altered excitability after chronic epilepsy in hippocampal slice cultures. *Proc. Natl. Acad. Sci. USA 90*, 257-261.

Multani, P. Myers, R.H., Blume, H.W., Schomer, D.L. and Sotrel, A. (1994) Neocortical dendritic pathology in human partial epilepsy: a quantitative Golgy study. *Epilepsia 35,* 728-736.

Murphy, S. and Pearce, B., 1987. Functional receptors for neurotransmitters on astroglial cells. *Neuroscience*, 2, 381-394.

Nakanishi, S. 1992. Molecular diversity of glutamate receptors and implications for brain function. *Science*, 258, 597-603.

Nicholson, C.H., 1995. Extracellular space as the pathway for neuron-glial cell interactions. In: Ketternmann, H and Ransom, B.R. (Eds) *Neuroglia*. New York, Oxford: Oxford University Press, pp: 387-397.

Obrenovitch, T.P., Urenjak, J., and Zilkha, E., 1996. Evidence disputing the link between seizure activity and high extracellular glutamate. *J.Neurochem.* 66, 2446-2454.

O'Brien, R.J., Kamboj, S., Ehlers, M.D., Rosen, K.R., Fischbach, G.D., and Huganir, R.L. (1998) Activity-dependent modulation of synaptic AMPA receptor accumulation. *Neuron 21,* 1067-1078.

O'Kane, R.L., Martínez-López, I., DeJoseph, M.R., Viña, J.R. and Hawkins, R.A. (1999) Na^+-dependent glutamate transporters (EAAT1, EAAT2, and EAAT3) of the blood-brain barrier. *J. Biol. Chem. 274*, 31891-31895.

Olsson, T., Rönnback, L. and Hansson, E., 1997. Astroglia and brain function. In: Hansson, E., Olsson, T., Rönnback, L. (Eds) *On astroglia and glutamate neurotransmission.* New York: Springer- R.G. Landes. pp: 1-14.

Pal, S., Sombati, S., Limbrick, D.D. and De-Lorenzo, R.J. (1999) *In vitro* status epilepticus causes sustained elevation of intracellular calcium levels in hippocampal neurons. *Brain Res. 851*, 20-31.

Pallen, C.J. and Wang, J.H. (1985) A multifunctional calmodulin-stimulated phosphatase. *Arch. Biochem. Biophys. 237*, 281-291.

Parent J.M., Yu, T.W., Leibowitz, R.T., Geschwind, D.H., Sloviter, R.S., Lowenstein, D.H. (1997) Dentate granule cells neurogenesis is increased by seizures and contributes to aberrant network reorganization in the adult rat hippocampus. *J. Neurosci. 17*, 3727-3728.

Parsons, J.T., Churn, S.B. and De-Lorenzo, R.J. (1997) Ischemia-induced inhibition of calcium uptake into rat brain microsomes mediated by Mg^{2+}/Ca^{2+} ATPase. *J. Neurochem. 68*, 1124-1134.

Parsons, J.T., Churn, S.B. and De-Lorenzo, R.J. (1999) Global ischemia-induced inhibition of the coupling ration of calcium uptake and ATP hydrolysis by rat whole brain microsomal Mg^{2+}/Ca^{2+} ATPase. *Brain Res. 834*, 32-41.

Parsons, J.T., Churn, S.B., Kochan, L.D. and De-Lorenzo, R.J. (2000) Pilocarpine-induced status epilépticus causes N-metyl-D-aspartate receptor-dependent inhibition of microsomal Mg^{2+}/Ca^{2+} ATPase mediated Ca^{2+} uptake. *J. Neurochem. 74*, 1209-1218.

Paxinos, G., and Watson, C., (1997) *The Rat Brain in Stereotaxic Coordinates* Academic Press, London.

Ponting, C. P., Phillips, C., Davies, K.E. and Blake, D.J. (1997) PDZ domains: targeting signalling molecules to sub-membranous sites. *BioEssays 19,* 469-479.

Peña, F and Tapia, R., 1999. Relation among seizures, extracellular amino acid changes and neurodegeneration induced by 4-aminopyridine in rat hippocampus: A microdialysis and electroencephalographic study. *J. Neurochem.*, 72, 2006-2014.

Rao, A. and Craig, A.M. (1997) Activity regulates the synaptic localization of the NMDA receptor in hippocampal neurons. *Neuron 19,* 801-812.

_____ . (2000) Signaling between the actin cytoskeleton and the postsynaptic density of dendritic spines. *Hippocampus.*10(5):527-41.

Riccio, A and Ginty, D.D. (2002) What a privilege to reside at the synapse: NMDA receptor signalling to CREB. *Nature Neurosci. 5*:5, 389-390.

Rice, A.C. and De-Lorenzo R.J. (1998) NMDA receptor activation during status epilepticus is required for the development of epilepsy. *Brain Res. 782*, 240-247.

Rogawski, M.A., 1995. Excitatory amino acids and seizures. In: Stone, T.W. (Ed) *CNS Neurotransmitters and Neuromodulators: Glutamate*. New York: CRC Press. pp: 219-37.

Rönnback, L. and Hansson, E., 1997. Does astroglial network perform qualitative modifications of neuronal messages? In: Hansson, E., Olsson, T., Rönnback, L. (Eds) *On astroglia and glutamate neurotransmission*. New York: Springer- R.G. Landes. pp: 155-187.

Rosenmund, C. and Westbrook, G.L. (1993) Calcium-induced actin depolymerization reduces NMDA channel activity. *Neuron 10,* 805– 814.

Roth, S.U., Sommer, C., Mundel, P and Kiessling, M. (2001) Expression of synaptopodin, an actin-associated protein, in the rat hippocampues after limbic epilepsy. *Brain Pathol. 11*, 169-181.

Rothstein, J.D., Dykes-Hobert, M., Pardo, C.A., Bristol, L.A., Jin. L., Kuncl. R.W., Kanai, Y., Hediger, M.A., Wang, Y.F., Schtelke, J.P. and Welty, D.F., 1996. Knockout of glutamate transporters reveals a major role from astroglial transport in exocitotoxicity and clearance of glutamate. *Neuron*, 16, 675-686.

Rzigalinski, B.A., Weber, J.T., Willoughby, K.A. and Ellis, E.F. (1998) Intracellular free calcium dynamics in strech-injured astrocytes. *J. Neurochem. 70*, 2377-2385.

Sattler, R., Xiong, Z., Lu, W.Y., MacDonald, J.F. and Tymianski, M. (2000) Distinct roles of synaptic and extrasynaptic NMDA receptors in excitotoxicity. *J. Neurosci. 20,* 22-33.

Seifert, G., and Steinhäuser, C., 1995. Glial cells in the mouse hippocampus express AMPA receptors with an intermediate Ca^{2+} permeability. *Eur. J. Neurosci.*, 7, 1872-1881.

Schab, R.S., 1967. Intravenous diazepam in the treatment of prolonged seizure activity. *N. Engl. J. Med.,* 276, 779-784.

Scharfman, H.E. (2002) Epilepsy as an example of neural plasticity. *The Neuroscientist*, *8* (2), 154-173.

Schousboe, A., 1981. Transport and metabolism of glutamate and GABA in neurons and glial cells. *Int. Rev. Neurobiol.*, 22, 1-45.

Schousboe, A., Svenneby, G. and Hertz, L., 1977. Uptake and metabolism of glutamate in astrocytes cultured from dissociated mouse brain hemispheres. *J. Neurochem.*, 29, 999-1005.

Sheibel, M.E. and Scheibel, A.B. (1973) Hipocampal pathology in temporal lobe epilepsy. A Golgi survey. In *Epilepsy: its phenomena on man* (M.A. Brazier, Ed.) pp. 311-337 Academic Press, New York.

Sheibel, M.E., Crandall, P.H. and Sheibel, A.B. (1974) The hippocampal-dentate complex in temporal lobe epilepsy. *Epilepsia 15,* 55-88.

Shi, S.H., Hayashi, Y., Petralia, R.S., Zaman, S.H., Wenthold, R.J., Svoboda, K., and Malinow, R. (1999) Rapid spine delivery and redistribution of AMPA receptors after synaptic NMDA receptor activation. *Science 284,* 1811-1816.

Shorte, S.L. (1997) N-methyl-D-aspartate evokes rapid net depolymerization of filamentous actin in cultured rat cerebellar granule cells. *J. Neurophysiol. 78,* 1135-1143.

Sierra, G., Acuña, C., Otero, J. and Docampo, G., 1969. Experimental study of a new psychosedative drug: HS-2314. *Int. J. Neuropharmac.,* 8, 153-160.

Sierra-Paredes, G., and Sierra-Marcuño, G., (1996) Microperfusion of picrotoxin in the hippocampus of freely moving rats through microdialysis probes: a new method of induce partial and secondary generalized seizures. *J. Neurosci. Meth. 67,* 113-120.

Sierra-Paredes, G., Cornes, J.M. and Sierra-Marcuño, G. (1999) Calpain inhibitor I reatards seizure arrest in the hippocampus of freely moving rats. *Neurosci. Lett. 263,* 165-168.

Sierra-Paredes, G., Galán-Valiente, J.,.Vazquez-Illanes, M. D., Aguilar-Veiga, E. and Sierra-Marcuño, G. (2000a) Effect of ionotropic glutamate receptors antagonists on the modifications in extracellular glutamate and aspartate levels during picrotoxin seizures: a microdialysis study in freely moving rats *Neurochem. Int. 37,* 377-386.

Sierra-Paredes, G., Oreiro-García, M.T., Cornes J.M., and Sierra-Marcuño, G. (2000b) Long-term decrease in picrotoxin seizure threshold induced by in vivo actin depolymerization in the rat hippocampus. *Epilepsia 41,* suppl 7, 23.

Sierra-Paredes, G., Senra-Vidal, A, and Sierra-Marcuño, G. (2001) Effect of extracellular high concentrations of glutamate and glycine on picrotoxin seizure thresholds: a microdialysis study in freely moving rats. *Brain Res. 888,* 19-25.

Sinor, J.D., Du, S., Venneti, S., Blitzblau, R.C., Leszkiewicz, D.N., Rosenberg, P-A. and Aizenman, E. (2000) NMDA and glutamate evoke excitotoxicity at distinct cellular locations in rat control neurons in vitro. *J. Neurosci. 20*: 8831-8837.

Smythies, J(2002). The Dynamic Neuron. Cambridge, Massachusetts. The MIT Press.

Soto-Otero, R., Méndez-Alvarez, E., Galán-Valiente, J., Aguilar-Veiga, E., and Sierra-Marcuño, G., 1994. Quantitative analysis of neuroactive amino acids in brain tissue by liquid chromatography using fluorescent pre-column labelling with *o* -phthalaldehide and *N*-acetyl-L-cysteine. *Biomed. Chromatogr.,* 8, 114-118.

Spector, I., Shochet, N.R., Kashman, Y. and Groweiss, A. (1983) Latrunculins: novel marine toxins that disrupt microfilament organization in cultured cells. *Science 219,* 493-495.

Spector, I., Shochet N.R., Blasberger, D. and Kashman, Y. (1989) Latrunculins– novel marine macrolides that disrupt microfilament organization and affect cell growth: I. Comparison with cytochalasin *D. Cell. Motil. Cytoskeleton 13,* 127–144.

Stasheff, S.F., Anderson, W.W., Clark, S. and Wilson, W.A. (1989) NMDA antagonists differentiate epileptogenesis from seizures in an in vitro model. *Science 245,* 648-651.

Steinhäuser, C and Gallo, V., 1996. News on glutamate receptors in glial cells. *Trend. Neurosci.,* 19, 339-345.

Steinhäuser, C., Jabs, R. and Kettenmann, H., 1994. Glutamate activates a cationic conductance and blocks potassium currents in identified glial cells of the mouse hippocampal slice. *Hippocampus* 4, 19-36.

Tallant, E.A., Brumley, L.M. and Wallace, R.W. (1988) Activation of a calmodulin-dependent phosphatase by a Ca^{2+} dependent protease. *Biochemestry 27*, 2205-2211.

Taylor, C.P. and Dudek, F.E., 1982. Synchronous neural afterdischarges in rat hippocampal slices without active chemical synapsis. *Science*, 218, 810-812.

Tehrani, M.H.J. and Barnes E.M. (1995) Reduced function of □-aminobutyric $acid_A$ receptors in tottering mouse brain: Role of cAMP-dependent protein kinase. *Epilepsy Res. 22*, 13-21.

Tovar, K.R. and Westbrook, G.L. (1999) The incorporation of NMDA receptors with a distinct subunit composition at nascent hippocampal synapses in vitro. *J. Neurosci. 19*, 4180-4188.

Traub, R.D. and Wong, R.K.S., 1982. Cellular mechanisms of neuronal synchronization in epilepsy. *Science*, 216, 745-747.

Ventura, R. and Harris, K.M. (1999) Three-dimensional relationships between hippocampal synapses and astrocytes. *J. Neurosci. 19*(16), 6897-6906.

Vizi, E.S. and Kiss, J.P., 1998. Neurochemistry and pharmacology of the major hippocampal transmitter systems: Synaptic and nonsynaptic interactions. *Hippocampus* 8, 566-607.

Vizi, E.S., Mike, A., and Tarnawa, I., (1996). 2,3-Benzodiazepines (GYKI 52466 and analogs): Negative allosteric modulators of AMPA receptors. *CNS Drug Reviews*, 2, 91-126.

Walaas, S.I. and Greengard P. (1991) Protein phosphorylation and neuronal function. *Pharmacol. Rev.*, *43*, 299-349.

Wang, K.K.W., Roufogalis, B.D. and Villalobo, A. (1989) Characterization of the fragmented forms of calcineurin produced by calpain I. *Biochem. Cell. Biol. 67*, 703-711.

Westphal, R.S., Tavalin, S.J., Lin, J.W., Alto, N.M., Fraser, I.D.C., Langeberg, L.K., Sheng, M. and Scott, J.D. (1999) Regulation of NMDA receptors by an associated phosphatase-kinase signaling complex. *Science 285*, 93-96.

Wheal, H.V., Simpson, L., Phelps, S. and Stockley, E., 1991. Excitatory amino acid receptor subtypes and their roles in epileptiform synaptic potentiation in the hippocampus. In: Wheal H. V. and Thomson A. (Eds), *Excitatory Amino Acids and Synaptic Transmission* New York: Academic Press. pp: 239-264.

Wilkin, G.P., Marriot, D.R. and Cholewinsky, A.J., 1990. Astrocyte heterogeneity. *Trends Neurosci.,* 13, 43-46.

Wyler, A.R., 1982. Neuronal activity during seizures in monkeys. *Exp. Neurol.*, 76, 574-585.

Wyneken, U., Smalla, K.H., Marengo J.J., Soto, D., de la Cerda, A., Tischmeyer, W., Grimm, R., Boeckers, T.M., Wolf, G., Orrego, F., Gundelfinger, E.D. (2001) Kainate induced seizures alter protein composition and N-methyl-D-aspartate receptor function of rat forebrain postsynaptic densities. *Neuroscience*, *102*:65-74.

Wyszynski, M., Lin, J., Rao, A., Nigh, E., Beggs, A.H., Craig, A.M., Weinberg, R.J., and Sheng, M. (1997) Competitive binding of alpha-actinin and calmodulin to the NMDA receptor. *Nature 385,* 439-442.

Yang, Y., Ge,W., Chen,Y., Zhang,Z., Shen,W., Wu,C., Poo,M. and Duan, S (2003) Contribution of astrocytes to hippocampal long-term potentiation through release of D-serine. Proc Natl Acad Sci U S A. 100(25): 15194–15199.

Yechikhov, S., Morenkov, E., Chulanova, T., Godukhin, O. and Shchipakina, T. (2001) Involvement of cAMP and Ca^{+2}/calmodulin-dependent neuronal protein phosphorylation in mechanisms underlying genetic predisposition to audiogenic seizures in rats. *Epilesy Res. 46*, 15-25.

Yuste, R. and Denk, W. (1995) Dendritic spines as basic fundamental units of neuronal integration. *Nature 375,* 682-684.

Zhang, S., Ehlers, M.D., Bernhardt, J.P., Su, C.T. and Huganir, R.L. (1998) Calmodulin mediates calcium-dependent inactivation of N-methyl-D-aspartate receptors. *Neuron 21,* 682-453.

In: Focus on Neurochemistry Research
Editor: Robert M. Coleman, pp.197-212
ISBN 1-59454-417-4

Chapter VII

Uptake of Amino Acid Neurotransmitters into Synaptic Vesicles: A Target for 5-aminolevulinic acid

Tatiana Emanuelli[1,2,*] Rejane G. Tavares[2], Candice E. S. Santos[2], Letícia B. Alves[2], Aníbal E. Vercesi[3] and Diogo O. Souza[2]
[1]Departamento de Tecnologia e Ciência de Alimentos, Centro de Ciências Rurais, Universidade Federal de Santa Maria, Santa Maria, RS;
[2]Departamento de Bioquímica, Instituto de Ciências Básicas da Saúde, Universidade Federal do Rio Grande do Sul, Porto Alegre, RS;
[3]Departamento de Patologia Clínica, Faculdade de Ciências Médicas, Universidade Estadual de Campinas, Campinas, SP, Brasil

Abstract

5-Aminolevulinic acid (ALA) is an haem precursor that accumulates in acute intermittent porphyria and lead poisoning. It has been proposed that ALA overload is involved in the neurological dysfunctions associated with those disorders. This chapter was aimed at investigating the effects of ALA on the uptake of amino acid neurotransmitters by rat brain synaptic vesicles. ALA significantly inhibited the uptake of the neurotransmitter amino acids [^{3}H]glycine, [^{3}H]γ-aminobutyric acid and [^{3}H]glutamate (IC_{50} values of 1.08 x 10^{-6}, 7.08 x 10^{-4} and 1.18 x 10^{-3} M, respectively). This effect was not due to the inhibition of the vesicular H^+-ATPase activity, which

* Corresponding Author: Tatiana Emanuelli, Departamento de Tecnologia e Ciência dos Alimentos, Centro de Ciências Rurais, Universidade Federal de Santa Maria, Campus – Camobi, 97105-900 - Santa Maria – RS, Brasil. FAX: +55 55 3220 8353; E-mail: tati@ccr.ufsm.br

generates the electrochemical gradient that drives the transport of neurotransmitters. ALA also had no effect on the transmembrane potential ($\Delta\psi$) measured across the synaptic vesicles. However, ALA dose-dependently inhibited the formation of ΔpH when added before gradient formation and dissipated ΔpH when added after gradient formation, indicating that its inhibitory effect is related to a selective dissipation of the proton gradient. These results provide evidence that ALA may disturb the vesicular uptake of amino acids, which is an important event in the storage and release of these neurotransmitters by synaptic vesicles.

Key Words: 5-Aminolevulinic acid; Synaptic vesicles; Vesicular uptake; Glycine; γ-Aminobutyric acid (GABA); Glutamate.

Introduction

Excitatory neurotransmission in the central nervous system is mediated mainly by the amino acid glutamate (Lipton & Rosenberg, 1994; Ozawa et al., 1998), which is involved in many physiological processes, such as neural development and plasticity, cognition, learning and memory (Ozawa et al., 1998). Conversely, the amino acids γ-aminobutyric acid (GABA) and glycine are responsible for mediating inhibitory neurotransmission in the brain and spinal cord, respectively, participating in a variety of motor and sensory functions (Olsen & DeLorey, 1999). Impairments of the inhibitory and/or excitatory neurotransmission have been suggested to be involved in several neurological and psychiatric disorders of humans (Lipton & Rosenberg, 1994; Olsen & DeLorey, 1999; Mattson, 2003; Pearl & Gibson, 2004).

5-Aminolevulinic acid (ALA) is an haem precursor that accumulates in tissues of patients suffering from acute intermittent porphyria (AIP), tyrosinosis or lead poisoning, due to a deficiency in the haem biosynthetic pathway (Kappas et al., 1995; Sakai & Morita, 1996; Nordman & Puy, 2002). Such disorders are characterized by acute attacks that start with abdominal pain (peripheral neuropathy) and may be accompanied by disturbances in the central nervous system (anxiety, psychosis, insomnia, disturbed consciousness and seizures; Kappas et al., 1995; Nordman & Puy, 2002). The pathogenesis of these manifestations is not completely elucidated but a significant amount of evidence suggests the direct involvement of ALA (Kappas et al., 1995; Demasi et al., 1996; Puy et al., 1996). There is experimental evidence that ALA may disturb the neurotransmission mediated by structurally related amino acids, such as GABA and glutamate, both *in vivo* and *in vitro*. Intrastriatal administration of ALA induces convulsions and body asymmetry in rats (Emanuelli et al., 2000; Prauchner et al., 2004). Such behavioral alterations were prevented by glutamate antagonists and by a $GABA_A$ agonist, suggesting the involvement of GABAergic and glutamatergic mechanisms (Emanuelli et al., 2000). In synaptosomal preparations, ALA activates GABA receptors, inhibits the K^+-stimulated release of GABA, and the Na^+-dependent uptake of GABA and glutamate, whereas it increases the basal release of GABA and glutamate (Dichter et al., 1977; Müller & Snyder, 1977; Brennan & Cantrill, 1979a; 1979b; 1979c; Emanuelli et al., 2001a; 2003). However, many of these effects were observed at ALA concentrations higher than those found in the extracellular fluids of patients (10^{-5} to 10^{-7} M in the cerebrospinal fluid and 10^{-3} to 10^{-5} M in the plasma; Yeung Laiwah, 1987).

Few studies have been conducted to investigate intracellular targets for the neurotoxic properties of ALA. The studies available have focused on the inhibitory effect of ALA on Na^+,K^+-ATPase activity (Russell et al., 1983), on adenylate cyclase activity (Emanuelli et al., 2001b), and on the prooxidant effects of this compound (Demasi et al., 1996).

One important intracellular step of synaptic transmission modulation is the uptake of neurotransmitters from the cytosol of the presynaptic terminal to the lumen of synaptic vesicles (Liu & Edwards, 1997). This transport is mediated by specific low affinity transporters present in the vesicle membrane (Fykse & Fonnum, 1996; Liu & Edwards, 1997) and is driven by a proton electrochemical gradient ($\Delta\mu_H^+$) generated by a vacuolar Mg^{2+}-dependent bafilomycin A_1-sensitive H^+-ATPase of the vesicle membrane (Cidon & Sihra, 1989; Nelson, 1991; Fykse & Fonnum, 1996). Three vesicular glutamate transporters (VGLUTs 1, 2, and 3) and one vesicular inhibitory amino acid transporter (VIAAT) that is responsible for GABA and probably glycine storage have been identified (Sagne et al., 1997; Bai et al., 2001; Shigeri et al., 2004). The vesicular transporters are distinct from the high affinity neurotransmitter transporters of the plasma membrane that use the Na^+ gradient across the plasma membrane to drive active transport (McGivan & Pastor-Anglada, 1994). The vesicular transport of neurotransmitters is coordinated with the cell membrane transport in order to decrease the neurotransmitter concentration in the synaptic cleft and effectively store these molecules inside the synaptic vesicles, for releasing when the presynaptic membranes are further depolarized (Liu & Edwards, 1997). It has been suggested that disturbances in the vesicular uptake of amino acids may be involved in the neurotoxic effects of some pesticides, heavy metals, and endogenous metabolites, as well as in the pathogenesis of epilepsy and neural degeneration (Lewis et al., 1997; Liu & Edwards, 1997; Vaccari et al., 1999; Reis et al., 2000; Tavares et al., 2000a; 2000b; 2001; Porciúncula et al., 2003).

Considering i) the role of synaptic vesicle uptake in the clearance of neurotransmitters from the synaptic cleft, ii) the relevance of the amino acid neurotransmitters in neurological dysfunctions, iii) the neurological symptoms of porphyries, and iv) the putative involvement of ALA in these symptoms, this chapter was aimed at investigating the effects of ALA on the uptake of the neurotransmitters glutamate, GABA, and glycine by synaptic vesicles isolated from rat brain.

Materials and Methods

Materials

ALA hydrochloride, ATP, acridine orange, bafilomycin A_1, EGTA, GABA, glutamic acid, glycine, Mops, nigericin, potassium gluconate, Trizma-base, and carbonyl cyanide p-(trifluoromethoxy)phenylhydrazone (FCCP) were purchased from Sigma (St. Louis, MO, USA). L-[^{3}H]glutamate, [^{3}H]GABA, and [^{3}H]glycine were from Amersham International (Buckinghamshire, UK). All other reagents were of analytical grade.

Preparation of Synaptic Vesicles

Synaptic vesicles were isolated from the brain of male Wistar rats (200-250g) as previously described (Fykse & Fonnum, 1988). The forebrains were homogenized (10% wt/vol) in a buffer containing 0.32 M sucrose, 10 mM Mops/Tris, pH 7.4, and 1 mM EGTA and centrifuged twice for 10 min at 1,000 *x g*. Both supernatants were pooled and centrifuged for 30 min at 20,000 *x g* to obtain the crude synaptosomal fraction (P_2). This fraction was osmotically lysed by resuspension in 10 mM Mops/Tris, pH 7.4, containing 0.1 mM EGTA (approximately 0.8 ml/g of fresh tissue) and centrifuged at 17,000 *x g* for 30 min. The supernatant containing synaptic vesicles was subjected to 0.4 M and 0.6 M sucrose density gradient centrifugation at 65,000 *x g* (SW 28 rotor, Beckman Instruments) for 2 h. The synaptic vesicle fraction was isolated from the 0.4 M sucrose band and stored at -70°C for up to four weeks, with no loss of activity. The purity of this preparation has been previously determined by other groups (Naito & Ueda, 1985). In the present study, we have performed controls with the addition of high sodium concentration in order to evaluate the possible presence of Na^+-dependent neurotransmitter uptake, which would indicate contamination with vesicles formed from plasma membranes. In all synaptic vesicle preparations assayed, the uptake was not stimulated by the addition of sodium (data not shown). The protein content of synaptic vesicle preparations was determined by the method of Lowry *et al.* (1951), using bovine serum albumine as the standard.

Vesicular Uptake of Neurotransmitters

Uptake experiments were performed in a standard medium (final volume 200 μl) containing 10 mM Mops/Tris, pH 7.4, 4 mM KCl, 140 mM potassium gluconate, 120 mM sucrose, 2 mM $MgCl_2$ and 2 mM ATP as previously described (Wolosker et al., 1996). Glutamate uptake was performed in standard medium with 50 μM L-[^{3}H]glutamate (3 μCi/ml), except for the saturation curve, where the concentrations used ranged from 0.025 to 4 mM. GABA and glycine uptake were performed in standard medium with 1 mM [^{3}H]GABA or [^{3}H]glycine (5 μCi/ml), except for the saturation curve, where the concentrations used ranged from 0.025 to 10 mM. ALA solutions had their pH adjusted to 7.4 with KOH, and were stored frozen until use, for no more than one week. The uptake was started by addition of synaptic vesicles (30-40 μg of protein/tube). The uptake of amino acids by our vesicular preparations attained the steady-state (achieved when the rate of efflux equals the rate of influx) at 10 min of incubation (data not shown). This incubation time was used in all experiments. Incubations were carried out at 35°C and the reaction was stopped by rapid filtration of the assay medium through 0.45 μm Millipore filters. The filters were quickly flushed three times with 4 ml of 10 mM Mops/Tris pH 7.4, at room temperature. Radioactivity was measured with a Wallac model 1409 liquid scintillation counter. Specific uptake was calculated by discounting the uptake measured in the absence of ATP. The glutamate vesicular transporter has a biphasic dependence on Cl^-, such that low concentrations (physiological, 1-8 mM) activate uptake, while high concentrations (15-20 mM) are inhibitory (Wolosker et al., 1996). Considering that the ALA salt used contains one chloride molecule per mol, additional controls (without ALA addition) were run with a Cl^- concentration similar to that obtained after the addition of ALA (hydrochloride) to the incubation medium. When compared to the uptake obtained in the standard medium (8 mM Cl^- – 100% uptake), glutamate uptake was not affected by 10 mM Cl^- (90%), but was significantly inhibited by 20 mM Cl^- (25%). Therefore, in the uptake experiments, we tested up to 1 mM ALA, which is expected to increase the Cl^- concentration of incubation medium to 9 mM. The same concentration limit was used in the uptake assays of GABA and glycine.

Measurement of Bafilomycin A_1-sensitive ATPase Activity

The activity was determined by measuring the rate of ATP hydrolysis by the synaptic vesicle preparation, in the same conditions used for neurotransmitter uptake, except that reaction was stopped by adding 200 μl 10% (wt/vol) trichloroacetic acid. ATP hydrolysis was colorimetrically determined by measuring the amount of inorganic phosphate released (Chan et al., 1986). Bafilomycin A_1-sensitive ATPase activity was calculated as the difference between ATP hydrolysis in the absence and in the presence of 50 nM bafilomycin A_1. This concentration of bafilomycin caused the maximum ATPase inhibition in our vesicle preparation (data not shown).

Measurement of Electrochemical Proton Gradient

The proton gradient (ΔpH) was determined by measuring the changes in the absorbance spectrum of acridine orange, using a SLM Aminco DW2000 spectrophotometer at the wavelength pair 493-530 nm (Palmgren, 1991). Reaction was started by adding 2 mM ATP to a 2 ml quartz cuvette, containing synaptic vesicles (70-100 μg protein), ALA, 3.0 μg acridine orange, 10 mM Mops-Tris (pH 7.4), 100 mM KCl, and 4 mM $MgCl_2$. The reaction was stopped by the addition of the protonophore nigericin (1.25 μg). The transmembrane potential ($\Delta\psi$) was determined by measuring the fluorescence quenching of oxonol V at 617 (excitation) and 643 nm (emission) (Maycox et al., 1988; Cidon & Sihra, 1989). Reaction was started by adding 2 mM ATP to a 2 ml quartz cuvette, containing synaptic vesicles (70-100 μg protein), ALA, 1.5 μM oxonol V, 10 mM Mops-Tris (pH 7.4), and 150 mM sucrose. The reaction was stopped by the addition of the protonophore FCCP (10 μM). Each experiment was repeated at least five times with different vesicle preparations, and the figures show representative experiments.

Statistical Analysis

Results were analysed by one-way ANOVA followed by Duncan's multiple range test when appropriate. Differences between means were considered significant when $p<0.05$.

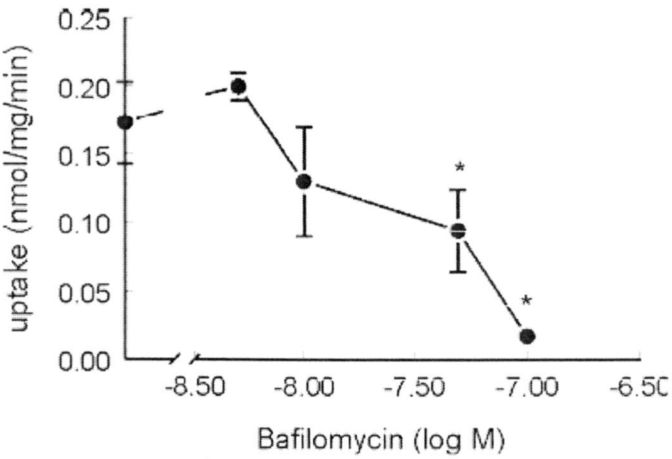

Figure 1. Effect of bafilomycin A_1 on [^{3}H]glutamate vesicular uptake. Glutamate uptake is expressed as mean ± S.E.M. (nmol mg protein^{-1} min^{-1}) of 5 independent experiments performed in triplicate. *Significantly different from control (absence of bafilomycin A_1, P<0.05).

Results

Vesicular uptake of amino acids is driven by an electrochemical gradient generated by a bafilomycin A_1-sensitive vacuolar H^+-ATPase (Cidon & Sihra, 1989; Nelson, 1991; Fykse & Fonnum, 1996). In order to assess the adequacy of our assay conditions we evaluated the effect of bafilomycin A_1 on glutamate uptake and on ΔpH measured across the synaptic vesicle preparation (Figures 1 and 2). Bafilomycin A_1 (≥ 50 nM) significantly inhibited glutamate uptake in a dose-dependent manner (Figure 1). As observed for glutamate uptake, ΔpH measured across our synaptic vesicle preparation was also completely dissipated by bafilomycin A_1 (Figure 2).

The kinetic parameters of the amino acid neurotransmitter uptake by synaptic vesicles are shown in Table 1. We used non-saturating concentrations of neurotransmitters (0.05 mM for glutamate and 1 mM for GABA and glycine) to investigate the effect of ALA on the uptake of these compounds by synaptic vesicles (Figure 3).

Table 1. Kinetic Parameters of [^{3}H]glutamate, [^{3}H]GABA, and [^{3}H]glycine Vesicular Uptake

Ligand	Vmax (nmol mg protein^{-1} min^{-1})	Km (mM)
[^{3}H]Glutamate	4.62±0.55	0.83±0.16
[^{3}H]GABA	25.90±3.10	2.92±0.81
[^{3}H]Glycine	20.30±2.60	3.40±0.70

Note: Results are expressed as mean ± S.E.M. of 6 independent experiments performed in triplicate. Vmax and Km values were determined by double-reciprocal plots of saturation curves.

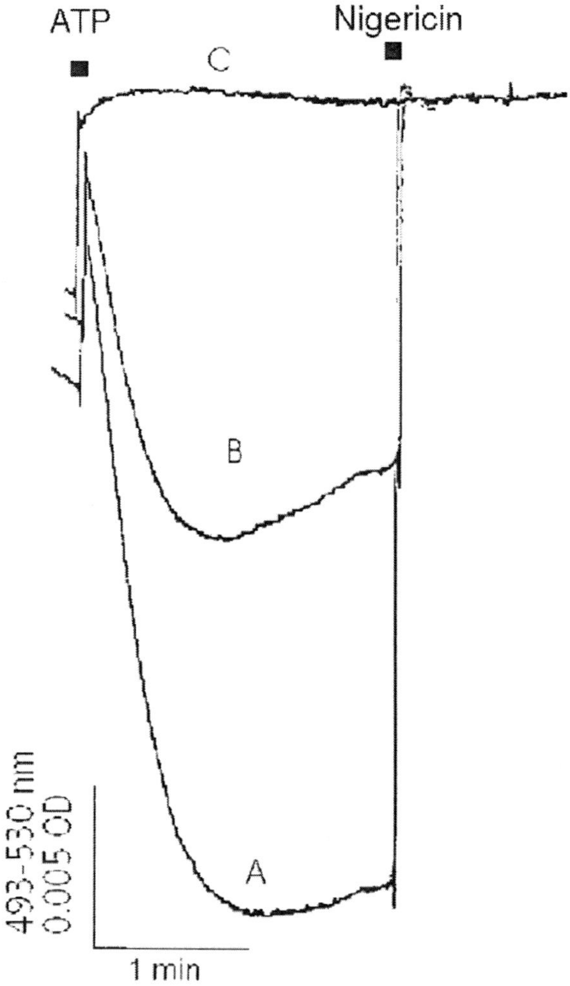

Figure 2. Effect of bafilomycin A_1 on ΔpH measured across rat brain synaptic vesicles. Assay was performed in the absence (control, A) or in the presence of 0.1 μM (B) or 0.5 μM (C) bafilomycin A_1. Traces are representative of five experiments performed with different synaptic vesicle preparations.

Figure 3 shows the effect of ALA on the uptake of amino acids into rat brain synaptic vesicles. ALA inhibited the uptake of [^{3}H]glutamate [$F(3,24)=3.78$, $p<0.05$, Figure 3a], [^{3}H]GABA [$F(3,17)=6.54$, $p<0.01$, Figure 3b], and [^{3}H]glycine [$F(6,45)=4.04$, $p<0.01$, Figure 3c], in a dose-dependent manner. Inhibition of glutamate and GABA uptake were statistically significant at ALA concentrations ≥ 100 μM, while for glycine uptake statistically significant inhibition was observed at ALA concentrations ≥ 0.1 μM. The maximum uptake inhibition observed in the presence of 1 mM ALA amounted to 50, 25 and 49% for glutamate, GABA, and glycine uptake. Higher ALA concentrations were not evaluated, since it would lead to a chloride concentration that could affect glutamate uptake. The estimated IC_{50} for ALA inhibition of glutamate, GABA, and glycine uptake were 1.18×10^{-3}, 7.08×10^{-4}, and 1.08×10^{-6} M, respectively.

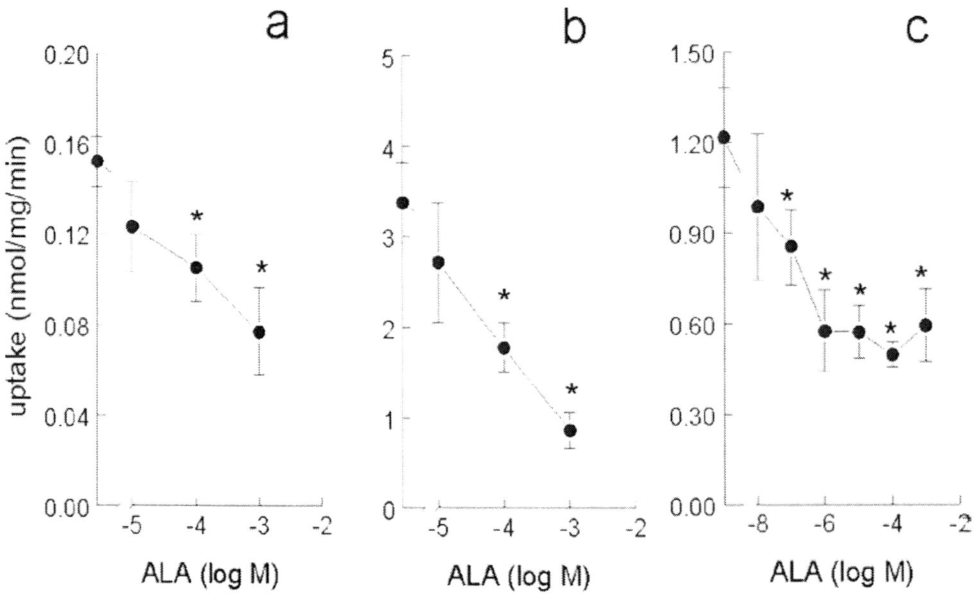

Figure 3. Effect of ALA on [^{3}H]glutamate (a), [^{3}H]GABA (b), and [^{3}H]glycine (c) uptake into rat brain synaptic vesicles. Results are expressed as mean ± S.E.M. (nmol mg protein^{-1} min^{-1}) of 5-7 independent experiments performed in triplicate. *Significantly different from control (absence of ALA) ($P<0.05$).

Previous reports have shown that ALA is a potent inhibitor of the plasma membrane Na^+,K^+-ATPase activity of neural cells (Russell et al., 1983). In order to determine if the inhibitory effect of ALA on the vesicular uptake of amino acid neurotransmitters could be attributed to an inhibition of the vesicular H^+-ATPase, the enzyme responsible for generating the electrochemical gradient across the vesicle membrane, we assessed the effect of ALA on bafilomycin A_1-sensitive ATPase activity (Figure 4). In the same assay conditions previously used to determine neurotransmitter uptake ALA had no effect on ATPase activity (Figure 4).

In order to further elucidate the mechanism responsible for the inhibitory effect of ALA on the vesicular uptake of amino acid neurotransmitters we assessed the effect of ALA on the $\Delta\mu_H^+$ formed by the ATPase pump across the vesicle membrane (Figure 5a and 5b). The $\Delta\psi$ measured across the synaptic vesicles was also unaffected by ALA (data not show). However, ALA (10-700 μM) dose-dependently inhibited the formation of ΔpH when added before gradient formation (Figure 5a). Besides, ALA (1 mM) also dissipated ΔpH when added after gradient formation (Figure 5b).

Discussion

ALA has been demonstrated to affect amino acid-mediated neurotransmission by interfering with neurotransmitter receptor binding and/or uptake by plasma membrane transporters (Muller & Snyder, 1977; Brennan & Cantrill, 1979a; McLoughlin & Cantrill, 1984; Emanuelli et al., 2001a; 2003).

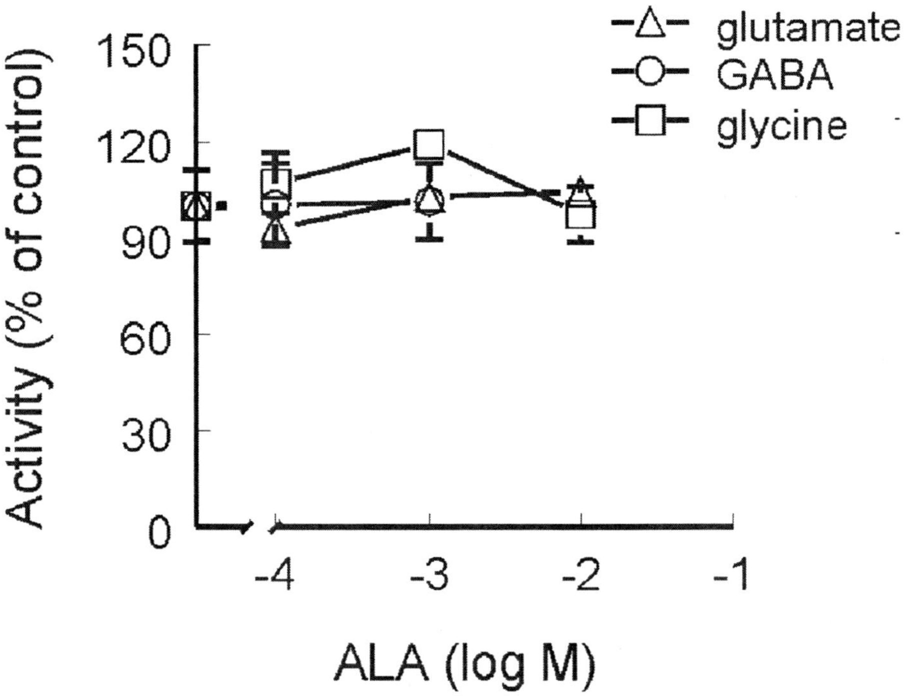

Figure 4. Effect of ALA on bafilomycin A_1-sensitive ATPase activity. ATPase activity (mean ± S.E.M., n=3) is expressed as percentage of control (35 nmol Pi mg protein^{-1} min^{-1}).

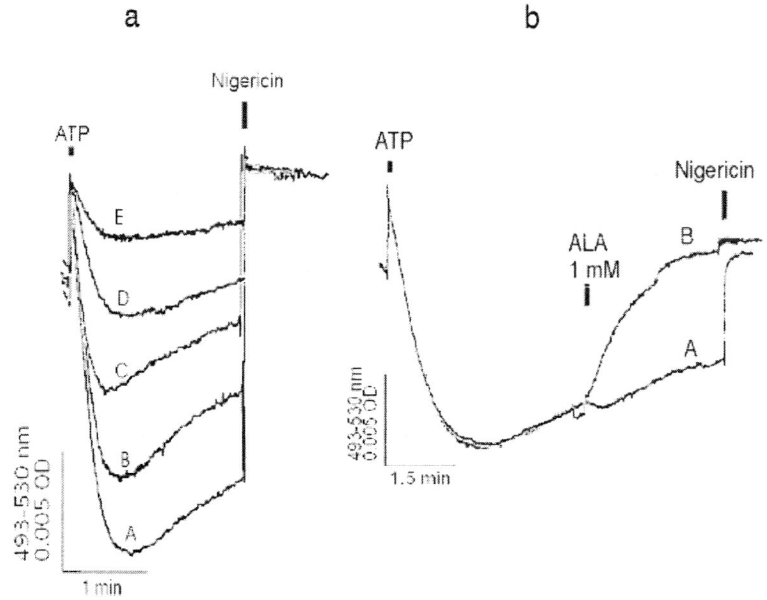

Figure 5. Effect of ALA on ΔpH measured across rat brain synaptic vesicles. ALA was added either before (a) or after (b) gradient formation. In (a), assay was performed in the absence (control, A) or in the presence of 10 (B), 100 (C), 500 (D) or 700 μM (E) ALA. In (b), assay was performed in the absence of ALA (control, A) or with 1 mM ALA addition after 4 min (B). Traces are representative of five experiments performed with different synaptic vesicle preparations.

In the present study we have demonstrated for the first time that ALA is an inhibitor of the vesicular uptake of amino acid neurotransmitters. The inhibitory potency varied in the following rank order: glycine>>GABA>glutamate vesicular uptake. GABA and glycine vesicular uptake seems to be mediated by the same transporter, while glutamate vesicular uptake is mediated by distinct transporters (Fykse & Fonnum, 1996; Liu & Edwards, 1997). However, the driving force responsible for the vesicular uptake is common to all amino acid transporters studied, and is provided by a proton electrochemical gradient ($\Delta\mu_{H}^{+}$) generated by a vacuolar Mg^{2+}-activated H^{+}-ATPase sensitive to bafilomycin A_1 (Fykse and Fonnum, 1996; Liu and Edwards, 1997). Although ALA inhibited the vesicular uptake of all amino acids tested, we have demonstrated that the inhibitory effect was not related to an inhibition of the bafilomycin A_1-sensitive H^{+}-ATPase activity present in our preparation.

Despite some controversy, it is generally accepted that the vesicular glutamate uptake relies mainly on the $\Delta\psi$ component (Fykse & Fonnum, 1996; Wolosker et al., 1996; Liu & Edwards, 1997; Shigeri et al., 2004). Conversely, the uptake of GABA and glycine is supported both by the electrical ($\Delta\psi$) and the chemical (ΔpH) components of the energy gradient. Therefore, the greater inhibitory potency of ALA towards GABA and glycine uptake could be related to a differential effect of ALA on the components of the energy gradient. Accordingly, evaluation of the effect of ALA on the electrochemical gradient revealed that ALA selectively dissipates ΔpH with no effect on $\Delta\psi$. Dissipation of ΔpH occurred regardless of ALA being added before or after gradient formation, with a complete dissipation being observed at 1 mM ALA. Nevertheless, the different inhibitory effect of ALA towards GABA and glycine uptake suggests the involvement of mechanisms other than ΔpH dissipation. Therefore, we cannot rule out that ALA could directly inhibit the vesicular transporter of amino acids, with a greater affinity for the glycine transporter.

The importance of vesicular transporters for synaptic transmission was experimentally demonstrated by the behavioral alterations observed in genetically manipulated animals presenting defects in the expression of vesicular transporters for monoamines and acetylcholine (Reimer et al., 1998). Recently, a loss of vesicular glutamate transport, consistent with a dysfunction in the recycling of this neurotransmitter has been observed in autopsies of Alzheimer disease patients (Westphalen et al., 2003). Moreover, the inhibition of glutamate vesicular uptake has been proposed to play a role in the neurotoxic effects of dithiocarbamates (Vaccari et al., 1999), quinolinic acid (Tavares et al., 2000a), organic acids (Reis et al., 2000; Tavares et al., 2000b; 2001), and methylmercury (Porciúncula et al., 2003).

The present results do not allow us to draw sound toxicological implications for the inhibition of vesicular amino acids uptake by ALA. However, it has been demonstrated that exposure to bafilomycin A_1, a specific inhibitor of vacuolar H^{+}-ATPase, which eliminates the driving force for the uptake of both glutamate and GABA, reduced inhibitory and excitatory postsynaptic currents in hippocampal neurones (Zhou et al., 2000). Besides, Rose Bengal, an inhibitor of vesicular glutamate uptake also reduced the amount of glutamate released upon synaptosomal depolarization (Ogita et al., 2001). Therefore, it is tempting to propose that ALA would decrease the physiological vesicular release of amino acid neurotransmitters. Additionally, it could increase the cytosolic concentrations of these compounds, possibly resulting in the non-vesicular release of these amino acids through the inversion of the plasma membrane transporters. A similar mechanism involving efflux of monoamines from storage vesicles, followed by reversal of plasma membrane transporters, has been previously proposed to explain the psychoactive effects and the neurotoxicity induced by amphetamines (Liu & Edwards, 1997). Hence, a disturbance in the vesicular cycling of amino acid neurotransmitters may have contributed to some of the previously reported neurotoxic effects observed in ALA-accumulating disorders, such as acute intermittent porphyria, tyrosinosis and lead poisoning (Yeung Laiwah, 1987; Kappas et al., 1995). Besides, oral administration of ALA has been recently used to delineate malignant gliomas (Stummer et al., 1998; 2000) and ALA or ALA-esters are being evaluated as photosensitizers in photodynamic therapy of human gliomas (Lilge & Wilson, 1998). Therefore, the observed ALA-induced inhibition of amino acid uptake into synaptic vesicles may also be of value in understanding the possible side-effects of ALA photodynamic therapy.

Although ALA inhibited the vesicular uptake of all neurotransmitters studied, our results indicated that at low micromolar concentrations, comparable to those found in the CSF of patients during the acute attacks (Yeung Laiwah, 1987), ALA is a selective inhibitor of glycine vesicular uptake. The main pathways responsible for glycine metabolization (oxidation to CO_2, conversion to lipids and incorporation into protein) saturate at 0.5 mM glycine in brain slices (Rotta et al., 2002). Considering that the physiological concentration of glycine in the CNS is around 1-2 mM (Turecký et al., 1980), an inhibition of the vesicular uptake of this amino acid could possibly increase its intracellular concentration. This observation may be of relevance for the pathogenesis of porphyria. Glycine is one of the substrates of ALA-synthase (ALA-S), which is the first enzyme in the haem biosynthetic pathway (Kappas et al., 1995). Data from the literature suggests that the activity of this enzyme, which regulates the biosynthesis of haem, may be modulated by physiological concentrations of glycine (1-2 mM; Turecký et al., 1980) that are 1/5 to 1/20 the Km of ALA-S for this substrate (Kappas et al., 1995). Therefore, the increase in the intracellular concentration of glycine, due to the inhibition of vesicular uptake, could positively modulate the biosynthesis of ALA, possibly increasing the intracellular concentration of this compound in neuronal tissue during an acute attack. However, experimental evidence must be provided before accepting this proposal.

In our study, the inhibition of GABA and glutamate vesicular uptake by ALA was observed at ALA concentrations higher than those previously reported in the CSF of patients with acute porphyrias (10^{-5} to 10^{-7} M; Yeung Laiwah et al., 1987). Although ALA exhibits low blood-brain barrier permeability (Ennis et al., 2003), which limits its transport from plasma (10^{-3} to 10^{-5} M), it is important to point out that cerebral particles have been demonstrated to accumulate ALA (Juknat et al., 1995) and that in these disorders ALA is actively produced within the brain (De Matteis et al., 1981; Kappas et al., 1995). These findings suggest that the intracellular cerebral ALA levels, which remain so far unknown, may be higher than those found in the CSF of patients. Therefore, considering the intracellular location of synaptic vesicles, an involvement of GABA and glutamate vesicular uptake inhibition in some of the neurological manifestations of acute porphyrias cannot be ruled out.

Conclusion

Results of the present study revealed that the haem precursor ALA inhibited the uptake of neurotransmitter amino acids by synaptic vesicles with an inhibitory potency order of: glycine >> GABA > glutamate. This effect was not due to the inhibition of vesicular bafilomycin A_1-sensitive H^+-ATPase activity, which is the common driver force to the three amino acid transport activities. However, ALA selectively dissipated ΔpH with no effect on Δψ measured across synaptic vesicles. This study provided evidence for an intracellular target to the neurotoxic effects of ALA, and may be of value to understand the role of this compound in the neurological dysfunctions observed in ALA-accumulating disorders, such as acute intermittent porphyria, tyrosinosis and lead poisoning. Besides, results may also be of value to understand the possible neurological side-effects associated with the use of ALA in tumor resection and in the photodynamic therapy of human gliomas.

Acknowledgements

Sunpported by CNPq (PRONEX grant 41960904-366/96 to D. O. Souza). T.E. was the recipient of CNPq research Fellowship (proc. 550691/02-2).

References

Bai, L.; Xu, H.; Collins, J. F. & Ghishan, F.K. (2001). Molecular and functional analysis of a novel neuronal vesicular glutamate transporter. *Journal of Biological Chemistry, 276,* 36764-36769.

Brennan, M. J. W. & Cantrill, R. C. (1979a). The effect of delta-aminolaevulinic acid on the uptake and efflux of amino acid neurotransmitters in rat brain synaptosomes. *Journal of Neurochemistry, 33,* 721-725.

Brennan, M. J. W. & Cantrill, R. C. (1979b). Delta-aminolevulinic acid is a potent agonist for GABA autoreceptors. Nature (London), 280, 514-515.

Brennan, M. J. W. & Cantrill, R.C. (1979c). The effect of delta-aminolaevulinic acid on the uptake and efflux of [^{3}H]GABA in rat brain synaptosomes. *Journal of Neurochemistry, 32,* 1781-1786.

Chan, K. M.; Delfert, D. & Junger, K. D. (1986). A direct colorimetric assay for Ca^{2+}-stimulated ATPase activity. *Analytical Biochemistry, 157,* 375-380.

Cidon, S. & Sihra, T. S. (1989). Characterization of a H^+-ATPase in rat brain synaptic vesicles: coupling to L-glutamate transport. *Journal of Biological Chemistry, 264,* 8281-8288.

De Matteis, F.; Zetterlund, P. & Wetterberg, L. (1981). Brain 5-aminolaevulinate synthase. *Biochemical Journal, 196,* 811-817.

Demasi, M.; Penatti, C. A. A.; DeLucia, R. & Bechara, E. J. H. (1996). The prooxidant effect of 5-aminolevulinic acid in the brain tissue of rats: implications in neuropsychiatric manifestations in porphyrias. *Free Radical Biology & Medicine, 20,* 291-299.

Dichter, H. N.; Tadeini, L.; Lin, S. & Ayala, G. F. (1977). Delta amino levulinic acid. Effect of a porphyrin precursor on an isolated neuronal preparation. *Brain Research, 126,* 189-195.

Emanuelli, T.; Pagel, F. W.; Alves, L. B.; Regner, A. & Souza, D. O. (2001a). 5-Aminolevulinic acid inhibits [^{3}H]muscimol binding to human and rat brain synaptic membranes. *Neurochemical Research, 26,* 101-105.

Emanuelli, T.; Pagel, F. W.; Alves, L. B.; Regner, A.; Souza, D. O. (2001b). Inhibition of adenylate cyclase activity by 5-aminolevulinic acid in rat and human brain. *Neurochemistry International, 38,* 213-218.

Emanuelli, T.; Pagel, F. W.; Porciúncula, L. O. & Souza, D. O. (2003). Effects of 5-aminolevulinic acid on the glutamatergic neurotransmission. *Neurochemistry International, 42,* 115-121.

Emanuelli, T.; Prauchner, C. A.; Dacanal, J.; Zeni, A.; Reis, E. C.; Mello, C. F. & Souza, D. O. (2000). Intrastriatal administration of 5-aminolevulinic acid induces convulsions and body asymmetry through glutamatergic mechanisms. *Brain Research, 868,* 88-94.

Ennis, S. R.; Novotny, A.; Xiang, J.; Shakui, P.; Masada, T.; Stummer, W.; Smith, D. E. & Keep, R. F. (2003). Transport of 5-aminolevulinic acid between blood and brain. *Brain Research, 959,* 226-234.

Fykse, E. M. & Fonnum F. (1996). Amino acid neurotransmission: dynamics of vesicular uptake. *Neurochemical Research, 21,* 1053-1060.

Fykse, E. M. & Fonnum, F. (1988). Uptake of γ-aminobutyric acid by synaptic vesicle fraction isolated from rat brain. *Journal of Neurochemistry, 50,* 1237-1242.

Juknat, A. A.; Kotler, M. L. & Batlle, A. M. C. (1995). High δ-aminolevulinic acid uptake in rat cerebral cortex: effect on pophyrin biosynthesis. *Comparative Biochemistry and Physiology, 111C,* 143-150.

Kappas, A.; Sassa, S.; Galbraith, R. A. & Nordmann, Y. (1995). The Porphyrias. In C.R. Scriver, A. L. Beaudet, W. S. Sly & D. Valle (Eds.), *The Metabolic Bases of Inherited Disease.* (7th edition, pp. 2103-2160). New York, NY: McGraw Hill.

Lewis, S. M.; Francis, S. L.; Todorova, M.; Seyfried, T. N. & Ueda, T. (1997). Synaptic vesicle glutamate uptake in epileptic (EL) mice. *Neurochemistry International, 31,* 581-585.

Lilge, L. & Wilson, B. C. (1998). Photodynamic therapy of intracranial tissues: a preclinical comparative study of four different photosensitizers. *Journal of Clinical Laser Medicine & Surgery, 16,* 81-92.

Lipton, S. A. & Rosenberg, P. A. (1994). Mechanisms of disease. Excitatory amino acids as a final common pathway for neurological disorders. *New England Journal of Medicine, 330,* 613-622.

Liu, Y. & Edwards, H. (1997). The role of vesicular transport proteins in synaptic transmission and neural degeneration. *Annual Review of Neuroscience, 20,* 125-156.

Lowry, O. H.; Rosebrough, N. J.; Farr, A. L. & Randall, R. J. (1951). Protein measurement with the Folin phenol reagent. *Journal of Biological Chemistry, 193,* 265-275.

Mattson, M.P. (2003). Excitotoxic and excitoprotective mechanisms: abundant targets for the prevention and treatment of neurodegenerative disorders. *Neuromolecular Medicine, 3,* 65-94.

Maycox, P. R.; Deckewerth, T.; Hell, J. W. & Jahn, R. (1988). Glutamate uptake by brain synaptic vesicles: energy dependence of transport and functional reconstitution in proteoliposomes. *Journal of Biological Chemistry, 263,* 15423-15428.

McGivan, J. D. & Pastor-Anglada, M. (1994). Regulatory and molecular aspects of mammalian amino acid transport. *Biochemical Journal, 299,* 321-334.

McLoughlin, J. L. & Cantrill, R. C. (1984). The effect of delta-aminolaevulinic acid on the high affinity uptake of aspartic acid by rat brain synaptosomes. *General Pharmacology, 15,* 553-555.

Muller, W. E. & Snyder, S. H. (1977). Delta-aminolevulinic acid: influences on synaptic GABA receptor binding may explain CNS symptoms of porphyria. *Annals of Neurology, 2,* 340-342.

Naito, S. & Ueda, T. J. (1985). Characterization of glutamate uptake into synaptic vesicles. *Journal of Neurochemistry, 44,* 99-109.

Nelson, N. (1991). Structure and pharmacology of the proton ATP-ases. *Trends in Pharmacological Sciences, 12,* 71-75

Nordmann, Y. & Puy, H. (2002). Human hereditary hepatic porphyrias. *Clinical Chimica Acta, 325,* 17-37.

Ogita, K.; Hirata, K.; Bole, D. G.; Yoshida, S.; Tamura, Y.; Leckenby, A. M. & Ueda, T. (2001). Inhibition of vesicular glutamate storage and exocytotic release by Rose Bengal. *Journal of Neurochemistry, 77,* 34-42.

Olsen, R. W. & DeLorey, T. M. (1999). GABA and Glycine. In G. J. Siegel, B. W. Agranoff, R. W. Albers, S. K. Fisher & M. D. Uhler (Eds.), *Basic Neurochemistry.* (6th edition, pp. 335-346) New York, NY: Lippincott-Raven.

Ozawa, S.; Kamiya, H. & Tsuzuki, K. (1998). Glutamate receptors in the mammalian central nervous system. *Progress in Neurobiology, 54,* 581-618.

Palmgren, M.G. (1991). Acridine orange as a probe for measuring pH gradients across membranes: mechanism and limitations. *Analytical Biochemistry, 192,* 316-321.

Pearl, P. L. & Gibson, K. M. (2004). Clinical aspects of the disorders of GABA metabolism in children. *Current Opinion in Neurology, 17,* 107-113.

Porciúncula, L. O.; Rocha, J. B.; Tavares, R. G.; Ghisleni, G.; Reis, M. & Souza, D. O. (2003). Methylmercury inhibits glutamate uptake by synaptic vesicles from rat brain. *Neuroreport, 14,* 577-580.

Prauchner, C. A.; Agostini, A. N.; Morimoto, A.; Augusti, P. R.; Unfer, T. C.; Zeni, G.; Mello, C. F.; Emanuelli, T. (2004). Dimethyl sulfoxide and ebselen prevent convulsions induced by 5-aminolevulinic acid. *Neurochemical Research, 29,* 1793-1800.

Puy, H.; Deybach, J. C.; Bogdan, A.; Callebert, J.; Baumgartner, M.; Voisin, P.; Nordmann, Y. & Touitou, Y. (1996). Increased δ-aminolevulinic acid and decreased pineal melatonin production: a common event in acute porphyria studies in the rat. *Journal of Clinical Investigation, 97,* 104-110.

Reimer, R. J.; Fon, E. A. & Edwards, R. H. (1998). Vesicular neurotransmitter transport and the presynaptic regulation of quantal size. *Current Opinion in Neurobiology, 8,* 405-412.

Reis, M.; Farage, M. & Wolosker, H. (2000). Chloride-dependent inhibition of vesicular glutamate uptake by α-keto acids accumulated in maple syrup disease. *Biochimica Biophysica Acta, 1475,* 114-118.

Rotta, L. N.; Valle, S. C.; Schweigert, I.; Ricardi, L. D.; Ferronatto, M. E.; Silva, L.; Souza, D. O. & Perry, M. L. S. (2002). Utilization of energy nutrients by cerebellar slices. *Neurochemical Research*, 27, 201-206.

Russell, V. A.; Lamm, M. C. L. & Taljaard, J. J. F. (1983). Inhibition of Na^+, K^+-ATPase activity by δ-aminolevulinic acid. *Neurochemical Research, 8,* 1407-1415.

Sagne, C.; El Mestikawy, S.; Isambert, M. –F.; Hamon, M.; Henry, J. –P.; Giros, B. & Gasnier, B. (1997). Cloning of a functional vesicular GABA and glycine transporter by screening of genome databases. *FEBS Letters, 417,* 177-183.

Sakai, T. & Morita, Y. (1996). δ -aminolevulinic acid in plasma or whole blood as a sensitive indicator of lead effects, and its relation to the other heme-related parameters. *International Archives of Occupational and Environmental Health, 68,* 126-132.

Shigeri, Y.; Seal, R. P. & Shimamoto, K. (2004). Molecular pharmacology of glutamate transporters, EAATs and VGLUTs. *Brain Research Reviews, 45,* 250-265.

Stummer, W.; Novotny, A.; Stepp, H.; Goetz, G.; Bise, K. & Reulen, H. J. (2000). Fluorescence-guided resection of glioblastoma multiforme by using 5-aminolevulinic acid-induced porphyrins: a prospective study in 52 consecutive patients. *Journal of Neurosurgery, 93,* 1003-1013.

Stummer, W.; Stocker, S.; Wagner, S.; Stepp, H.; Fritsch, C.; Goetz, C.; Goetz, A.; Kiefmann, R. & Reulen, H. J. (1998). Intraoperative detection of malignant gliomas by 5-aminolevulinic acid-induced porphyrin fluorescence. *Neurosurgery, 42,* 518-526.

Tavares, R. G.; Santos, C. E. S.; Tasca, C. I.; Wajner, M.; Souza, D. O. & Dutra-Filho, C. S. (2001). Inhibition of glutamate uptake into synaptic vesicles from rat brain by 3-nitropropionic acid in vitro. *Experimental Neurology, 172,* 250-254.

Tavares, R. G.; Santos, C. E.; Tasca, C. I.; Wajner, M.; Souza, D. O. & Dutra-Filho, C. S. (2000b). Inhibition of glutamate uptake into synaptic vesicles of rat brain by the metabolites accumulating in maple syrup urine disease. *Journal of Neurological Science, 181,* 44-49.

Tavares, R. G.; Tasca, C. I.; Santos, C. E. S.; Wajner, M.; Souza, D. O. & Dutra-Filho, C. S. (2000a). Quinolinic acid inhibits glutamate uptake into synaptic vesicles from rat brain. *NeuroReport, 11,* 249-253.

Turecký, L.; Líska, B. & Pechán, I. (1980). Glycine in the central nervous system in experimental allergic encephalomyelitis. *Journal of Neurochemistry, 35,* 735-738.

Vaccari, A.; Saba, P.; Mocci, I. & Ruiu, S. (1999). Dithiocarbamate pesticides affect glutamate transport in brain synaptic vesicles. *Journal of Pharmacology and Experimental Therapeutics, 288,* 1-5.

Westphalen, R. I.; Scott, H. L. & Dodd, P. R. (2003). Synaptic vesicle transport and synaptic membrane transporter sites in excitatory amino acid nerve terminals in Alzheimer disease. *Journal of Neural Transmission, 110,* 1013-1027.

Wolosker, H.; Souza, D. O. & De Meis, L. (1996). Regulation of glutamate transport into synaptic vesicles by chloride and proton gradient. *Journal of Biological Chemistry, 271,* 11726-11731.

Yeung Laiwah, A. C.; Moore, M. R. & Goldberg, A. (1987). Pathogenesis of acute porphyria. *Quarterly Journal of Medicine, 63,* 377-392.

Zhou, Q.; Petersen, C. C. & Nicoll, R. A. (2000). Effects of reduced vesicular filling on synaptic transmission in rat hippocampal neurones. *Journal of Physiology, 525(Pt 1),* 195-206.

In: Focus on Neurochemistry Research
Editor: Robert M. Coleman, pp.213-228
ISBN 1-59454-417-4

Chapter VIII

Potential Mechanisms of Methylmercury (MeHg)-Induced Neurotoxicity: The Role of Glutamate and Reactive Oxygen Species (ROS)

Gouri Shanker[1], Tore Syversen[2], Judy L. Aschner[3], Ursula Sonnewald[2], and Michael Aschner[3,4,5*]

[1]Department of Physiology and Pharmacology, Wake Forest University School of Medicine, Medical Center Boulevard, Winston-Salem, NC
[2]Department of Clinical Neuroscience, Norwegian University of Science and Technology, Trondheim, Norway, and [3]Departments of Pediatrics, [4]Pharmacology, and the [5]Kennedy Center, Vanderbilt University Medical Center, Nashville, TN

Abstract

Methylmercury (MeHg) is a significant environmental contaminant that will continue to pose great risk to human health. Considerable attention in the scientific and health policy fora has focused on the question of whether MeHg intake from a diet high in fish is associated with aberrant CNS function. A number of studies suggest that fetal exposure at levels attained by mothers regularly eating fish during pregnancy are associated with neurological deficits in their offspring. Astrocytes play a key role in MeHg-induced excitotoxicity. [1] MeHg preferentially accumulates in astrocytes. [2] MeHg potently and specifically inhibits glutamate uptake in astrocytes. [3] Neuronal dysfunction is secondary to disturbances in astrocytes. [4] Co-application of nontoxic

* Please send request for reprints to Michael Aschner, Ph.D., Department of Pediatrics, B-3307 Medical Center North, Vanderbilt University School of Medicine, Nashville, TN 37232-2495; Phone: 615 322 8024; Fax: 615-322-6541; E-mail: michael.aschner@vanderbilt.edu

concentrations of MeHg and glutamate leads to the typical appearance of neuronal lesions associated with excitotoxic stimulation. [5] MeHg induces swelling of astrocytes. These observations are fully consistent with MeHg-induced dysregulation of excitatory amino acid homeostasis, and indicate that a glutamate-mediated excitotoxic mechanism is involved. In the present review we present data in support of the hypotheses that activation of the cytosolic phospholipase A_2 ($cPLA_2$) and the ensuing hydrolysis and release of arachidonic acid (AA) are mediators for glutamate release upon exposure to MeHg. We also provide evidence that MeHg alters the uptake of amino acids comprising the tripeptide, glutathione (GSH), a major antioxidant system. Finally, we present data on the role of astrocytes and oxidative stress in modulating MeHg neurotoxicity.

Introduction

Methylmercury (MeHg) is a major neurotoxicant that continues to pose appreciable risk to human health as evidenced by the tragic epidemics of MeHg poisoning in Japan and Iraq (Takeuchi et al., 1972; Bakir et al., 1973). Anthropogenic sources resulting in the acidification of freshwater streams and lakes, and the impoundment of water for large hydroelectric schemes have led to increasingly higher MeHg concentrations in fish. Given that contaminated fish represent a common source for human MeHg exposure, considerable attention in the scientific and health policy fora is focused on the question of whether MeHg intake from a diet high in fish is associated with aberrant CNS function. All sources of environmental mercury represent a potential risk for poisoning in humans given the ability of methanogenic bacteria to methylate inorganic mercury to MeHg in waterways, ultimately resulting in its accumulation and bioconcentration in sea food chain. Fish at the upper hierarchy of the marine biota, namely predators, such as shark, swordfish, tuna and alike, contain the highest MeHg concentrations. MeHg-adulterated fish represent the most prevalent source for human exposure to this metal.

It is well established that excessive MeHg ingestion from a diet high in fish is associated with aberrant central nervous system (CNS) functions (Kjelstrom et al., 1989; Grandjean et al., 1997, 1999). The developing nervous system is the most vulnerable and sensitive to MeHg toxicity. Recent studies in human populations support the earlier findings that the maternal mercury exposure during pregnancy is associated with neurological as well as neuropsychological deficits detectable at 6 to 7 years of age (Steuerwald et al., 2000; Cordier et al., 2002; Grandjean et al., 2003). Another recent study has pointed to MeHg's selective detrimental effects on neurogenesis (Faustmann et al., 2002). However, despite these observations, the ability of MeHg derived from food consumption to affect cognitive and behavioral functions children remains controversial as exemplified by other studies (Myers et al., 1997; Davidson et al., 1998) in which no association was noted between MeHg and neurodevelopmental outcomes in children at 66 months of age. Thus we still do not know the effect of MeHg in the human population and clearly there remains a need to understand the mechanisms and consequences of MeHg exposure on CNS function. The current literature suggests that no single mechanism can explain the multitude of effects observed in MeHg-induced neurotoxicity.

MeHg –Mediated Cellular Damage

Affectd Cell Types

The last 2 decades have seen a tremendous insurgence in the understanding of the role of astrocytes in brain function. Although these cells do not conduct action potentials, nor do they form synapses with neuronal cells, it is clear that they interact amongst themselves and with neurons in a complex fashion, assuring optimal structural, metabolic and trophic support throughout life. With the advent of novel technology, recent breakthroughs in astrocyte physiology have established dynamic and reciprocal signaling networks between astrocytes and with neurons. Proper CNS function involves interactive signaling between astrocytes and neurons. Accordingly, defining the nature of aberrant interactions between the 2 cell types might open new perspective for understanding the pathogenesis of brain diseases. As illustrated below, recent studies suggest that astrocytes (as well as other glial cells, and particularly, microglia) may directly mediate neuropathological outcome, both during CNS development and senescence.

Among various cell types, astrocytes represent a major cell type occupying approximately 50% of the CNS volume (Chen and Swenson, 2003; Aschner et al., 1999) and their dysfunction following toxicant exposure is suggested to be the main cause of the observed neurotoxicity (Chen et al., 2000). The "foot" processes of these cells are known to be closely associated with synapses, axonal tracts, nodes of Ranvier and capillaries. Astrocytes perform important duties and some of their functions include the maintenance of normal extracellular ion concentrations, the uptake of potassium (K^+) and the control of extracellular pH (Aschner et al., 1996). They express a pantheon of receptors and uptake systems for neurotransmitters, properties that were formerly thought to be exclusively neuronal. Thus, astrocytes can respond to and prevent the build up of potentially dangerous levels of neurotransmitters (e.g. glutamate) in the extracellular fluid (Aschner and Kimelberg, 1996). Important roles of astrocytes during early brain development include the synthesis and elaboration of cues for neuronal migration and the production of neurotrophic factors important for neuronal division and differentiation. Astrocytes also induce the high electrical resistance (tightness) of the blood-brain barrier (BBB) that limits the transport of noxious substances entering the brain, modulating optimal transport of nutrients and metabolites (Aschner and Kimelberg, 1996; Aschner et al., 1999).

Mechanisms Associated with MeHg-induced CNS Damage

Although not the only cell type to be adversely affected by MeHg (Atchison and Hare, 1994), a number of studies have established a key role for astrocytes in mediating MeHg neurotoxicity.

(a) Astrocytes and microglia represent a preferential cellular site for Hg accumulation after MeHg exposure (Garman et al., 1975; Charleston et al., 1996; Charleston et al.,

1996; Aschner, 1996; Aschner et al., 1997; Aschner et al., 2000). However, some of the morphological methods used to visualize mercury will only detect inorganic mercury and not MeHg in its intact form. Thus, evaluation of cellular distribution of mercury should be cognizant of the particular method that is used, and consideration should be given to the fact that adequate methodologies and studies to determine to determine the precise cellular localization of MeHg have yet to be performed.

(b) MeHg selectively inhibits astrocytic transport of cystine and cysteine (Figure 1), thereby adversely affecting their redox status and attenuating glutathione (GSH) content (Allen et al., 2001; Shanker et al., 2001; Shanker and Aschner, 2001a; Shanker and Aschner, 2001b; Shanker and Aschner, 2001c; Shanker et al., 2003). Astrocytes have recently been proposed to play an essential role in providing GSH precursors to neurons. The symbiotic relationship between neurons and glia is exemplified by the neuronal dependence on astrocyte-derived thiols for the maintenance of stable GSH concentrations. GSH (γ-glutamylcysteinylglycine) is synthesized in two steps by the actions of γ-glutamylcysteine synthetase forming γ-glutamylcysteine from cysteine and glutamate, followed by the addition of glycine by GSH synthetase. It is a major antioxidant in mammalian cell systems, constituting approximately 90% of the intracellular non-protein thiols, and it is crucial in conjugation and elimination of various toxic molecules, and the maintenance of intracellular redox status. In general, GSH levels are lower in neurons than in astrocytes (Sagara et al., 1993), rendering the former more susceptible to increased intracellular levels of reactive oxygen species (ROS). Recent studies establish that cysteine and GSH are unstable and oxidized to their disulfide forms under aerobic conditions, and that constant release of cysteine GSH by astrocytes is essential for the maintenance of stable levels of thiols in the CNS (Shanker et al., 2001a; Shanker et al., 2001b; Shanker et al., 2001c; Dringen et al., 1999).

We have accordingly characterized cystine transport (Allen et al., 2001), a prerequisite to GSH production. Astrocytes were shown to possess a temperature dependent transport systems for cystine uptake. Astrocytes accumulated cystine by Na^+-independent (X_C-) and-dependent (X_{AG}-), and inhibition of the X_{AG}^- transport system decreased cystine transport to levels equivalent to those in sodium-depleted conditions, suggesting that cystine is carried by a glutamate/aspartate transporter in astrocytes. Inhibition of the multifunction ectoenzyme/amino acid transporter γ-glutamyltranspeptidase (GGT) decreased astrocytic cystine transport (Allen et al., 2001). Inhibition of System X_C- with quisqualate also decreased cystine uptake in these cells, demonstrate that cultured astrocytes accumulate cystine via three independent mechanisms, system X_{AG}-, system X_C-, and GGT. Notably, inhibition of cystine uptake in astrocytes by MeHg appears to be due to actions on the System X_{AG}- transporter (Allen et al., 2001; Figure 1).

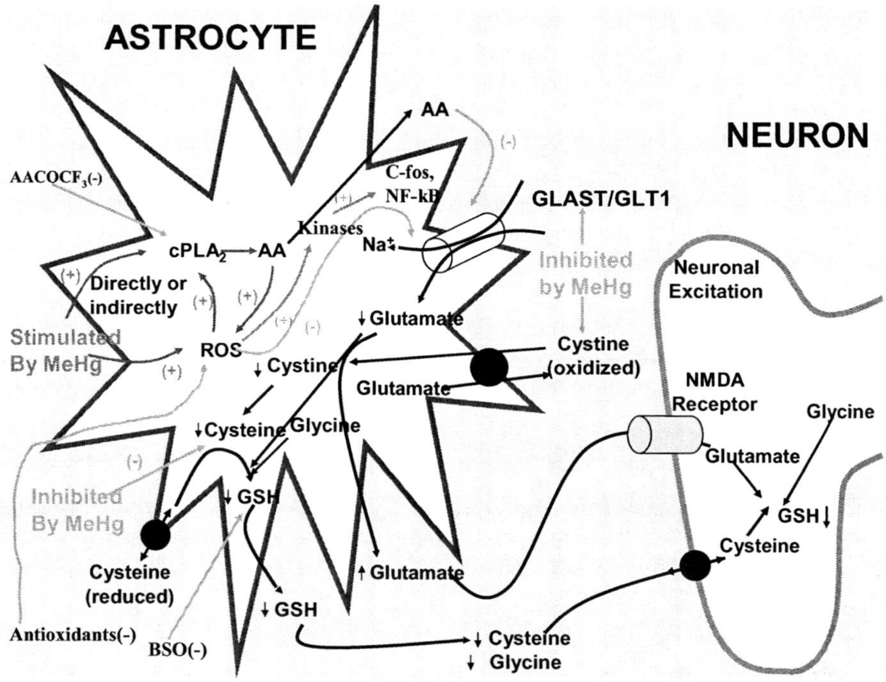

Figure 1. A schematic model of some of the currently proposed as well as some of our previously studied processes resulting in MeHg-dependent neurotoxicity

(c) Since astrocytes are a major site of glutamate inactivation, their dysfunction may indirectly contribute to MeHg-induced damage to juxtaposed neurons. MeHg inhibits astrocytic glutamate (and aspartate) uptake (Figure 1) and stimulates its efflux, thereby increasing glutamate concentrations in the extracellular fluid and sensitizing neurons to excitotoxic injury (Brooks and Kristt, 1989; Aschner et al., 1993; Dave et al., 1994; Mullaney et al., 1994; Allen et al., 2001).

(d) MeHg-induced neuronal dysfunction is secondary to disturbances in astrocytes (Brookes, 1992), and the in vitro co-application of non-toxic concentrations of mercury with glutamate results in the appearance of typical neuronal lesions found with excitotoxic stimulation (Matyja and Albrecht, 1993). Furthermore, MeHg increases glutamate extracellular levels in frontal cortex of awake rats as measured by microdialysis (Juarez et al., 2002).

(e) Cytosolic phospholipase A_2 ($cPLA_2$) plays an important role in the maintenance of cell membrane phospholipids. It catalyzes the production of arachidonic acid (AA) from the sn-2 position of membrane phospholipids to yield lysophospholipids. AA release occurs upon the activation of $cPLA_2$ (by phosphorylation of Ser-505 and calcium-dependent translocation to membranes), allowing $cPLA_2$ access to its substrate. $cPLA_2$ expression is strongly increased during cerebral ischemia and trauma, Alzheimer's disease, and epilepsy. MeHg causes the activation of cytosolic phospholipase A_2 ($cPLA_2$) leading to AA release and further inhibition of glutamate transporter (Figure 1), feeding forward on an unimpeded cytotoxic cycle (Shanker et al., 2002; Shanker et al., 2003; Shanker et al., 2004).

MeHg and Oxisative Stress

Oxidative stress has been implicated in a variety of neurodegenerative diseases (Coyle and Puttafarken, 1993), such as Alzheimer's disease, Parkinson's disease and amyotropic lateral sclerosis (Kaltschmidt et al., 1993; Hunot et al., 1997; Kitamura et al., 1997; Lukiw and Bazan, 2000). It also plays an important role in other degenerative conditions such as autoimmune and inflammatory diseases (i.e. ischemia and rheumatoid arthritis), cancer, diabetes mellitus and arteriosclerosis (Munthe et al., 1986; Kehrer et al., 1994; Halliwell, 1994; Betteridge, 2000), as well as in metal induced toxicity (LaBel et al., 1992; Yee and Choi, 1996; Britton, -1996; Yiins et al., 1999; Almazan et al., 2000; Bush, 2000; Chen et al., 2000; Galaris and Evangelou, 2002). The balance between oxidative and reductive cellular processes is known to be adversely affected in these various disorders (Betteridge, 2000; Sayre et al., 1999). Oxidative stress is associated with the accumulation of high levels of toxic reactive species, such as, reactive oxygen species (ROS), reactive nitrogen species (RNS), reactive nitrogen oxygen species (RNOS), as well as unbound metal ions (Betteridge, 2000; Bush, 2000; Davis et al., 2001). Typical ROS include oxygen radicals such as superoxide radical ($O_2{}^{*-}$), hydroxyl radical (*OH), as well as non-radical derivatives of oxygen including hydrogen peroxide, H_2O_2). Examples of RNS include nitric oxide (NO) radical, and that of RNOS include a highly reactive oxidant species peroxynitrite ($ONOO^-$), which is a product of reaction between NO and the superoxide radical. These reactive species are highly oxidizing and potently damaging to cellular redox-sensitive proteins, enzymes and DNA, and they also cause peroxidation of membranes.

ROS are known to mediate MeHg-induced neurotoxicity in multiple experimental models. For example, MeHg induces ROS formation in vivo (rodent cerebellum), and in vitro (isolated rat brain synaptosomes) (Ali et el., 1992), as well as in cerebellar neuronal cultures, a hypothalamic neuronal cell line and in mixed reaggregating cell cultures (Sarafian, 1994; Park et al., 1996; Sorg et al., 1998; Sarafian et al., 1999; Gasso et al., 2001). In addition, an increase in ROS has been observed in (a) mitochondria isolated from MeHg-injected rat brains (Yee and Choi, 1996); (b) mitochondria isolated in vitro from rat brain and then exposed to MeHg (Myhre and Fonnum, 2001); (c) mitochondria from Hg and glutamate exposed astrocytes and neurons (Dugan et al., 1995; Brawer et al., 1998). This ROS formation resulting from MeHg and glutamate induced oxidative stress contributes to mitochondrial dysfunction. Overproduction of ROS is mediated, at least in part, by glutamate, since this toxicity can be attenuated by N-methyl D-aspartate (NMDA) receptor antagonists (Park et al., 1996). The source of glutamate is likely to be astrocytic given the effect of MeHg on glutamate uptake and release (Figure 1) from astrocytes (Aschner et al., 1994; Allen et al., 2001). Therefore, although, the toxic damage of MeHg might be most prevalent in neurons, a large body of literature suggests that neuronal damage in response to MeHg, most likely represents aberrant control by astrocytes of the extracellular milieu.

A number of reports indicate oxidative stress-induced activation of a number of signaling molecules [i.e. distinct kinases/ transcription factors (nuclear factor kappa B, NF-kB), (activator protein-1, AP-1, etc.), early response genes (c-fos, c-jun etc.)] (See, Figure 2). This activation leads to induction of various target genes [i.e. inducible nitric oxide synthase (iNOS), cyclooxygenase II (Cox II), manganese superoxide dismutase (Mn-SOD), inducible form (HSP-72), cytokines etc.], which contribute to cell damage (Kumagai et al., 1997; Hsieh et al., 1998; Goering et al., 2000; Sano et al., 2001). Studies have also shown that distinct kinases mediate the metal/toxicant-induced toxicity both downstream as well as upstream of generated ROS (Myhre and Fonnum, 2001; El Bekay et al., 2003). For example, ROS generating nicotinamide-adenine dinucleotide phosphate (NADPH) Oxidase enzyme is stimulated by zinc in astrocytes and neurons in a protein kinase C (PKC)-dependent manner. Furthermore, the upstream involvement of tyrosine kinase, PKC, and MAPK pathway kinases in MeHg-induced generation of ROS in synaptosomes has been demonstrated with selective inhibitors (Myhre and Fonnum, 2001). The generated ROS resulting from oxidative stress (induced by a variety of agents like copper, arsenic, chromium, H_2O_2, angiotensin II, and from degenerative conditions) causes the activation downstream of a variety of kinases (p38MAPK, ERK and JNK), as well as transcription factors such as NF-kB, leading to final cytotoxic response (Floyd, 1999; Chen et al., 2000; Rosenberger et al., 2001; Chen and Shi, 2002; El Bekay et al., 2003; Zafarullah et al., 2003).

Recent studies in our laboratory examined the ability of MeHg to induce oxidative stress in primary cultures of cerebrocortical astrocytes from neonatal rat brains. The studies were carried out in 4-6 weeks old cultures. We examined the effect of 60 minutes exposure to MeHg on astrocytic ROS generation (Figure 3A). Compared to controls (100 %), there was a concentration-dependent increase in ROS generation in astrocytes treated with 5 uM (158 ± 24 % of control), 10 μM (197 ± 22 % of control) and 20 μM MeHg (301 ± 57 % of control), with a significant treatment effect at 10 uM ($p<0.01$) and 20 uM MeHg ($p<0.001$).

If ROS generation contributes to MeHg-induced (10 μM) astrocytic dysfunction, then antioxidants should attenuate this effect. As shown in Figure 3B, pre-incubation (10 minutes) of astrocytes with PBN inhibited ROS generation by MeHg (217 ± 22 % of control; $p<0.001$) to levels statistically indistinguishable from control cultures (0.05 mM PBN, 141 ± 10 % of control; 0.1 mM PBN, 150 ± 17 % of control; 0.5 mM PBN, 123 ± 21 % of control).

To substantiate MeHg-induced ROS generation in astrocytes, additional studies were carried out with laser scanning confocal microscopy (LSCM). Figure 4 shows confocal microscopic images of astrocytes treated with 10 uM MeHg. Cells were exposed to 10 uM MeHg for 1 h, then 20 μM hydroethidine (HE) fluorescent dye was added and the indicated confocal images were obtained at 0 min (left) and 10 min (right), after HE addition. The results support the earlier observations on MeHg-induced ROS formation (Figure 3), as reflected by the increased formation of fluorescent oxidized form of HE from the dihydro non-fluorescent molecule. Taken together, these results demonstrate that MeHg exposure leads to increased formation of ROS in astrocytes. This increased ROS could be damaging to astrocytes as it may compromise their ability to maintain control over the extracellular milieu thus damaging juxtaposed neurons.

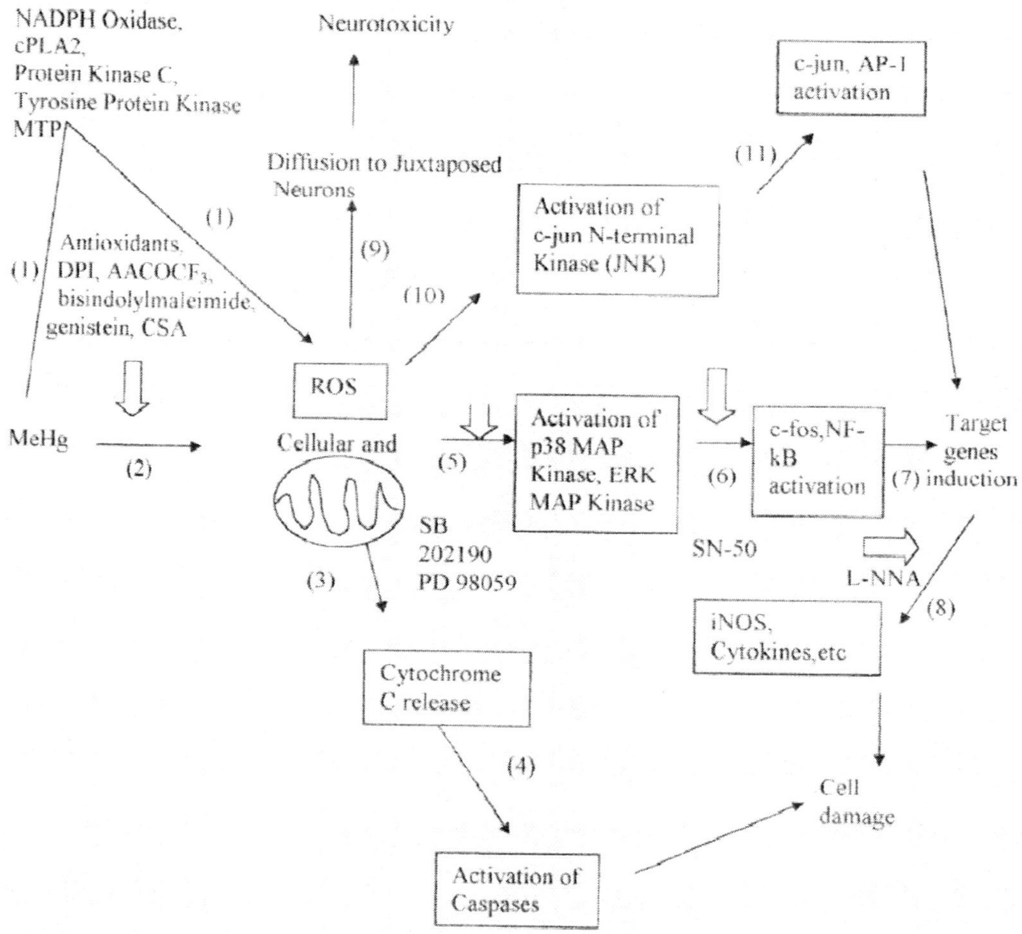

Figure 2. A model on the role of oxidative stress in MeHg-induced neurotoxicity. The numbers indicate the following steps/pathways: 1, involvement of various enzymes and MTP in MeHg-induced ROS formation; 2, possible blocking of increased ROS by pretreatment with a) indicated antioxidants, b) NADPH Oxidase inhibitor (DPI), c) $cPLA_2$ inhibitor ($AACOCF_3$), d) protein kinase C inhibitor (bisindolylmaleimide), e) tyrosine protein kinase inhibitor (genistein) and e) MTP inhibitor (CSA); 3,4, release of cytochrome C into cytosol from mitochondria; 5, ROS formed leads to activation of p38 MAP Kinase and ERK MAP Kinase (blocked by the inhibitors SB 202190 and PD 98059); 6, stimulated kinases mediates the activation of c-fos and NF-kB (blocked by the NF-kB activation inhibitor SN-50, kinase inhibitors as well as the antioxidants NAC and catalase); 7,8, possible induction of target proteins like iNOS (inhibitable with L-NNA, an iNOS inhibitor); 9, the diffusion of ROS from astrocytes to adjacent neurons leading to mitochondrial damage in the latter; 10,11, involvement of other signaling (i.e. JNK, c-jun and AP-) pathways in modulating MeHg cytotoxicity.

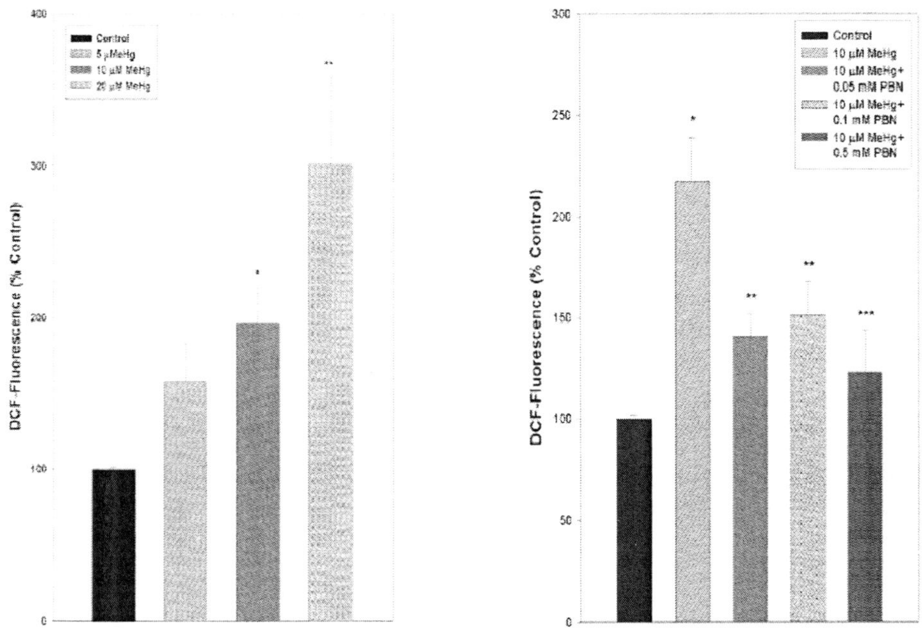

Figure 3. A (Left). MeHg (60 minute exposure) induces a concentration-dependent increase in ROS formation in rat neonatal primary cerebral astrocytes (*p<0.01 versus control, **p<0.001 versus control, n = 6-10; mean ± SEM). **B** (Right). Compared with 10 μM MeHg alone (*p<0.001 versus control), pre-incubation (10 minutes) of astrocytes with PBN was found to attenuate ROS generation to levels statistically indistinguishable from control cultures (**p<0.05, PBN plus 10 uM MeHg versus 10 uM MeHg alone; ***p<0.01, PBN plus 10 uM MeHg versus 10 uM MeHg alone) (n= 4-8; mean ± SEM).

Figure 4. Confocal images of astrocytes. The cells were treated for 1 h at 37°C with10 μM MeHg. Then after addition of 20 μM HE, the images were taken at 0 min (left) and 10 min (right)

Summary

There have not been detailed and systematic studies to examine how MeHg-induced ROS formation in astrocytes leads ultimately to cytotoxicity, and what, if any, is the role of various signal transduction pathways in this process. Based on the existing literature, it would seem reasonable to suggest that at least in part, the generation of ROS and may involve activation of signaling pathways involving distinct kinases [i.e. protein kinase C, tyrosine protein kinase, p38 MAP kinase and ERK MAP kinase], phospholipase A_2, as well as immediate early genes (c-fos) and transcription factor NF-kB (please refer to Figure 1B). MeHg-dependent astrocytic ROS/RNS increase could be damaging to mitochondria in GSH compromised astrocytic cells, as well as to juxtaposed neurons with relative paucity of GSH (Stewart et al., 2000; Heales and Bolanos, 2002). These studies should be undertaken to further advance our knowledge of CNS mediated injury, with particular emphasis on the role afforded by astrocytes in this process.

Acknowledgments

This study was supported by Public Health Service Grant ES07331 to MA.

References

[1] Ali, SF; LeBel CP; Bondy SC. Reactive oxygen species formation as a biomarker of methylmercury and trimethyltin neurotoxicity. *Neurotoxicology* 1992, 13, 633-648.

[2] Allen, JW; Mutkus, LA; Aschner, M. Isolation of neonatal rat cortical astrocytes for primary cultures, in: M. Maines, L.G. Costa, G.I. Sipes (Eds.), *Current Protocols in Toxicology,* Wiley, New York, 2000, pp 12.4.1-12.4.15.

[3] Allen, JW; Mutkus, LA; Aschner, M. Methylmercury mediated inhibition of 3H-D-aspartate transporter in cultured astrocytes is reversed by the antioxidant catalase. *Brain Res.* 2001, 902, 92-100.

[4] Allen, JW; Shanker, G; Aschner, M. Methylmercury inhibits the in vitro uptake of the glutathione precursor, cystine, in astrocytes, but not in neurons. *Brain Res.* 2000, 894,131-140.

[5] Almazan, G; Liu, HN; Khorchild, A; Sunderrajan, S; Martinez-Bermudez, AK; Chemtoz, S. Exposure of developing oligodendrocytes to cadmium causes HSP 72 induction, free radical generation, reduction in glutathione levels, and cell death. *Free Rad. Biol. Med.* 2000, 29, 858-869.

[6] Aschner, M. Methylmercury in astrocytes- what possible significance? *Neurotoxicology* 1997, 17, 93-106.

[7] Aschner, M. Astrocytic swelling, phospholipase A_2, glutathione and glutamate: interactions in methylmercury induced neurotoxicity. *Cell. Mol. Biol.* 2000, 46, 843-854.

[8] Aschner, M; Allen, JW; Kimelberg, HK; LoPachin, RM; Streit, WJ. Glial cells in neurotoxicity development. *Annul. Rev. Pharmacol. Toxicol.* 1999, 39, 151-173.

[9] Aschner, M; Bennett, BA. Astrocyte and neuron co-culturing method. In Methods in *Molecular Medicine, vol 22: Neurodegeneration Methods and Protocols.* (J. Harry and H.A. Tilson, eds.), 1998, pp. 133-144. Humana Press, Totowa, NJ.

[10] Aschner, M; Du, YL; Gannon, M; Kimelberg, HK. Methylmercury-induced alterations in excitatory amino acid transporter in rat primary astrocytic cultures. *Brain Res.* 1993, 602, 181-186.

[11] Aschner, M; Kimelberg, HK eds. In: The role of glia in neurotoxicity, 1996, CRC Press, Boca Raton, FL.

[12] Aschner, M; Kimelberg, HK. Metallothionein in brain-the role in physiology and pathology. *Toxicol. Appl. Pharmacol.* 1997, 142, 229-242.

[13] Aschner, M; Mullaney, KJ; Wagoner, D; Lash, LH; Kimelberg, HK. Intracellular glutathione (GSH) levels modulate mercuric chloride (MC)- and methylmercuric chloride (MeHgCl)- induced amino acid release from neonatal rat primary astrocytic cultures. *Brain Res.* 1994, 664, 133-140.

[14] Aschner, M; Yao, CP; Allen, JW; Tan, KH. Methylmercury alters glutamate transport in astrocytes. *Neurochem. Intl.* 37, 199-206, 2000.

[15] Atchison, WD; Hare, MF. Mechanism of methylmercury-induced neurotoxicity. *FASEB. J.* 1994, 8, 622-629.

[16] Bakir, F; Damluji, SF; Amin-Zaki, L; Murthada, M; Khalidi, A; Al-Rawi, NY. Tikriti, S; Dhahrir, HI; Clarkson, TW; Smith, JC; Doherty, RA. Methylmercury poisoning in Iraq. *Science* 1973, 181, 230-242.

[17] Betteridge, DJ. What is oxidative stress? *Metabolism* 2000, 49, 3-8.

[18] Brawer, JR; McCarthy, GF; Gornitsky, M; Frankel, D; Mehindale, M; Schipper, HM. Mercuric chloride induces a stress response in cultured astrocytes characterized by mitochondrial uptake of iron. *Neurotoxicology* 1998, 19, 767-776.

[19] Britton, RS. Metal-induced hepatotoxicity. *Sem. Liver Dis.* 1996,16, 3-12.

[20] Brookes, N. In Vitro evidence for the role of glutamate in CNS toxicity of mercury. *Toxicology* 1992, 76, 245-256.

[21] Brookes, N; Kristt, DA. Inhibition of amino acid transport and protein synthesis by $HgCl_2$ and methylmercury in astrocytes, selectivity and reversibility. *J. Neurochem.* 1989, 53, 1228-1237.

[22] Bush, AI. Metals and neuroscience. *Cur. Opin. Chem. Biol.* 2000, 4, 184-191.

[23] Charleston, JS; Body, RC; Bolander, RP; Mottet, NK; Vahter, ME; Burbacher, TM. Changes in the number of astrocytes and microglia in the thalamus of the monkey Macaca Fascicularis following long-term subclinical methylmercury exposure. *Neurotoxicology* 1996, 17, 27-38.

[24] Charleston, JS; Bolander, RP; Mottet, NK; Body, RL; Vahter, ME; Burbacher, TM. Increases in the number of reactive glia in the visual cortex of Macaca Fascicularis following subclinical long-term methylmercury exposure. *Toxicol. Appl. Pharmacol.* 1994, 129, 196-206.

[25] Chen, SH; Liu, LH; Liang, Y-C; Lin, J.K; Lin-Shiau, S. Death signaling pathway induced by pyrrolidine dithiocarbamate-Cu^{2+} complex in the cultured rat cortical astrocytes. *Glia* 2000, 31, 249-261.

[26] Chen, F; Shi, X. Signaling from toxic metals to NF-kB and beyond: not just a matter of reactive oxygen species. *Environ. Health Perspect.* 2002, 110, 807-811.

[27] Chen, Y; Swanson, RA. Astrocytes and brain injury. *J. Cereb. Blood Flow Metab.* 2003, 23, 137-149.

[28] Cordier, S; Garel, M; Mandereau, L; Morcel, H; Doineau, P; Gosme-Segure, S; Josse, D; White, R; Amiel-Tison, C. Neurodevelopmental investigations among methylmercury-exposed children in French Guiana. *Environ. Res.* 2002, 89, 1-11.

[29] Coyle, JT; Puttafarken, P. Oxidative stress, glutamate and neurodegenerative disorders. *Science* 1993, 262, 689-695.

[30] Dave, D; Mullaney, KJ; Goderie, S; Kimelberg, HK; Aschner, M. Astrocytes as mediators of methylmercury neurotoxicity: effects on D-aspartate and serotonin uptake. *Dev. Neurosci.* 1994, 16, 222-231.

[31] Davidson, PW; Myers, GJ; Cox, C; Axtell, C; Shamlaye, C; Sloane-Reeves, J; Cernichiari, E; Needham, L; Choi, A; Wang, Y; Berlin, M; Clarkson, TW. Effects of prenatal and postnatal methylmercury exposure from fish consumption on neurodevelopment: outcomes at 66 months of age in the Seychelles child development study. *JAMA* 1998, 280, 701-707.

[32] Davis, KL; Martin, E; Turko, IV; Murad, F. Novel effects of nitric oxide. *Ann. Rev. Pharmacol. Toxicol.* 2001, 41, 203-236.

[33] Dringen, R; Pfeiffer, B; Hamprecht, B. Synthesis of the antioxidant glutathione in neurons: supply by astrocytes of CysGly as precursor for neuronal glutathione. *J Neurosci* 1999, 19, 562-569.

[34] Dugan, LL; Sensi, SL; Canzoniero, LM; Haudran, SD; Rothman, SM; Lin, TS; Goldberg, MP; Choi, DW. Mitochondrial production of reactive oxygen species in cortical neurons following exposure to N-methyl-D-aspartate. J. *Neurosci.* 1995, 151, 6377-6388.

[35] El Bekay, RE; Alvarez, M; Monteserin, J; Alba, G; Chacon, P; Vega, A; Martin-Nieto, J; Jimenez, J; Pintado, E; Bedoya, FJ; Sobrino F. Oxidative stress is a critical mediator of the angiotensin II signal in human neutrophils: involvement of mitogen-activated protein kinase, calcineurin, and the transcription factor NF-kB. *Blood* 2003, 102, 662-671.

[36] Faustmann, EM; Ponce, RA; Ou, YC; Mendoza, MAC; Lewandowski, T; Kavanagh, T. Investigations of Methylmercury-induced alterations in neurogenesis. *Environ. Health Perspect.* 2002, 110 (suppl. 5), 859-864.

[37] Floyd, RA. Antioxidants, oxidative stress, and degenerative neurological disorders. *Proc. Soc. Exp. Biol. Med.* 1999, 222, 236-245.

[38] Garman, RH; Weiss, B; Evans, HL. Alkylmercurial encephalopathy in the monkey: a histopathologic and autoradiographic study. *Acta Neuropathol.* (Berlin), 1975, 32, 61-74.

[39] Gasso, S; Cristofol, RM; Selena, G; Rosa, R; Rodriguez-Farre, E; Sanfeliu, C. Antioxidant compounds and Ca^{2+} pathway blockers differentially protect against

methylmercury and mercuric chloride neurotoxicity. *J. Neurosci. Res.* 2001, 66, 135-145.

[40] Goering, PL; Fisher, BR; Noren, BT; Papaconsantinou, A; Rojko, JL; Marler, RJ. Mercury induces regional and cell-specific stress protein expression in rat kidney. *Toxicol Sci.* 2000, 53, 447-457.

[41] Grandjean, P; Weihe, P; White, RF; Debes, F; Araki, S; Yokoyama, S; Murata, K; Sorenson, N; Dahl, R; Jorgensen, PJ. Cognitive deficit in 7-year old children with prenatal exposure to methylmercury. Neurotoxicol. Teratol. 1997, 19, 417-428.

[42] Grandjean, P; Budiz-Jorgensen, E; White, RF; Jorgensen, PJ; Weihe, P; Debes, F; Keiding, N. Methylmercury exposure biomarkers as indicators of neurotoxicity in children aged 7 years. *Am. J. Epidemiol.* 1999, 150, 301-305.

[43] Grandjean, P; White, RF; Weihe, P; Jorgensen, PJ. Neurotoxic risk caused by stable and variable exposure to methylmercury from seafood. *Ambul. Pediat.* 2003, 3, 18-23.

[44] Halliwell, B. Free radicals and antioxidants: a personal view. *Nutr. Rev.* 1994, 52, 253-265.

[45] Heales, SJ; Bolanos, JP. Impairment of brain mitochondrial function by reactive nitrogen species: the role of glutathione in dictating susceptibility. *Neurochem. Intl.* 2002, 40, 469-474.

[46] Hsieh, HJ; Cheng, CC; Wu, ST; Chiu, JJ; Wang, BS; Wang, DL. Increase of reactive oxygen species (ROS) in endothelial cells by shear flow and involvement of ROS in shear-induced c-fos expression. *J. Cellular Physiol.* 1998, 175, 156-162.

[47] Hunot, S; Brugg, B; Ricard, D; Michel, PP; Muriel, MP; Ruberg, M; Faucheux, BA; Agid, Y; Hirsch, EC. Nuclear translocation of NF-kappa B is increased in dopaminergic neurons of patients with Parkinson disease. *Proc. Natl. Acad. Sci.* USA 1997, 94, 7531-7536.

[48] Juarez, BJ; Martinez, ML; Montante, M; Dufour, L; Garcia, E; Jimenez-Capdeville, ME. Methylmercury increases glutamate extracellular levels in frontal cortex of awake rats. *Neurotoxicol. Teratol.* 2002, 24, 767-771.

[49] Kaltschmidt, B; Baeuerle, PA; Kaltschmidt, C. Potential involvement of the transcription factor NF-kappa B in neurological disorders. *Mol. Aspects Med.* 1993, 14, 171-190.

[50] Kehrer, JP; Smith, CV. Free radicals in biology: sources, reactivities, and roles in the etiology of human diseases. In: Frei, B., Ed. Natural antioxidants in human health and disease. San Diego: Academic Press: 1994: 25-62.

[51] Kitamura, Y; Shimohama, S; Ota, T; Matsuoka, Y; Nobura, Y; Taniguchi, T. Alterations of transcription factor NF-kappa B and STAT1 in Alzheimer's disease brains. *Neurosci. Lett.* 1997, 237, 17-20.

[52] Kjellstrom, T; Kennedy, P; Wallis, S; Stewart, A; Friberg, L; Lind, B; Witherspoon, T; Manteal, C. Physical and mental development of children with prenatal exposure to mercury from fish. Stage II: Interviews and psychological tests at age 6. Solna, Sweden. 1989, National Swedish Environmental Protection Board Report 3642.

[53] Kumagai, Y; Mizukado, S; Nayafune, J; Shinyashiki, M; Homma-Takeda, S; Shimojo, N. Post-transcriptional elevation of mouse brain Mn-SOD protein by mercuric chloride. *Brain Res.* 1997, 769, 178-182.

[54] C.P. LeBel, H. Ishiropoulos, S.C. Bondy, Evaluation of the probe 2′, 7′-dichlorofluorescin as an indicator of reactive oxygen species formation and oxidative stress, Chem. Res. Toxicol. 5 (1992) 227-231.

[55] Lukiw, WJ; Bazan, NG. Neuroinflammatory signaling Upregulation in Alzheimer's disease. *Neurochem. Res.* 2000, 25, 1173-1184.

[56] Matyja, E; Albrecht, J. Ultrastructural evidence that mercuric chloride lowers the threshold for glutamate neurotoxicity in an organotypic culture of rat cerebellum. *Neurosci. Lett.* 1993, 158, 155-158.

[57] Mullaney, KJ; Fehm, MN; Vitarella, D; Wagoner, DE Jr.; Aschner, M. The role of –SH groups in methylmercuric chloride induced D-aspartate and rubidium release from rat primary astrocytic cultures. *Brain Res.* 1994, 641, 1-9.

[58] Munthe, E; Aaseth, J; Jellum, E. Trace elements and rheumatoid arthritis (RA): pathogenetic and therapeutic aspects. *Acta Pharmacol. Toxicol.* 1986, 59, 365-373.

[59] Myers, GJ; Davidson, PW; Shamlaye, CF; Axtell, CD; Cernichiari, E; Choisy, O; Choi, A; Cox, C; Clarkson, TW. Effects of prenatal methylmercury exposure from a high fish diet on developmental milestones in the Seychelles Child Development Study. *Neurotoxicology* 1997, 18, 819-829.

[60] Myhre, O; Fonnum, F. The effect of aliphatic, naphthenic, and aromatic hydrocarbons, on production of reactive oxygen species and reactive nitrogen species in rat brain synaptosome fraction: the involvement of calcium, nitric oxide synthase, mitochondria, and phospholipase A_2. *Biochem. Pharmacol.* 2001, 62, 119-128.

[61] Nose, K; Shibanuma, M; Kikuchi, K; Kageyama, H; Sakiyama, S; Kuroki, T. Transcriptional activation of early response genes by hydrogen peroxide in a mouse osteoblastic cell line. *Eur. J. Biochem.* 1991, 201, 99-106.

[62] Park, ST; Lim, KT; Chung, YT; Kim, SU. Methylmercury- induced neurotoxicity in cerebral neuron culture is blocked by antioxidants and NMDA receptor antagonists. Neurotoxicology 1996, 17, 37-46.

[63] Rosenberger, J; Petrovics, G; Buzas, B. Oxidative stress induces proorphanin FQ and proenkephalin gene expression in astrocytes through p38- and ERK-MAP kinases and NF-kB. *J. Neurochem.* 2001, 79, 35-44.

[64] Sano, M; Fukuda, K; Sato, T; Kawaguchi, H; Suematsu, M; Matsuda, S; Koyasu, S; Matsui, H; Yamauchi-Takihara, K; Harada, M; Saito, Y; Ogawa, S. ERK and p38 MAPK, but not NF-kB, are critically involved in reactive oxygen species–mediated induction of IL-6 by angiotensin II in cardiac fibroblasts. *Circ. Res.* 2001, 89, 661-669.

[65] Sarafian, TA. Methylmercury-induced generation of free radicals: Biological implications. *Metal Ions Biol. Sys.* 1999, 36, 415-444.

[66] Sarafian, TA; Vartavarian, L; Kane, DJ; Bredesen, DE; Verity, MA. Bcl2 expression decreases methylmercury induced free-radical generation and cell killing in a neural cell line, *Toxicol. Lett.* 1994, 74, 149-155.

[67] Sayre, LM; Perry, G; Smith, MA. Redox metals and neurodegenerative disease. *Curr. Opin. Chem. Biol.* 1999, 3, 220-225.

[68] Shanker, G; Allen, JW; Mutkus, LA; Aschner, M. The uptake of cysteine in cultured primary astrocytes and neurons. *Brain Res.* 2001a, 902, 156-163

[69] Shanker, G; Allen, JW; Mutkus, LA; Aschner, M. Methylmercury inhibits cysteine uptake in cultured primary astrocytes, but not in neurons. *Brain Res.* 2001b 914, 159-165.

[70] Shanker, G; Aschner, M. Identification and characterization of uptake systems for cystine and cysteine in cultured astrocytes and neurons: Evidence for methylmercury-targeted disruption of astrocytic transport. *J. Neurosci. Res.* 2001c, 66, 998-1002.

[71] Shanker, G. Mutkus, S.J. Walker, M. Aschner. Methylmercury enhances arachidonic acid release and cytosolic phospholipase A_2 expression in primary cultures of neonatal astrocytes. *Mol. Brain Res.* 2002, 106, 1-11.

[72] Shanker, G; Aschner, M. Methylmercury-induced reactive oxygen species formation in neonatal cerebral astrocytic cultures is attenuated by antioxidants. *Mol. Brain Res.* 2003, 110, 85-91.

[73] Shanker, G; Syversen, T; Aschner, M. Astrocyte-mediated methylmercury neurotoxicity. *Biol. Trace Elem. Res.* 2003, 95, 1-10.

[74] Sorg, O; Schilter, B; Honnegger, P; Monnet-Tschudi, F. Increased vulnerability of neurons and glial cells to low concentrations of methylmercury in a prooxidant situation. *Acta Neuropathol.* (Berlin) 1998, 96, 621-627.

[75] Steuerwald, U; Weihe, P; Jorgensen, PJ; Bjerve, K; Brock, J; Heinzow, B; Budtz-Jorgenson, E; Grandjean, P. Maternal seafood diet, methylmercury exposure, and neonatal neurologic function. *J. Pediat.* 2000, 136, 599-605.

[76] Stewart, VC; Sharpe, MA; Clark, JB; Heales, SJ. Astrocyte-derived nitric oxide causes both reversible and irreversible damage to the neuronal mitochondrial respiratory chain. *J. Neurochem.* 2000, 75, 694-700.

[77] Takeuchi, T. Biological reactions and pathological changes in human beings and animals caused by organic mercury contamination. In R. Hartung and B.D. Dinman (Eds.), Environmental Mercury contamination, Ann Arbor Science, Ann Arbor, 1972, pp 247-289.

[78] Yee, S; Choi, BH. Oxidative stress in neurotoxic effects of methylmercury poisoning. *Neurotoxicology* 1996, 17, 17-26.

[79] M. Zafarullah, W.Q. Li, J. Sylvester, M. Ahmad, Molecular mechanisms of N-acetylcysteine actions, Cell Mol. Life Sci. 60 (2003) 6-20.

In: Focus on Neurochemistry Research
Editor: Robert M. Coleman, pp.229-247
ISBN 1-59454-417-4

Chapter IX

Purine Nucleoside Mediated Protection of Primary Cerebellar Granule Cells after Hypoxic Insult

Veronika Heftberger, Bettina Tomaselli, Valerie Podhraski and Gabriele Baier-Bitterlich*

Institute for Med. Chemistry and Biochemistry, Dept. of Neurobiochemistry,
University of Innsbruck, Fritz Pregl Str.3, A-6020 Innsbruck, Austria

Abstract

Neurons strongly depend on oxygen supply and ATP production. Short periods of hypoxia or anoxia occurring e.g. in stroke or trauma can result in necrosis/apoptosis and damaged axons. Attendant with energy depletion is the formation of purine nucleosides adenosine, inosine, and guanosine. In this study, hypoxia was induced with rotenone, a mitochondrial complex I inhibitor, which leads to loss of viable cells and neurite formation. Purine nucleosides partially restored cell survival and neurite outgrowth of primary cerebellar granule cells, whereby inosine had the most striking effect. Nucleosides may influence cellular signaling either by entering cells via nucleoside transporters or alternatively by binding to adenosine receptors. Pharmacological inhibition of nucleoside transport led to an almost complete abolishment of neuroprotective effects. In addition, immune fluorescence analysis revealed a strong surface expression of the adenosine A1 receptor. In our study, neuroprotective effects of adenosine and inosine were apparently supported by A1 receptor-mediated signaltransduction, whereas guanosine-mediated protection seemed to be A1 receptor

* A.Univ.Prof.Dr.Gabriele Baier-Bitterlich; Institute for Med. Chemistry and Biochemistry, Dept. of Neurobiochemistry, University of Innsbruck, Fritz Pregl Str.3, A-6020 Innsbruck, Austria; Tel.: ++43/0 512-507-3272, e-mail: gabriele.baier-bitterlich@uibk.ac.at

independent. Our data support current hope of the possible use of purine nucleosides for the neuronal regeneration of primary brain cells following hypoxic insult.

Introduction

In acute neurological conditions such as stroke, trauma, and infection severe, usually focal injuries to the central nervous system occur. Stroke caused by cerebral ischemia is the third leading cause of death and the leading cause of neurological disability. Since neurons strongly depend on oxygen supply and ATP production, such serious injuries to the brain due to hypoxic or ischemic insult generally cause necrosis, although in most cases there exists a process of delayed injury in regions surrounding an area of most severe damage. Some of this 'delayed type' appears to be more of an apoptotic type [1]. Programmed cell death (apoptosis) is an active, evolving process that involves new protein synthesis ending in internucleosomal DNA fragmentation. As a result of hypoxia, and in severe cases ischemia, ATP production is diminished and the energy status of neurons and glia is compromised. Neurons in the adult mammalian central nervous system, which are injured by stroke or trauma, normally fail to regenerate axons, which results in long lasting disabilities in sensory, motor, or cognitive functions.

To reduce apoptosis in the brain is central to functional recovery after stroke. Present search is focused on the development of drugs that block the apoptotic process in the hope to improve clinical outcome [2]. Numerous authors (see reviews by [3-5]) have proposed adenosine and its receptors as targets for therapeutic approaches to stroke and related disorders. Exogenous application of adenosine, a cell-permeant purine nucleoside, has been shown in a variety of *in vivo* models to protect neural cell viability (for a review see: [6, 7]).

Neuroprotective Mechanism

Adenosine acts as a powerful endogenous neuroprotectant during ischemia-induced energy failure, by decreasing neuronal metabolism and increasing cerebral blood flow, and by playing a variety of different roles as an intercellular messenger. This is particularly the case in the brain, which expresses high concentrations of adenosine receptors. These effects are mediated through interaction of adenosine with specific receptors, and stimulation of adenosine receptors was hypothesized to result in an effective treatment of stroke (for a review, see: [8]).

There is reduction in the release of excitotoxic neurotransmitters, attenuation of NMDA receptors, vasorelaxation, and anti-inflammatory effects. But most importantly, there is, at least in animal models, an impressive reduction of neuronal damage and mortality hardly attained by any other drug (for a review, see: [5]).

Adenosine acts at four receptor subtypes: A1, A2A, A2B and A3 [9]. All subtypes belong to the super family of G-protein coupled receptors, with A1 and A3 interacting with Gi/ Go proteins, and A2A and A2B with Gs [10]. In similarity with the majority of other G-protein coupled receptors, all adenosine receptor types are characterized by seven transmembrane helices [11]. The stimulation of adenosine A2 receptors elevates, while activation of both adenosine A1 and A3 receptors reduces, the concentration of intracellular cAMP ([12, 13]. In addition to their effects on adenylate cyclase, adenosine receptors are also characterized by their stimulatory effect on phospholipase C [14, 15] (for a review, see: [16]).

Evaluation of the effects of adenosine receptor agonists and antagonists in stroke models indicate that adenosine acting through A1 receptors has neuroprotective effects [3-5], probably by inhibiting excitatory synaptic neurotransmission in the brain during hypoxia [17]. In contrast, activation of A2A receptors may enhance neuronal damage, as mice lacking these receptors exhibited reduced damage following focal ischemia [18]. On the other hand, recent results indicate that hypoxia-induced membrane responses of PC12 cells are likely to be mediated via activation of the A2A receptor [19]. Adenosine A3 agonists are cytoprotective, but at higher concentrations may also enhance ischemia-induced neuronal damage [5, 15, 20], although these receptors are of low abundance in brain.

Adenosine is the main agonist of these receptors, but the adenosine metabolite inosine has been shown to bind and activate adenosine A3 receptors [21], and yet another group [22] found that A1 and A2 receptor antagonists partially reversed the suppressive effect of inosine on proinflammatory cytokine production. Inosine, an adenosine derivative, was shown to protect astrocytes, and the effects exerted by adenosine were apparently the result of its conversion to inosine by adenosine deaminase [23]. Inosine has multiple anti-inflammatory effects. These findings, coupled with the fact that inosine has very low toxicity, suggested that this agent may be useful in the treatment of inflammatory/ischemic diseases [22]. In addition, it was shown that also guanosine, which is derived from GTP, was able to preserve glial cell viability to the same extent as inosine [24]. The effects of purine nucleosides were blocked with purine transport inhibitors [25]. One of the targets, which has been found to be stimulated by inosine and guanosine, may coincide with a serine-threonine kinase, protein kinase N [26, 27].

In this study, we wanted to investigate the mechanism by which purine nucleosides promote viability and neurite outgrowth of ischemic brain cells. In our test model, A1 receptor expressing primary rat cerebellar granule cells [28] were subjected to chemical hypoxia using rotenone, a mitochondrial complex I inhibitor [29]. Many authors employ chemical hypoxia models for their studies [6], and recently results were published, showing that rotenone mimicked the effect of hypoxia on endothelial cells [30].

Material and Methods

Separation and Cultivation of Cerebellar Granule Cells from Rats

Cerebellar granule cells were prepared from 7-day-old Sprague-Dawley rats according to a protocol based on [31]. Briefly, cells were dissociated from freshly dissected cerebella by mechanical disruption in the presence of trypsin (Gibco, Scotland) and DNase (Deoxyribonuclease 1, Roche, Austria). Cerebellar granule neurons were further enriched by a preplating step and were finally seeded on poly-L-ornithine (Sigma, Austria) coated culture plates at a density of 5 x 104 cells/cover slip. The deployed medium was Neurobasal TM medium 1x (without glutamine), supplemented with 2% B-27-Supplement 50x (both from Gibco, U.S.A), as well as with 1% L- glutamine and (10 U/ml) penicillin- (10 µg/ml) streptomycin (both from PAA Laboratories, Austria). The preparation of cerebellar granule cells is well characterized and is reported to contain predominantly granule neurons [31]. The cells were cultivated 2-4 days at 37 °C and 5% CO2. After two days in culture, the cerebellar neurons were provided with fresh neurobasal medium (with the above mentioned supplements).

Evaluation of Hypoxic Insult by Fluorescence Microscopy

On the first day in vitro (DIV) -2 hours after seeding-, chemical hypoxia was induced with rotenone, an inhibitor of mitochondrial complex I (1 µM, Calbiochem, Germany). Cell death of cerebellar granule cells was analyzed 16-24 hours later after double staining with Hoechst 33342 (Molecular Probes, U.S.A) and propidium iodide (Sigma, Austria). First, cells were incubated for 15 min with Hoechst 33342 (10 µg/ml), then for another five minutes with propidium iodide (5 µg/ml), both at 37 °C and 5% CO2. After these incubation periods with DNA dyes, cells were washed twice with phosphate buffer (1xPBS, pH 7.0; 10xPBS= 19 mM NaH2PO4, 81 mM NaH2PO4, 1,54 M NaCl, diluted with destilled water 1:10). Cerebellar granule cells were studied on a Zeiss Axioplan Fluorescence Microscope. Digital images were taken using MetaVue, and corresponding images were overlaid by means of Adobe Photoshop 7.0. While all cells are stained with Hoechst (blue), only dead cells (late apoptosis and necrosis) allow propidium iodide to penetrate their membranes (pink).

Quantification of Decreased Cell Death (=Rescue) by Application of Purine Nucleosides

Cerebellar granule cells (first day in vitro, DIV 1) were incubated with intermediary concentrations of purine nucleosides (Sigma, Austria) adenosine, guanosine, or inosine (500 µM) within 5 min [6] after inducing hypoxia with rotenone (1 µM). After 16-24 hours, effects of purine nucleosides were visualized after double staining with Hoechst 33342 and propidium iodide. Rescue was defined as decrease of dead cells after treatment with purine nucleosides as compared to control cells:

Rescue(%) = 100 P/R (deadP + R) x 100R (dead)

In addition to the 500 µM concentration of purine nucleosides that was chosen based upon earlier data [6, 32] and used throughout the studies, two more concentrations of adenosine, guanosine, and inosine were tested.

Pharmacological Inhibitors

Specific antagonists, such as the adenosine A1 receptor antagonist 8-cyclopentyl-1, 3-dipropylxanthine (DPCPX, 10 μM, Sigma, Austria) and the nucleoside transport inhibitor S-(4-Nitrobenzyl)-6-thioinosine (NBTI, 10 μM, Sigma, Austria) were used to study the contribution of the A1 receptor- and nucleoside transport -mediated signaling. Another pharmacological inhibitor, the adenosine A2A receptor antagonist 8-(3-Chlorostyryl) caffeine (CSC, 10 μM, Sigma, Austria) was employed to investigate the contribution of the A2A receptor signaling. The adenosine deaminase inhibitor erythro-9- (2-hydroxy-3-nonyl) adenosine (EHNA, 10 μM, Sigma, Austria) was used to study, whether adenosine effects are mediated via its metabolite inosine [19, 25]. On the first DIV, these four inhibitors were added to cerebellar granule cells 30 min prior stimulation with purine nucleosides.

Measurement of Neurite Outgrowth

The neurite outgrowth of cerebellar granule cells after incubation with purine nucleosides adenosine, guanosine, and inosine was measured after a period of 24 hours. The effects of rotenone and purine nucleosides were visible in the increased or decreased expression of processes. A few hours after seeding (DIV 1), cerebellar granule cells were incubated with adenosine, guanosine, and inosine (500 μM). To evaluate the effects of purines on *hypoxic* cells, chemical hypoxia was induced with rotenone, and within 5 min, purine nucleosides (500 μM) were added. After 16-24 hours, neurite outgrowth was measured according to a method by [33]. Only cells with intact cell shapes and with neurites longer than two times the cell diameter were counted. Lengths of neurites from either normoxic or hypoxic cerebellar granule neurons were measured using Scion Image.

Adenosine Receptor Staining

Cerebellar granule cells (second and fourth DIV, respectively) were fixed in 4% PFA for 10 min at 4 °C. The cells were washed 3 x 10 min with phosphate buffer saline (PBS), pH 7.4, followed by a treatment with blocking solution, consisting of 10% goat serum and 3% BSA (Calbiochem, Canada) in PBS (30 min). Receptor staining using anti A1-adenosine receptor (Ab-1) rabbit IgG (1:100, Oncogene, U.S.A.) and anti A2a-adenosine receptor rabbit IgG (1:1000, Chemicon, U.S.A.) was carried out for 1 hour at room temperature (or overnight at 4 °C). After a series of washes with PBS, cells were treated with the secondary antibody Alexa Fluor 488 goat ant-rabbit IgG H+L(1:4000, Eubio, Austria) for 1 hour at room temperature, followed by another series of washes. Both primary and secondary antibodies were diluted with blocking solution (3% Goat and 1% BSA in PBS). The cerebellar granule cells were mounted in antifade mounting medium (Dako, Austria) and studied on an Axioplan Zeiss Fluorescence Microscope. Two different exposure times were used: 2000 milliseconds (ms) for cerebellar granule cells on DIV 2, and 500 ms for cells on DIV 4. Digital images were assembled by means of MetaVue.

Results

Effect Of Hypoxia On The Viability Of Cerebellar Granule Cells

Viability of cerebellar granule cells following hypoxic insult was studied after double staining of cerebellar granule cells with Hoechst 33342 and propidium iodide (Fig.1A-D). Whereas the Hoechst dye can penetrate all cells (Fig.1B), propidium iodide enters only dead cells (Fig.1C). In order to be able to quantify the effect of hypoxia and purine nucleosides, the three corresponding levels were overlaid (Fig.1D). Rotenone, an inhibitor of mitochondrial complex I, induced the loss of 40% cells as compared to a spontaneous cell death of cerebellar granule cells, which amounted to 10-20%. Neurite outgrowth after 24 hours was reduced from 45-55% of controls cells (every 2nd normoxic cell had 1-2 lengthy processes) as compared to 3-4% of cells after treatment with rotenone. Hypoxic cells (first DIV, 24 hours incubation) can form only very short neurites (up to 2-3 times the cell diameter), whereas control cells had processes, which ranged from 20 µm up to 70 µm.

Effect of Purine Nucleosides on Viability and Neurite Outgrowth of Normoxic Cerebellar Granule Cells

Cerebellar granule cells were incubated with adenosine, guanosine, and inosine (500 µM) under normoxic conditions. Whereas adenosine appeared ineffective, guanosine and inosine improved viability of normoxic cells by 9,88% and 14,11%, respectively. Interestingly, adenosine (1,3 fold of control) in contrast to guanosine and inosine (same as control) more effectively enhanced neurite outgrowth of normoxic cerebellar granule cells. In addition, the variance of neurite-lengths was studied after 24 hours incubation. The positive effect of purine nucleosides was most obvious in cells with neurites of 40-60 µm length (Fig.2C-E). 5% of cerebellar granule cells stimulated with adenosine and inosine even formed neurites with lengths up to 140-150 µm.

Effect of Various Concentrations of Purine Nucleosides on Hypoxic Cerebellar Granule Cells

As mentioned above, hypoxia induced a 40% loss of cerebellar granule cells. In a next step, cells were simultaneously incubated with various concentrations of purine nucleosides and the rescue was calculated as described in material and methods. Adenosine and inosine appeared most effective in the highest concentrations (500 µM and 1000 µM) used, whereas guanosine already reached the maximum capacity at a lower concentration (500 µM) (Fig.3A). For the following experiments, we therefore always used purine nucleosides at 500 µM and found that up to 22% hypoxic cells may be rescued upon addition of purine nucleosides. Concomitant to results under normoxia, inosine was the most advantageous purine nucleoside (Fig.3B). Interestingly, again, adenosine was most effective in rescuing neurite outgrowth of cerebellar granule cells following hypoxic insult (Fig.3C).

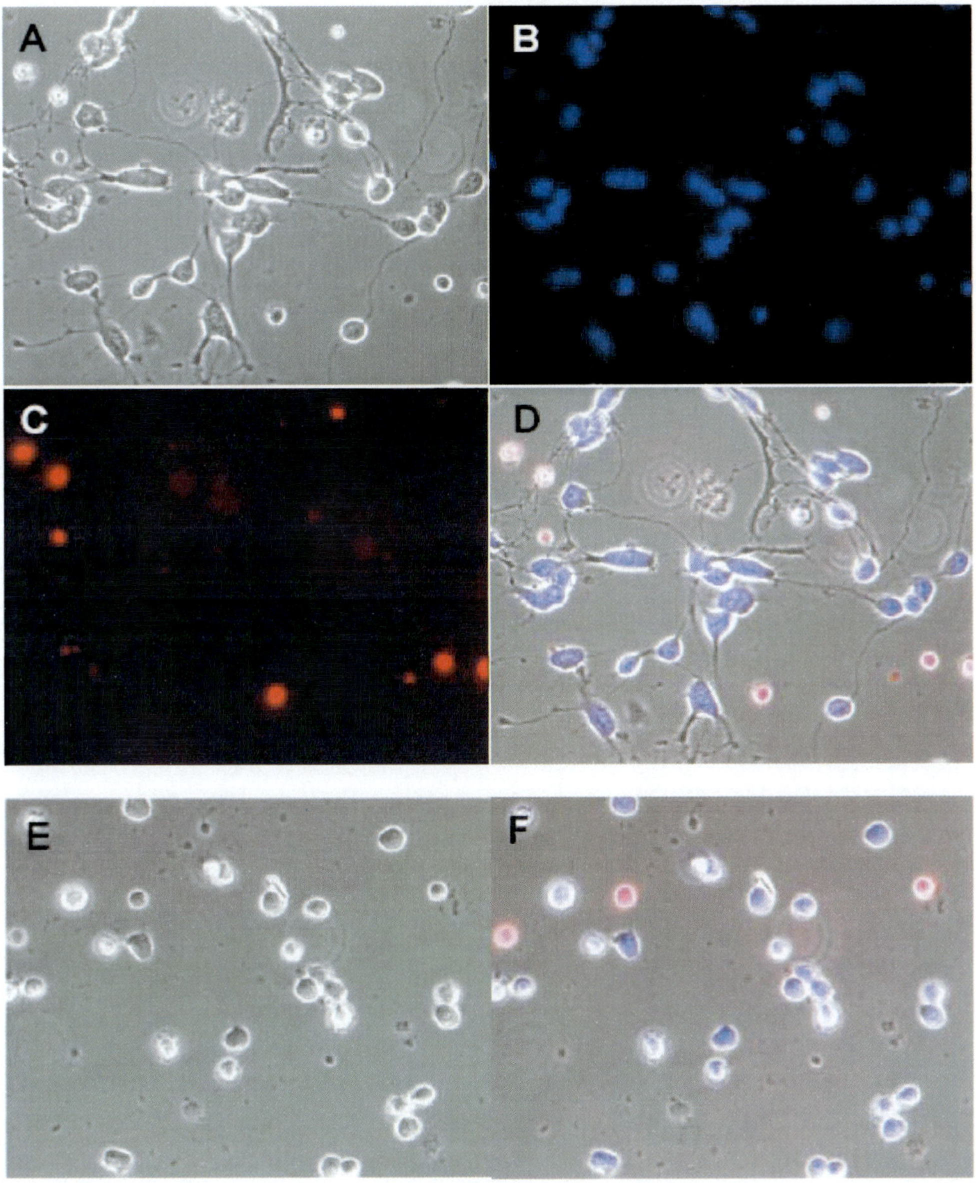

Figure 1. Hypoxic insult of cerebellar granule cells. Rat cerebellar granule cells P7 (5x104/cover slip) were cultured on poly-L-ornithine-coated culture plates at 37 °C and 5% CO2 for 24 hours. Two hours after seeding (first DIV), chemical hypoxia was induced with rotenone (1 μM). Cell death of cerebellar granule cells was visualized after double staining with Hoechst 33342 and propidium iodide, and digital images [A-C] were taken on a Zeiss Axioplan Fluorescence Microscope, and all three corresponding images were overlaid using Adobe Photoshop 7.0 [D]. Double stained cerebellar granule cells (pink) are dead cells. [A-D] Hoechst [B]/propidium iodide [C] staining of control cells. [E, F] Hoechst/propidium iodide staining of rotenone-treated cells.

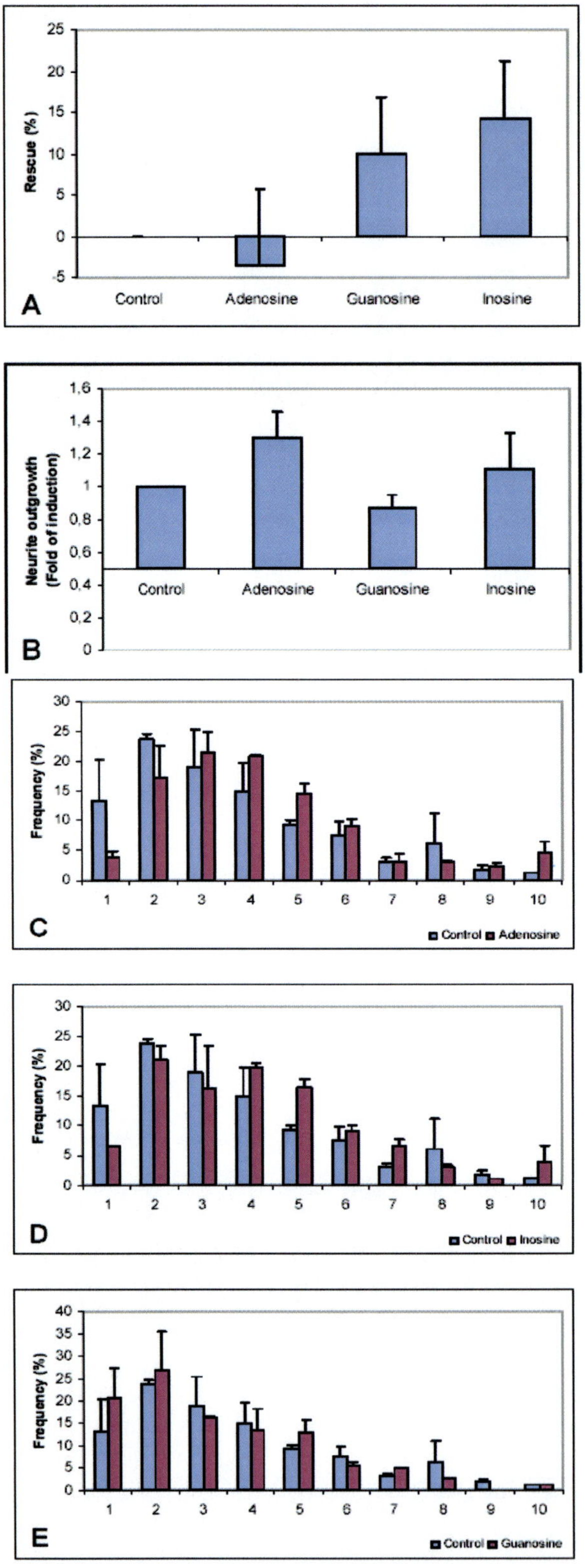

Figure 2. Purine nucleosides improve viability and neurite outgrowth of cerebellar granule cells. On the first DIV, cerebellar granule cells were incubated with purine nucleosides adenosine, guanosine, or inosine (500 μM) for 16-24 hours. Spontaneous cell death (due to handling of the cells, etc) was visualized after double staining with Hoechst 33342 and propidium iodide. For each sample three microscopic fields were counted. Cells with neurites longer than two times the diameter were considered to be cells with neurite outgrowth. Cells with more than two neurites were neglected. Length of neurites was measured using Scion Image. [A] Quantification of Hoechst/propidium iodide staining of control and purine nucleoside - treated cells. [B] Neurite outgrowth of healthy cerebellar granule cells; both control and purine nucleoside-treated cells. [C-E] Neurite lengths of healthy cerebellar granule cells; direct comparison of control cells and cells stimulated with either adenosine, guanosine, or inosine (bar 1= 10-20 μm, bar 2= 20-30 μm, bar 3= 30-40 μm, bar 4= 40-50 μm, bar 5= 50-60 μm, bar 6= 60-70 μm, bar 7= 70-80 μm, bar 8= 80-90 μm, bar 9= 90-100 μm, bar 10= >100 μm). Values represent the means and S.E.M. of nine [A] and six [B] independent experiments.

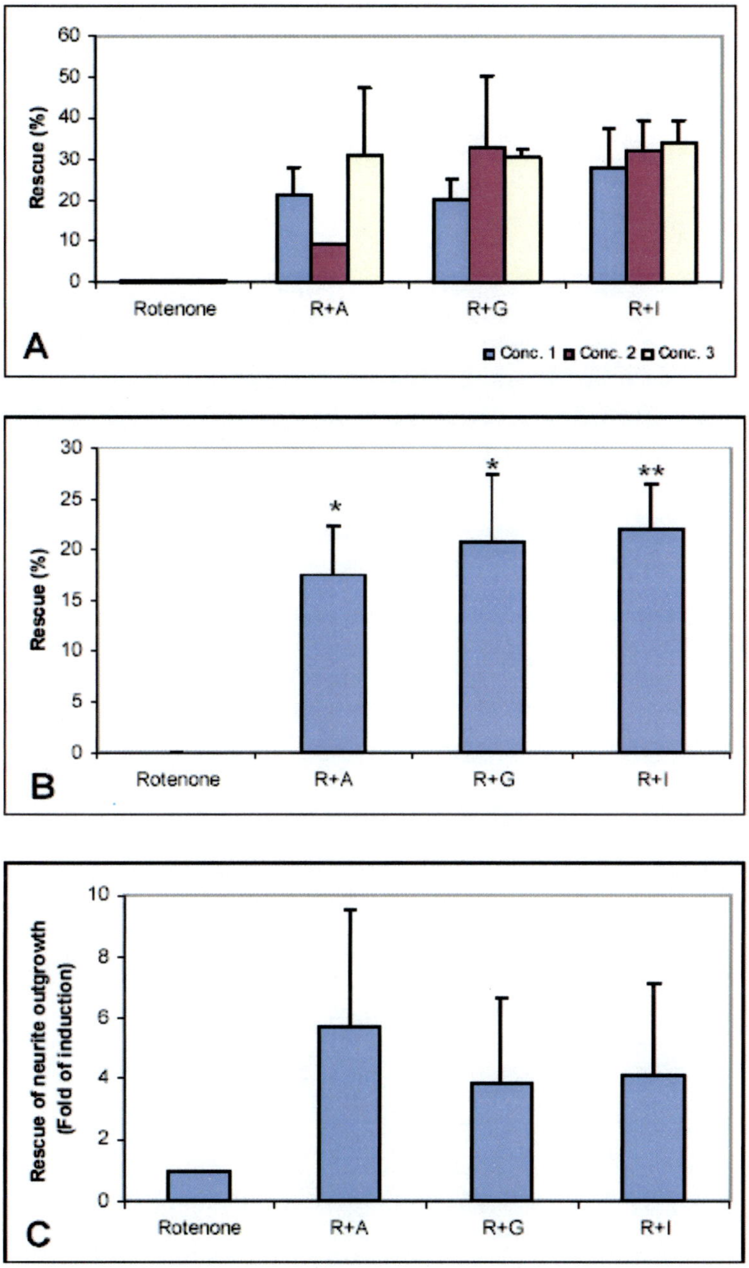

Figure 3. Neuroprotection of hypoxic cerebellar granule cells by purine nucleosides. Cerebellar granule cells (first DIV) were treated with rotenone (1 μM). Within five minutes, purine nucleosides adenosine, guanosine, or inosine (500 μM) were added for 16-24 hours. Cell death was visualized after double staining with Hoechst 33342 and propidium iodide. For each experiment three microscopic fields were counted. Cells with neurites longer than two times the diameter were considered to be cells with neurite outgrowth. Cells with more than two neurites were neglected [C]. [A, B] Quantification of purine-mediated rescue of the viability of rotenone-treated cerebellar granule cells (bar 1: R rotenone, bar 2: R+A rotenone + adenosine, bar 3: R+G rotenone + guanosine, bar 4: R+I rotenone + inosine). Other concentrations of purine nucleosides [A] were investigated as well: concentration 1: 10 μM (all three purine nucleosides), concentration 2: 100 μM (adenosine) and 500 μM (guanosine and inosine), respectively, and concentration 3: 500 μM (adenosine) and 1000 μM (guanosine and inosine), respectively. [C] Purine–mediated rescue of cerebellar granule cells calculated as percentage of outgrowth in rotenone-treated cultures. (Fig.3B, bars 2-4, P values for R vs. R+A = 0,0153, P for R vs. R+G = 0,0432, and P for R vs. R+I = 0,0050 in a paired t-test of mean values). Values represent the means and S.E.M. of nine [B] and four [C] independent experiments.

Expression of Adenosine Receptor A1 in Cerebellar Granule Cells

Cerebellar granule cells were stained with anti A1-Adenosine-receptor-antibody (1:100) and the secondary antibody Alexa Fluor 488 (1:4000) (Fig.4A,B,E,F). Next, cells were also stained for A2A receptor, which is also abundantly expressed in brain. As may be seen in Fig.4B, cells abundantly express A1 as compared to A2A (Fig.4C) and control (Fig.4D).

We also studied the expression of A1 receptor on day 4 in vitro and found that cells express increased amounts of A1 after being in culture for three days (Fig.4E,F).

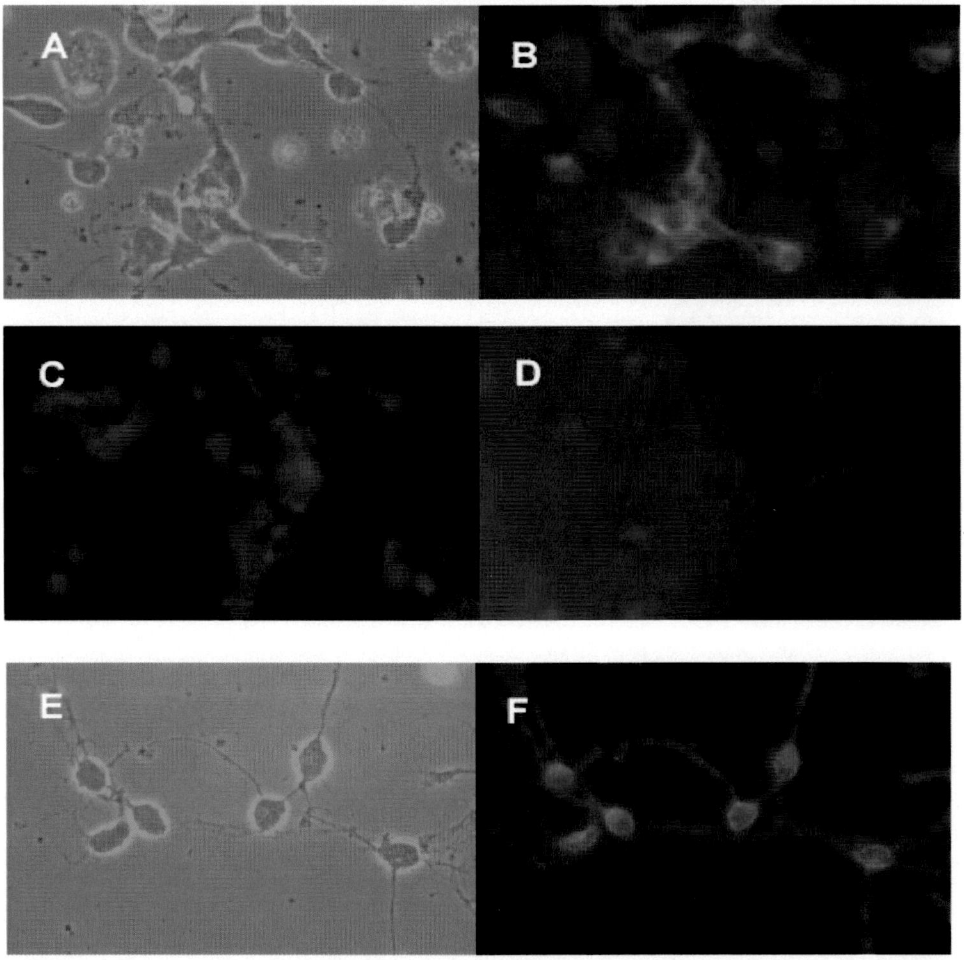

Figure 4. Cerebellar granule cells express A1 adenosine receptor. Adenosine receptor staining (first antibodies anti A1 adenosine receptor (Ab-1) rabbit IgG 1:100 and anti A2A adenosine receptor rabbit IgG antibody 1:1000), with Alexa Fluor 488 goat anti-rabbit IgG H+L as second antibody (1:4000). Digital images were taken at two different exposure times: 2000 ms (second DIV) or 500 ms (fourth DIV), using a Zeiss Axioplan Fluorescence Microscope (40-times magnified, oil objective) and MetaVue. [A, B] A1 adenosine receptor-staining; exposure time: 2000 ms (second DIV). [B-D] Comparison of A1 [B] and A2A [C] receptor expression of cerebellar granule cells, with a positive-control [D] (no first antibody); exposure time: 2000 ms (second DIV). [E, F] Expression of A1 adenosine receptor of cerebellar granule cells on a later stage of development; exposure time: 500 ms (fourth DIV).

Effect of Pharmacological Inhibitors on Neuroprotective Effect of Purine Nucleosides

Effects of the purine nucleosides adenosine, guanosine, and inosine on hypoxic cerebellar granule cells were studied with or without the addition of CSC, an inhibitor of the A2 receptor, and no remarkable effect detected (Fig.5A). In contrast, addition of the A1 antagonist DPCPX considerably inhibited the neuroprotective effect of adenosine and inosine while it did not effect guanosine-mediated protection (Fig.5B). The incubation of hypoxic cerebellar granule cells with the nucleoside transport inhibitor NBTI completely abolished rescue of all three nucleosides (Fig.5C). Last, the effect of EHNA, an antagonist of the adenosine deaminase, was tested. In our hands no effect was observed (Fig.5D).

Discussion

The adult brain is extremely vulnerable to various injuries. The recent discovery of neural progenitors in adult mammals, however, raises the possibility of repairing damaged tissue by employing their latent regenerative potential [34]. The effect of hypoxia on the viability of cerebellar granule cells was studied after double staining of cerebellar granule cells with Hoechst 33342 and propidium iodide. Upon addition of rotenone, many propidium iodide positive cells were observed, indicating a marked hypoxia-induced increase of dead cerebellar granule cells in rotenone-treated cells as compared to control cells. Also neurite outgrowth was diminished drastically after treatment with rotenone. Moreover, we observed that hypoxic cells can form only very short neurites as compared to control cells. Our data are in line with earlier findings by other authors employing chemical hypoxia models for their studies [6].

At first, cerebellar granule cells were incubated with adenosine, guanosine, and inosine (500 μM) under *normoxic* conditions. Whereas adenosine appeared ineffective, guanosine and inosine improved viability of cells. Interestingly, adenosine was more effective in enhancing neurite outgrowth of *normoxic* cerebellar granule cells. The positive effect of nucleosides was most obvious in cells with an average length of neurites, but cerebellar granule cells stimulated with adenosine and inosine even formed neurites with lengths up to 140-150 μm. Next, the effect of purine nucleosides on the viability of rotenone-treated cells was studied and quantified. As mentioned above, hypoxia induced a 40% loss of cerebellar granule cells. In a next step, cells were simultaneously incubated with various concentrations of purine nucleosides and the rescue was calculated. The loss of viability after the addition of rotenone was inhibited by all three purine nucleosides. Adenosine and inosine appeared most effective in the highest concentration (1000 μM) used, whereas guanosine had already reached its maximum capacity at a lower concentration (500 μM). For the following experiments, we always used purine nucleosides at 500 μM and found that up to 22% hypoxic cells may be rescued by the addition of purine nucleosides. In parallel to results with cells under normoxia, inosine was most effective. Interestingly, similar to normoxic cells,

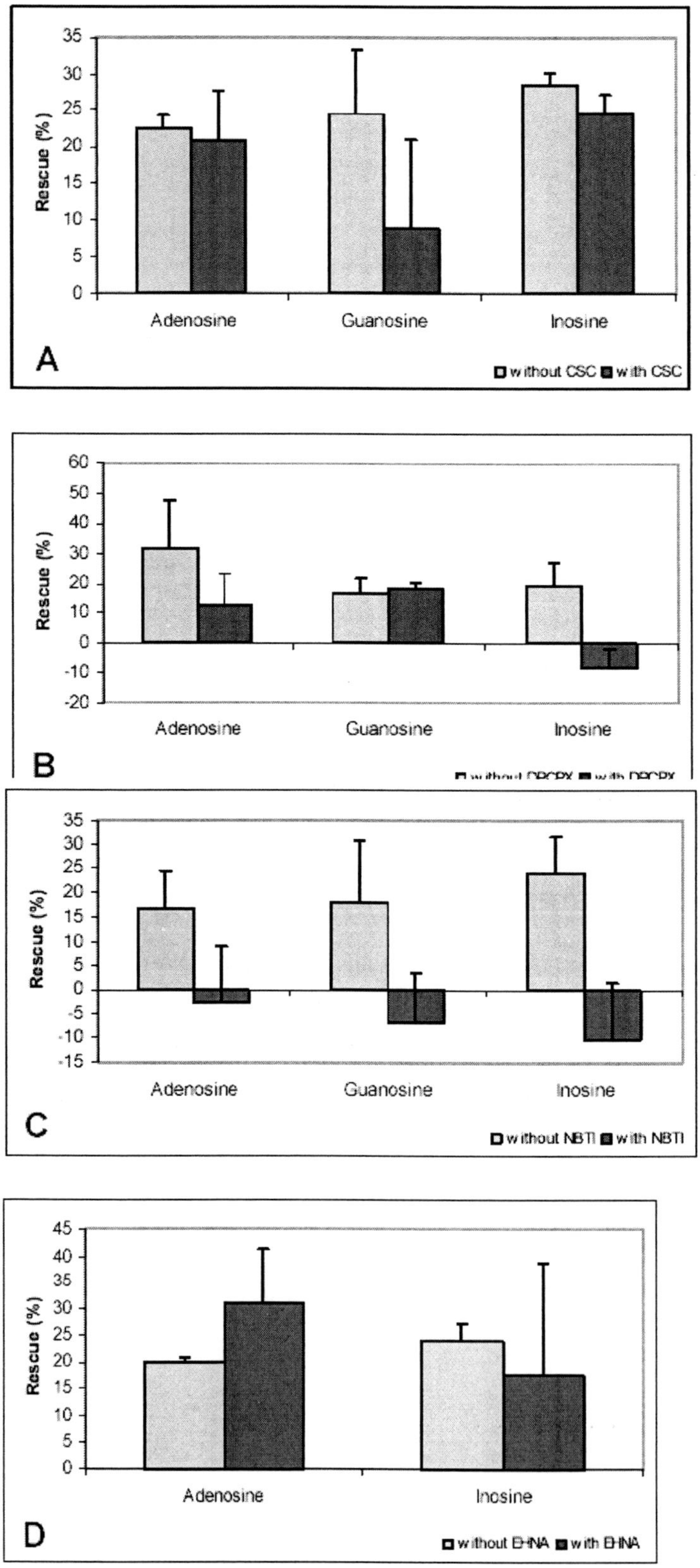

Figure 5. Effect of pharmacological inhibitors on protection of hypoxic cerebellar granule cells by purine nucleosides. Cerebellar granule cells (first DIV) were treated with pharmacological inhibitors (10 μM) 30 minutes before chemical hypoxia was induced with rotenone (1 μM). Within five minutes, cells were incubated with purine nucleosides adenosine, guanosine, or inosine (500 μM) for 16-24 hours. Cell death was visualized after double staining with Hoechst 33342 and propidium iodide. For each experiment three microscopic fields were counted. [A-D] Quantification of Hoechst/propidium iodide staining of rotenone/purine nucleoside-treated cells with and without addition of pharmacological inhibitors CSC [A], DPCPX [B], NBTI [C], and EHNA [D], respectively. Values represent the means and S.E.M. of three [A, D] and five [B, C] independent experiments.

adenosine was most effective in rescuing neurite outgrowth of cerebellar granule cells following hypoxic insult. Purine nucleosides have been shown to protect neural cell viability during culture under hypoxic conditions, the exact protective mechanism however varies depending on the cell type. Therefore, we wanted to study the role of adenosine receptor-mediated signaltransduction in purine-mediated effects.

Adenosine receptors are divided into four subtypes, A1, A2A, A2B, and A3 receptors, whereby brain cells, especially cerebellar granule cells predominantly express the A1 receptor [28]. This was also seen in our cultures. Cerebellar granule cells expressed increasing amounts of A1 receptor over the days in vitro, but only negligible amounts of A2A receptor. In contrast, in rat PC12 cells, adenosine exerts its effects via adenosine receptor A2A [19], while astrocytes are apparently protected via an intracellular-receptor independent mechanism [23].

In addition to adenosine receptors, both glia and neurons have nucleoside transport systems that play an important role in regulating the concentrations of purine nucleosides (for a review, see: [35]). We have used the transport inhibitor NBTI, and the A1 and A2A receptor antagonists DPCPX and CSC to study the mechanism of purine nucleoside – mediated signaling. As expected, the A2A receptor antagonist CSC had no remarkable effect. On the contrary, addition of the A1 antagonist DPCPX considerably inhibited the neuroprotective effect of adenosine and inosine, while it did not effect guanosine-mediated protection. These data suggest that the A1 adenosine receptor mediates the effect of adenosine and inosine, while guanosine exerts its effect probably via nucleoside transport and other not yet defined mechanisms. Concomitant with this hypothesis, the incubation of hypoxic cerebellar granule cells with the nucleoside transport inhibitor NBTI completely abolished rescue of all three nucleosides, suggesting that the effect of purine nucleosides was supported by nucleoside transport into the cell. Our results are in support of data by other authors [25], showing that the effects of the purine nucleosides were blocked with purine transport inhibitors. They claimed that exogenously applied inosine and guanosine act directly upon an intracellular target, which may coincide with protein kinase N [26, 36], Other data [37] showed that the action of guanosine is independent of A1 receptor and nucleoside transport. The way by which guanosine effects viability and neurite outgrowth is indeed very controversial. Whereas recent studies confirmed the hypothesis of the existence of specific G-coupled receptors for guanosine [38], others [39], however, observed that the mitogenic activity of guanosine was partly inhibited by antagonists of Ado A1 and A2B receptors, and claimed that guanosine-mediated effects may be partly based on an interaction with adenosine receptors or via the stimulation of adenosine release [35]. Others, [6], even suggested that neuronal protection by purine nucleosides depends on bystanding glia cells. Amongst the purine nucleosides tested in this study, inosine showed the best protection of cerebellar granule cells. According to other authors [25] who showed that the addition of adenosine or guanosine induced goldfish retinal ganglion cells to extend lengthy neuritis, and that these effects were highly specific and did not reflect conversion of the nucleosides to their nucleotide derivatives. They also claimed that the action of inosine was not due to a hydrolysis to hypoxanthine, since hypoxanthine was inactive in retinal ganglion cells. On the other hand, in certain cell types adenosine owes its effects actually to a conversion to inosine [23, 25]. Our own results however showed that the activity of adenosine is not blocked by

EHNA, an inhibitor of adenosine deaminase, suggesting an independent effect of adenosine on cerebellar granule cells.

Our own data on inosine-enhanced neurite outgrowth of hypoxia-treated cerebellar granule cells are also in line with recent findings [40], reviewed in: [41], showing that inosine may stimulate significant axonal reorganization after stroke. According to [40], however, '*in vivo*' inosine apparently does not exert neuroprotective effects as evidenced by its failure to diminish stroke volume, but contributes to improved outcome due to the induction of axonal rewiring. Inosine exerts its receptor-independent effects probably via protein kinase N that has been implicated in mediating NGF-induced differentiation [26, 36], and induction of expression of genes associated with axon growth, including GAP-43, L1, and □-1 tubulin [25, 42].

Results showing the positive impact of purine nucleosides inosine, adenosine, and guanosine on viability and neurite outgrowth of ischemic cerebellar granule cells support current studies on the possibility of novel neuronal cell regeneration therapies, and confirm the important role of purine nucleosides in neuronal pathologies involving hypoxic insult.

Acknowledgments

This work was supported by a grant of the Austrian FWF (P15823). We are grateful to Dr's H. Dietrich and A. Hanni for animal house keeping.

Correspondence should be addressed to G.B. (e-mail: gabriele.baier-bitterlich@uibk.ac.at).

References

[1] Honig, L. S. & Rosenberg, R. N. (2000) Apoptosis and neurologic disease, *Am J Med. 108*, 317-30.

[2] Raff, M. (1998) Cell suicide for beginners, *Nature. 396*, 119-22.

[3] Rudolphi, K. A., Schubert, P., Parkinson, F. E. & Fredholm, B. B. (1992) Neuroprotective role of adenosine in cerebral ischaemia, *Trends Pharmacol Sci. 13*, 439-45.

[4] Sweeney, M. I. (1997) Neuroprotective effects of adenosine in cerebral ischemia: window of opportunity, *Neurosci Biobehav Rev. 21*, 207-17.

[5] von Lubitz, D. K. (1999) Adenosine and cerebral ischemia: therapeutic future or death of a brave concept?, *Eur J Pharmacol. 371*, 85-102.

[6] Litsky, M. L., Hohl, C. M., Lucas, J. H. & Jurkowitz, M. S. (1999) Inosine and guanosine preserve neuronal and glial cell viability in mouse spinal cord cultures during chemical hypoxia, *Brain Res. 821*, 426-32.

[7] Rodriguez-Nunez, A., Camina, F., Lojo, S., Rodriguez-Segade, S. & Castro-Gago, M. (1993) Concentrations of nucleotides, nucleosides, purine bases and urate in cerebrospinal fluid of children with meningitis, *Acta Paediatr. 82*, 849-52.

[8] Dunwiddie, T. V. & Masino, S. A. (2001) The role and regulation of adenosine in the central nervous system, *Annu Rev Neurosci. 24*, 31-55.

[9] Fredholm, B. B., Abbracchio, M. P., Burnstock, G., Daly, J. W., Harden, T. K., Jacobson, K. A., Leff, P. & Williams, M. (1994) Nomenclature and classification of purinoceptors, *Pharmacol Rev. 46*, 143-56.

[10]Palmer, T. M. & Stiles, G. L. (1997) Structure-function analysis of inhibitory adenosine receptor regulation, *Neuropharmacology. 36*, 1141-7.

[11]Palmer, T. M. & Stiles, G. L. (1995) Adenosine receptors, *Neuropharmacology. 34*, 683-94.

[12]van Calker, D., Muller, M. & Hamprecht, B. (1979) Adenosine regulates via two different types of receptors, the accumulation of cyclic AMP in cultured brain cells, *J Neurochem. 33*, 999-1005.

[13]Zhou, Q. Y., Li, C., Olah, M. E., Johnson, R. A., Stiles, G. L. & Civelli, O. (1992) Molecular cloning and characterization of an adenosine receptor: the A3 adenosine receptor, *Proc Natl Acad Sci U S A. 89*, 7432-6.

[14]Linden, J. (1991) Structure and function of A1 adenosine receptors, *Faseb J. 5*, 2668-76.

[15]Abbracchio, M. P., Brambilla, R., Ceruti, S., Kim, H. O., von Lubitz, D. K., Jacobson, K. A. & Cattabeni, F. (1995) G protein-dependent activation of phospholipase C by adenosine A3 receptors in rat brain, *Mol Pharmacol. 48*, 1038-45.

[16]Fredholm, B. B., AP, I. J., Jacobson, K. A., Klotz, K. N. & Linden, J. (2001) International Union of Pharmacology. XXV. Nomenclature and classification of adenosine receptors, *Pharmacol Rev. 53*, 527-52.

[17]Scanziani, M., Capogna, M., Gahwiler, B. H. & Thompson, S. M. (1992) Presynaptic inhibition of miniature excitatory synaptic currents by baclofen and adenosine in the hippocampus, *Neuron. 9*, 919-27.

[18]Chen, J. F., Huang, Z., Ma, J., Zhu, J., Moratalla, R., Standaert, D., Moskowitz, M. A., Fink, J. S. & Schwarzschild, M. A. (1999) A(2A) adenosine receptor deficiency attenuates brain injury induced by transient focal ischemia in mice, *J Neurosci. 19*, 9192-200.

[19]Kobayashi, S., Conforti, L., Pun, R. Y. & Millhorn, D. E. (1998) Adenosine modulates hypoxia-induced responses in rat PC12 cells via the A2A receptor, *J Physiol. 508 (Pt 1)*, 95-107.

[20]Yao, Y., Sei, Y., Abbracchio, M. P., Jiang, J. L., Kim, Y. C. & Jacobson, K. A. (1997) Adenosine A3 receptor agonists protect HL-60 and U-937 cells from apoptosis induced by A3 antagonists, *Biochem Biophys Res Commun. 232*, 317-22.

[21]Jin, X., Shepherd, R. K., Duling, B. R. & Linden, J. (1997) Inosine binds to A3 adenosine receptors and stimulates mast cell degranulation, *J Clin Invest. 100*, 2849-57.

[22]Hasko, G., Kuhel, D. G., Nemeth, Z. H., Mabley, J. G., Stachlewitz, R. F., Virag, L., Lohinai, Z., Southan, G. J., Salzman, A. L. & Szabo, C. (2000) Inosine inhibits inflammatory cytokine production by a posttranscriptional mechanism and protects against endotoxin-induced shock, *J Immunol. 164*, 1013-9.

[23]Haun, S. E., Segeleon, J. E., Trapp, V. L., Clotz, M. A. & Horrocks, L. A. (1996) Inosine mediates the protective effect of adenosine in rat astrocyte cultures subjected to combined glucose-oxygen deprivation, *J Neurochem. 67*, 2051-9.

[24]Jurkowitz, M. S., Litsky, M. L., Browning, M. J. & Hohl, C. M. (1998) Adenosine, inosine, and guanosine protect glial cells during glucose deprivation and mitochondrial inhibition: correlation between protection and ATP preservation, *J Neurochem. 71*, 535-48.

[25]Benowitz, L. I., Jing, Y., Tabibiazar, R., Jo, S. A., Petrausch, B., Stuermer, C. A., Rosenberg, P. A. & Irwin, N. (1998) Axon outgrowth is regulated by an intracellular purine-sensitive mechanism in retinal ganglion cells, *J Biol Chem. 273*, 29626-34.

[26]Volonte, C., Rukenstein, A., Loeb, D. M. & Greene, L. A. (1989) Differential inhibition of nerve growth factor responses by purine analogues: correlation with inhibition of a nerve growth factor-activated protein kinase, *J Cell Biol. 109*, 2395-403.

[27]Batistatou, A., Volonte, C. & Greene, L. A. (1992) Nerve growth factor employs multiple pathways to induce primary response genes in PC12 cells, *Mol Biol Cell. 3*, 363-71.

[28]Wojcik, W. J. & Neff, N. H. (1983) Adenosine A1 receptors are associated with cerebellar granule cells, *J Neurochem. 41*, 759-63.

[29]Tai, K. K. & Truong, D. D. (2002) Activation of adenosine triphosphate-sensitive potassium channels confers protection against rotenone-induced cell death: therapeutic implications for Parkinson's disease, *J Neurosci Res. 69*, 559-66.

[30]Schafer, M., Ewald, N., Schafer, C., Stapler, A., Piper, H. M. & Noll, T. (2003) Signaling of hypoxia-induced autonomous proliferation of endothelial cells, *Faseb J. 17*, 449-51.

[31]Hatten, M. E. (1985) Neuronal regulation of astroglial morphology and proliferation in vitro, *J Cell Biol. 100*, 384-96.

[32]Gysbers, J. W. & Rathbone, M. P. (1996) Neurite outgrowth in PC12 cells is enhanced by guanosine through both cAMP-dependent and -independent mechanisms, *Neurosci Lett. 220*, 175-8.

[33]Bhatt, R. S., Tomoda, T., Fang, Y. & Hatten, M. E. (2000) Discoidin domain receptor 1 functions in axon extension of cerebellar granule neurons, *Genes Dev. 14*, 2216-28.

[34]Nakatomi, H., Kuriu, T., Okabe, S., Yamamoto, S., Hatano, O., Kawahara, N., Tamura, A., Kirino, T. & Nakafuku, M. (2002) Regeneration of hippocampal pyramidal neurons after ischemic brain injury by recruitment of endogenous neural progenitors, *Cell. 110*, 429-41.

[35]Rathbone, M. P., Middlemiss, P. J., Gysbers, J. W., Andrew, C., Herman, M. A., Reed, J. K., Ciccarelli, R., Di Iorio, P. & Caciagli, F. (1999) Trophic effects of purines in neurons and glial cells, *Prog Neurobiol. 59*, 663-90.

[36]Greene, L. A., Volonte, C. & Chalazonitis, A. (1990) Purine analogs inhibit nerve growth factor-promoted neurite outgrowth by sympathetic and sensory neurons, *J Neurosci. 10*, 1479-85.

[37]Gysbers, J. W. & Rathbone, M. P. (1996) GTP and guanosine synergistically enhance NGF-induced neurite outgrowth from PC12 cells, *Int J Dev Neurosci. 14*, 19-34.

[38]Traversa, U., Bombi, G., Camaioni, E., Macchiarulo, A., Costantino, G., Palmieri, C., Caciagli, F. & Pellicciari, R. (2003) Rat brain guanosine binding site. Biological studies and pseudo-receptor construction, *Bioorg Med Chem. 11*, 5417-25.

[39]Ciccarelli, R., Di Iorio, P., D'Alimonte, I., Giuliani, P., Florio, T., Caciagli, F., Middlemiss, P. J. & Rathbone, M. P. (2000) Cultured astrocyte proliferation induced by

extracellular guanosine involves endogenous adenosine and is raised by the co-presence of microglia, *Glia. 29*, 202-11.

[40]Chen, P., Goldberg, D. E., Kolb, B., Lanser, M. & Benowitz, L. I. (2002) Inosine induces axonal rewiring and improves behavioral outcome after stroke, *Proc Natl Acad Sci U S A. 99*, 9031-6.

[41]Benowitz, L. I., Goldberg, D. E. & Irwin, N. (2002) Inosine stimulates axon growth in vitro and in the adult CNS, *Prog Brain Res. 137*, 389-99.

[42]Petrausch, B., Tabibiazar, R., Roser, T., Jing, Y., Goldman, D., Stuermer, C. A., Irwin, N. & Benowitz, L. I. (2000) A purine-sensitive pathway regulates multiple genes involved in axon regeneration in goldfish retinal ganglion cells, *J Neurosci. 20*, 8031-41.

Index

A

B

C

F

G

H

M

N

O

P

Q

R

S